•A15042 573307

OCCUPATIONAL SAFETY & HEALTH:
7 Critical Issues for the 1990s

A BNA Special Report

Copyright © 1989
The Bureau of National Affairs, Inc.
1231 25th Street, N.W.
Washington, D.C. 20037

International Standard Book Number: 1-55871-128-7

Extra copies of this special report (BSP-136) are $95.00 each and may be obtained by mail from BNA's Customer Service Center, 9435 Key West Ave., Rockville, Md. 20850, or by telephone from BNA's Response Center, toll free (800) 372-1033. The following quantity discounts are available: 6-10 copies, 10%; 11-25, 15%; 26-50, 20%; 51-500, 25%; 501-1,000, 30%; and more than 1,000 copies, 35%.

Table of Contents

INTRODUCTION 1
    INITIATIVES IN SEVEN AREAS 1
    STRONGER PENALTIES URGED 2
        Industry Sees 'Administrative Inertia' 3
        Unions Push for Generic Standards 3
        Employers Seek Flexible Standards 4
    BEYOND THE REAGAN ERA 4
        'Old Days are Gone' 5
    HIGHLIGHTS 5
    ACKNOWLEDGEMENTS 6

CHAPTER I: Criminal Prosecutions 7
    CHAPTER OVERVIEW 8
    'THE TOLL IS FAR TOO HIGH' 10
        Legal Barriers Fall, But Slowly 10
        Green Light for State Prosecutions 11
        Prosecuting Workplace Deaths and Injuries 12
        Industry Cites Potential for Abuse 13
    MOST WORKER'S COMPENSATION LAWS BLOCK LAWSUITS 13
        The 'Third-Party' Lawsuit Option 15
    PROBLEMS IN THE OSHA SYSTEM 16
        Reagan 'Self-Policing' Policy Criticized 16
        Osha, State Fines Seen as Too Small 17
    CONGRESS ALLEGES 'INSTITUTIONAL RELUCTANCE' 18
    'EVEN IF YOU GET A CONVICTION, SO WHAT?' 18
        Longer Jail Sentences Needed, Critics Say 19
    LEGISLATION PLANNED 19
        'Beyond a Reasonable Doubt' 20
        Factors in Trench Cases 20
    STATE PROSECUTOR SEES LESS DIFFICULTY 21
    CASES 'WASHED OUT' ON FIRST REVIEW 22
    1990 BUDGET WOULD CUT PROSECUTION FUNDS 23
    STATE PROSECUTORS: 'WE GIVE THIS PRIORITY' 24
        'When the Feds Drop the Ball' 24
        Stronger State Laws Sought 25
        Bills Proposed by District Attorney 26
        Push from Attorney General in Minnesota 27
        The 'Barking Dog' of Deterrence 27
        Prosecutorial Discretion 28
    A LEGAL POINT-COUNTERPOINT: 29
    Preemption: A Union Lawyer's View by George H. Cohen 30
        Specter in the Boardroom 31

Statutory Shortcomings		32
High Visibility Cases		33

STAGE SET FOR LEGAL SHOWDOWN 33
STATES NOT PREEMPTED 35
    An Overview and Summary 35
    State's Right Expressly Preserved 35
    Heavy Burden for Proponents 36
    Strained Interpretations 37
    Integral Part of Tort Remedies 38
    Narrow Criminal Provisions 39
NO INTENT TO NULLIFY PROTECTION 40
    Section 18 Not Relevant to Preemption Issue 41
COMPREHENSIVE REGULATORY SCHEME 43
CONCLUSION 44
ENDNOTES 44
Use of General Criminal Laws by Robert C. Gombar, Willis J. Goldsmith,
Frank A. White, and Arthur G. Sapper 48
    A Red Herring: Getting Away With Murder 48
    Section 18 Means What It Says 49
    OSHA Regulation In Disguise 49
    The Preemption Decisions 51
CONGRESS MEANT TO PREEMPT STATE CRIMINAL LAWS 52
    Conscious Rejection 53
    The Act's Savings Clause—Section 4(b)(4) 54
    Much Ado About Little 55
CONCLUSION 56
ENDNOTES 56

CHAPTER II: High-Risk Notification 61
    'THOUSANDS BEING MADE SICK' 61
        Government Urged to Reveal Risks 61
        State Initiatives 62
    SOME PROGRAMS ALREADY PLANNED 62
        Reasons for DOE Concern 63
    LEGISLATIVE OUTLOOK DIM 63
        No Position Yet by Bush 63
        'Unnecessary, Costly Bureaucracy' 64
        Existing Standards Said Sufficient 64
        Unions See Gaps in Current Regulations 65
    'PIECE' WORK PROGRAM 65
        'Different Priority' by Labor 66
    PROVISIONS OF METZENBAUM BILL 66
        Elements of Notification 66
        Board Membership 67
    MEETING CONCERNS ABOUT LAWSUITS, COSTS 67
        Exemptions for Small Employers 68

ISSUE 'HAS NOT GONE AWAY' 68
    Likelihood of Floor Vote Questioned 68
    House Measure Expected 69
NIOSH RISK NOTIFICATION 69
    34 Candidate Studies 69
    Means for Notifying Workers 70
    Reasons for Not Notifying Workers 71
GOING TO COURT WITH NOTIFICATION 71
    'An Adversarial Relationship' 72
    State Authorities Intervene 73
    Basis for Court Action 73
    'Delay ... Doesn't Wash With This Court' 74
    Mesothelioma Found 75
    'How Much Do You Spend to Inform?' 76
    Six Lawsuits Filed, More Expected 76
    'Possibility of Going Back to Court' 77

CHAPTER III: The Construction Industry 79
    NEW CONSTRUCTION OFFICE 80
        Workings of the New Office 80
        'After the Fact' 81
        Better Inspection Targeting Urged 82
    NEW OSHA CONSTRUCTION UNIT PROPOSED 83
        Recommendation for 'Safety Specialist' 84
        Stronger Reporting Requirements, Penalties 85
    UNIONS 'PLEASED,' BUT SEPARATE AGENCY URGED 85
    EMPLOYERS SAY WORKER ACCOUNTABILITY NEEDED 86
        Industry Opposes Safety Specialist Idea 87
    OTHER LEGISLATIVE PLANS 87
    HAZARD COMMUNICATION 88
        Requirements for Employers 88
        Training Requirement Problems Seen 89
        Data Sheets: 'Regulatory Twilight Zone' 89
        What Inspectors Will Seek 90

CHAPTER IV: New Chemical Exposure Standards 93
    GENERIC STANDARDS 94
        Generic Efforts To Continue 94
    THE PEL PROJECT 95
        Short-Term Limits Imposed 95
        Skin Notations Added 96
        Deferred Decisions 96
        'Most Commonly Used Substances' 96
        Limits Selected 97
        Generally Accepted Limits 97
        160 Substances Not Addressed 98

   Completion In 'Near Future'? 98
   Industries Affected 99
   Health Benefits 99
   1989 Compliance Date 100
   Agency Assistance Available 100
   Transitional Limits Enforced 101
   Legal Challenges 101
   New Limits 102
   Early Reaction 103
   'Rubber-Stamping' Industry Levels 103
   Controversial Substances 104
   Mixed Compliance Success Expected 108
   'Arbitrary Decision' 108
   'It Will Put Some ... Out of Business' 109
   Unions To Press For More Standards 109
   Labor Priorities 110
  EXPOSURE MONITORING, MEDICAL SURVEILLANCE 110
   Potential Model Provisions 111
   Doubts About Schedule 111
   Small Businesses at a Disadvantage? 112
   'More Potential Harm Than Benefit' 112
   Unions: 'Key to Protecting Workers' 113

CHAPTER V. Recordkeeping 115
  INDUSTRY, LABOR COMPLAINTS 115
  THE KEYSTONE REPORT 116
   Pilot Studies on Forms 117
   Avoiding 'Future Shock' 117
  PREPARATIONS ON NEW GUIDELINES 118
   'Restricted Work Activity' Criticized 119
  'OUTREACH' PLANNED 119
  CHECKING RECORDS' ACCURACY 120
  MODELS FOR NEW LEGISLATION 121

CHAPTER VI: Regulating Blood-Borne Diseases 123
  'NO KNOWN CURE' 124
   'A Panic About AIDS' 124
  OSHA INSPECTION POLICY 125
   Protective Clothing 126
   Housekeeping Requirements 127
   Tags and Warning Signs 127
   Citing Section 5(a)(1) 128
   Testing After Possible Exposure 129
   Training for Workers 129
  PROPOSED FINES TOTAL $152,173 129
  OSHA'S BLOOD-BORNE DISEASE PROPOSAL 130

INDUSTRY QUESTIONS OSHA ROLE 131
    Paperwork, Liability Concerns 132
UNIONS SEEK MORE PROTECTION 133
    Additional Union Funding 133
    Workers To Monitor Employers' Programs 134
    Contract Provisions Recommended 135

CHAPTER VII: The Right to Act 137
BUILDING ON RIGHT-TO-KNOW 137
    Current UAW/Ford Program 138
    Safety Committee Requirements Abroad 139
WHISTLEBLOWER PROTECTION 140
    Disrupt Existing Protections? 140
'MANAGEMENT HAS THIS RESPONSIBILITY' 141
CONNECTICUT BILL STALLED 141
    Inspection, Hearing Process 142
    'Existing Laws Protect Employees' 142
    Plans for 1990 143
    Building Grassroots Support 143

APPENDIX A "Getting Away With Murder in the Workplace: OSHA's Nonuse of
Criminal Penalties for Safety Violations," Report by the
House Government Operations Committee, Oct. 4, 1988 . . . . A-1

APPENDIX B Letter from Thomas M. Boyd, Assistant Attorney General for Legislative and
Intergovernmental Affairs, U.S. Department of Justice, to Rep. Tom Lantos (D-Calif) on
DOJ's Position on Federal Preemption of State Prosecutions, Dec. 9, 1988 . . . . B-1

APPENDIX C *State of Illinois v. Chicago Magnet Wire Corp.*, Feb. 2, 1989 . . . . B-1

APPENDIX D State Prosecution Contacts D-1

APPENDIX E Memorandum from John A. Pendergrass, Assistant Secretary, Occupational
Safety & Health Administration, Regarding Cooperation in
State or Local Criminal Prosecutions E-1

APPENDIX F High Risk Occupational Disease Notification and Prevention Act (S 582) F-1

APPENDIX G Bill to Amend the OSH Act of 1970 to Improve Regulation of the
Construction Industry (S 930) G-1

APPENDIX H Proposed OSHA Rules:
Generic Standard for Exposure Monitoring [53 FR 37591, Sept. 27, 1988] H-1
Medical Surveillance Programs for Employees [53 FR 37595 Sept. 27, 1988] . . . . -H-7

APPENDIX I Bureau of Labor Statistics Guide to Recordkeeping Requirements for
Occupational Injuries and Illnesses, June 1986 . . . . I-1

APPENDIX J OSHA AIDS Inspection Directive, August 15, 1988 . . . . J-1

APPENDIX K Employee Health and Safety Whistleblower Protection Act (S 436) K-1

APPENDIX L Connnecticut Right To Act (House Bill No. 7346) L-1

INTRODUCTION

When the Occupational Safety and Health Act (OSH Act) was signed into law on Dec. 29, 1970, it was widely hailed as a landmark measure that would ensure safe working conditions for millions of Americans by replacing a patchwork of largely ineffective state laws with a strong, consistent federal regulatory program.

In mid-1989, as the 20th anniversary of the act approaches, many experts say that early promise has not been fulfilled.

Bureaucratic red tape, insufficient funding, a lack of consistent direction, and enforcement cutbacks under the Reagan administration have seriously hindered efforts by the Occupational Safety and Health Administration (OSHA) to set and enforce effective regulations, according to union officials, public interest groups, legislators, state and local authorities, and many legal and medical experts.

These problems are compounded by basic shortcomings in the act itself, including weak penalties for violations and the absence of provisions that would give workers more say in running workplace safety programs, the critics also say.

Not surprisingly, the issue has divided business and labor. The latter says the federal government should enact new laws to correct current inadequacies. Also needed as U.S. industry enters the 1990s, these critics say, are new OSHA standards for chemical hazards and emerging problems in the growing high-technology and service industries.

But industry spokesmen note that U.S. employers spend millions of dollars every year on protective equipment for workers, controls for hazardous processes and machinery, education and training, and other safety and health measures. For the most part, these officials disagree that major legislative changes are needed to meet the concerns voiced by union critics. In many cases, they say, OSHA and other agencies already have the necessary power to take administrative steps to address these concerns.

Business officials also say they are aware of the need to address emerging health and safety concerns. Already, companies are investing money to research these problems and come up with solutions, they note. But new OSHA standards could create more problems than they solve, these observers say.

INITIATIVES IN SEVEN AREAS

As America enters the 1990s, debate over these issues will center on a series of bills and proposed standards already under consideration by Congress, OSHA, and some states, observers say. In the 1990s, these measures would impose significant new responsibilites on industry and the government in seven areas:

- Criminal prosecution. Bills have been introduced that would significantly increase penalties for criminal violations of the OSH Act.

Copyright © 1989 by The Bureau of National Affairs, Inc.

- High-risk notification. A current Senate bill would require the federal government to notify workers determined to be at high risk of serious disease as a result of past exposure to hazardous substances in the workplace.
- Construction safety. Measures pending before the Senate and House would intensify OSHA's regulation of construction sites.
- Chemical exposure. OSHA is considering new regulations that would impose broad requirements for monitoring worker exposure to hundreds of hazardous substances.
- Recordkeeping. Concerns by labor and industry alike have resulted in plans to overhaul current requirements for reporting work-related injuries and illnesses.
- Blood-borne diseases. OSHA is proposing new regulations to protect health care employees from the risk of infection from acquired immune deficiency syndrome and other blood-borne infectious diseases.
- Right-to-act. Labor unions are seeking federal and state legislation that would give workers the right to refuse to perform jobs they think would expose them to the risk of serious injury.

Advocates of these measures say current job-related injury and illness statistics bolster their position. According to the National Institute for Occupational Safety and Health (NIOSH), some 7,000 people die every year in the United States in job-related accidents.

Other studies indicate the total may be as high as 20,000 deaths a year, Margaret Seminario, associate director of occupational safety, health, and social security for the AFL-CIO, told BNA.

"Most of these fatalities could have been prevented," Seminario maintained.

No reliable figures on occupational illnesses exist because such disorders often are not reported, most experts agree. However, a 1989 study by Mt. Sinai School of Medicine in New York, extrapolating from disease rates in New York state, estimated that job-related diseases may account for as many as 50,000 to 70,000 deaths annually.

Such figures show that "not only can we do better [in controlling workplace hazards], we have to do better," says Irving J. Selikoff, director emeritus of the Division of Occupational and Environmental Disease at Mt. Sinai and an expert on asbestos disease.

STRONGER PENALTIES URGED

Endorsing legislation to set stronger penalties for health and safety violations, labor officials and some legal experts note that the OSH Act currently sets maximum civil and criminal fines of $10,000. The maximum jail sentence for a criminal violation is six months.

These penalties are too low to deter violations and to give prosecutors an incentive for aggressively pursuing criminal charges for accidents involving worker deaths, critics say.

Because there are too few OSHA compliance officers to inspect every U.S. workplace, right-to-act legislation is needed to give workers more authority to protect themselves against workplace hazards, union officials say. A high-risk notification law is necessary because workers generally may be unaware of long-term disease risks from workplace chemical exposures that occurred years or decades ago, these officials also contend.

Industry Sees 'Administrative Inertia'

For the most part, industry officials say these measures are not needed because channels already exist for addressing such concerns.

"It's very possible to say that certain things in OSHA are not working as well as they could, but I'm not persuaded that translates into a need to revise the act," Kyle Olson, a health and safety specialist with the Chemical Manufacturers Association, told BNA.

In many cases, the complaints raised by unions are generated more by "administrative inertia" on the part of OSHA rather than by inherent problems in the OSH Act itself, Olson said.

Scott Railton, a Washington, D.C., attorney who has represented numerous industry clients in safety and health cases, agrees that OSHA can do more administratively if enough pressure is applied. For example, in 1986, OSHA began to get around the cap on penalties under the act by using more creative means to calculate fines, he says.

"The statutory mechanism is there," Railton said May 19, 1989, at a BNA conference on occupational safety and health, noting that OSHA has sought penalties of more than a million dollars against some employers in the past three years.

Similarly, many industry officials say, concerns about high-risk notification are satisfied by current OSHA regulations requiring employers to inform workers about the hazards of workplace chemicals. Business officials also point to an existing provision in the OSH Act that make it illegal for an employer to fire or otherwise reprimand an employee for filing a safety complaint with OSHA. This provision makes further right-to-act legislation unnecessary, they say.

Unions Push for Generic Standards

But union officials say that beyond legislative change, improvements are needed in OSHA's standards-setting program. For example, they say, OSHA needs to find ways to more quickly regulate the growing number of chemicals used in the workplace.

Rather than continue regulating most chemicals on a substance-by-substance basis, OSHA should make increasing use of "generic" standards that set broad requirements for groups of chemicals, these critics say.

OSHA also should begin to regulate hazards like cumulative trauma disorder that will pose risks for increasing numbers of workers in the next decade, as more employees enter clerical and service professions, according to some union officials.

Similarly, with the number of cases of AIDS in the general population expected to more than double by the end of 1991, health care workers need a standard to protect them from a growing risk of AIDS infection, unions say.

Employers Seek Flexible Standards

Industry officials say they already are trying to address emerging hazards such as carpal tunnel syndrome and blood-borne infectious diseases by funding research and implementing voluntary safety programs. "Generally speaking, most employers realize that safety and health in the workplace contributes to the bottom line," Donna Costlow, associate director of risk management for the National Association of Manufacturers, told BNA.

New OSHA standards to regulate emerging hazards may be premature because the agency may not be equipped yet to deal properly with hazards that fall outside its traditional realm of expertise, Costlow and other industry representatives say.

These officials also question the usefulness of generic standards in some instances. For example, a generic rule requiring employers to monitor workers' exposure to certain chemicals may impose unnecessary costs on companies where those substances are present in low concentrations that pose no risks to workers, according to Edward Burkeen, a spokesman for the American Iron and Steel Institute.

Needed instead, according to industry officials, are more flexible standards to give employers greater leeway to address particular hazards in the individual workplace.

BEYOND THE REAGAN ERA

Despite industry opposition, advocates of stricter safety regulation say such initiatives can be enacted by Congress or adopted by OSHA now because the public has become highly skeptical about the ability and will of industry to control workplace hazards effectively without strong oversight by government agencies accompanied by labor union and citizen group monitoring.

Although Ronald Reagan won the presidency in 1980 partly on the basis of his promise to reduce federal regulation, developments in the mid- and late 1980s made it too often clear that approach jeopardized the safety of workers, according to critics.

These developments included the December 1984 toxic chemical leak in Bhopal, India, that killed 3,000 people; the April 1987 building collapse in Bridgeport, Conn., that killed 28 construction workers; and widely publicized reports alleging serious hazards in some manufacturing and service industries.

"De-regulation means de-control," Selikoff said Jan. 21, 1988, at a meeting in Washington, D.C., on occupational health in the 1990s, sponsored by the New York

INTRODUCTION

Academy of Sciences. "[Without government regulation], we do not have the social organization necessary to address [health and safety] problems."

'Old Days are Gone'

Industry spokesmen say that such criticisms are too simplistic, and that corporate executives generally want to maintain safe workplaces regardless of whether regulatory pressures exist. But many agree that public concern about safety has grown.

They note that after the Bhopal disaster, federal regulators and state and local governments faced a groundswell of pressure from unions and community groups to enact right-to-know regulations requiring companies to notify workers and local residents about the hazards of chemicals used at work sites.

This concern will carry over into the 1990s and the presidency of George Bush, according to observers like Richard F. Boggs, vice president of Organization Resources Counselors Inc., a Washington, D.C., consulting firm that provides technical advice on health and safety matters to some 60 leading U.S. corporations.

"Every citizen is so much more concerned now," Boggs said. "The old days are gone."

HIGHLIGHTS

This Special Report examines the seven issues that experts say will dominate debates over occupational safety and health regulation in the coming year and beyond. The report includes these highlights:

- Over the next three years, the number of criminal cases filed against companies and business executives for fatal workplace accidents may climb into the hundreds as a result of recent legal and administrative actions giving federal and state prosecutors greater incentives to pursue such charges, according to experts.

- As Senate advocates of high-risk notification re-introduce a controversial bill on the issue, some notification programs already have been launched or are being planned by federal and state agencies. A Minnesota state court has ordered a company to turn over records needed by the state Health Department for a local worker notification effort, and two federal agencies plan to proceed with their own programs.

- In the wake of the Bridgeport disaster, Connecticut's congressional delegations have introduced bills to require employers to develop formal safety programs for construction sites and to give workers added authority to demand that hazards they find be corrected.

- Having issued an unprecedented regulation setting more-stringent exposure limits for more than 400 chemicals, OSHA is considering additional rules that would require employers to monitor workers' exposure to the chemicals and to conduct medical surveillance programs to detect harmful effects from exposure.

- Fines totaling $12 million have been proposed against employers in the past three years for alleged OSHA recordkeeping violations. The government now plans to impose new requirements in the next two years through major changes in recordkeeping forms and regulations.

- OSHA is developing a standard that would require hospitals, medical laboratories, companies in general industry, and other employers to protect health-care employees from workplace exposure to AIDS and other blood-borne diseases, and the agency has proposed $152,173 in penalties against employers for alleged failure to follow existing infection-control guidelines.

- Building on the right-to-know movement and some current collective bargaining programs that mandate joint labor-management safety committees in the workplace, unions are lobbying in favor of a bill in Congress and legislative proposals in two states to give workers the right to refuse to perform tasks they believe would put them at risk of serious injury.

ACKNOWLEDGEMENTS

This report was prepared by the Special Projects unit of The Bureau of National Affairs Inc., under the direction of Drew Douglas, managing editor. BNA senior staff editor Fred Blosser served as editorial coordinator.

Senior staff editor Roger Feinthel served as copy editor. Staff editor Gwen Moulton served as production editor. Researcher Loretta Kotzin helped prepare the appendices.

Chapter I was written by Blosser and Special Projects staff editor Steven Teske. BNA staff correspondents Martha W. Kessler, Terry Hammond-Smith, and Mark Wolski also contributed, as did BNA legal staff editor Marcia Mjoseth, and free-lance editor Ann Allen. Graphics staff Virginia Wheaton and Robert T. Savidge of BNA's Surveys Unit designed the graphics. Also contributing were the BNA Library staff.

The principal authors of the legal discussion in Chapter I were attorneys George H. Cohen of Bredhoff & Kaiser in Washington, D.C., and Robert C. Gombar of the Washington, D.C., office of Jones, Day, Reavis & Pogue.

Chapter II was written by Teske and BNA staff correspondents Bebe Raupe and Wolski. Chapter III was written by Pamela M. Prah, staff editor of BNA's *Occupational Safety & Health Reporter (OSHR)*, and Blosser. Chapter IV was written by *OSHR* staff editor Patricia A. Logan. Chapter V was written by Blosser; Chapter VI by *OSHR* staff editor Lisa L. Myrick; and Chapter VII by Blosser and BNA staff correspondents Mary Chapman and Kessler.

I. Criminal Prosecutions

It was sleeting in Minneapolis in the early spring of 1988 when Jana Studelska's telephone rang. A trembling voice said her husband had been in an accident at the concrete block plant where he worked, but the caller said no more.

"I pretty much knew at that point that he was dead," Studelska told BNA.

The rest of that April 26—when John L. "Butch" Studelska, 32, died on the shop floor under an avalanche of tons of sand and gravel from a collapsed storage hopper—is little more than a blur, the 25-year-old widow recalled.

She said she vaguely remembers arriving at the plant and noticing several police cruisers nearby; helicopters from several news organizations whirled overhead. "I've blocked out a lot of it," she said.

But she recalled handing her three-month-old son, Sam, to a bystander while she tried to enter the factory, then panicking later when she forgot what she had done with the child.

Today, more than a year after her husband became Minnesota's seventh job-related fatality of that year, Jana Studelska's panic and pain persist. And now she carries yet another burden: An increasing sense of outrage with her husband's employer, with the state and federal governments, and with a national occupational safety system she feels is so weak that it fails to protect workers adequately and a judicial system so tilted toward business that it serves as no deterrent.

"As far as I'm concerned, Butch was basically murdered. He did nothing wrong except go to work, punch in, and do his job."

The accident that killed John Studelska could have been prevented, according to state investigators. The two-story, 30-year-old bin at Oscar Roberts Concrete Products Co. was in disrepair long before it collapsed under a load of more than 250 tons, its steel structure eaten away by calcium chloride and steam used to keep its contents from freezing, they said.

When an official from the Minnesota Occupational Safety and Health Division arrived at the plant later that day, he also questioned whether the bin had been overloaded, according to the official accident report.

"It is my opinion this accident could have been prevented with routine maintenance, periodic tests of structural integrity, and establishment of loading and operating procedures," the inspector, Sherrill Benjamin, wrote. He also "strongly recommended" that the company develop better safety, training, and loading procedures.

Benjamin originally cited the company for willfully not maintaining the bin "free from recognizable hazards associated with deterioration and corrosion," according to the accident report. Under Minnesota law, a willful citation means a company knows a condition is dangerous but does nothing to correct it, Roy Miner, senior safety engineer at MOSHD, told BNA.

Five months after the fatality, the concrete company, without acknowledging any wrongdoing, paid a negotiated fine of $750 to settle the state's citation against it. The state had reduced the citation to the "serious" category, meaning the condition existed but the company was not aware of the danger it posed, Miner said.

The firm, which has not been charged criminally, strongly asserts it did nothing wrong. "We're totally not at fault. [The accident was] one of those things, an act of God," Richard W. Smith, safety director for the company, told BNA in a telephone interview. He declined to discuss the case further, citing a lawsuit brought by Jana Studelska against the alleged owner of the building, Oscar Roberts Co.

Throughout the country, an increasing number of ordinary people, families of victims killed on the job, members of Congress and state legislatures, public policy and labor activists, and state prosecutors are beginning to share at least part of Studelska's anger and concern.

Experts predict state and local officials, armed with new court rulings and their own outrage at perceived federal inaction, dramatically will increase the number of criminal cases brought against employers and corporate officials.

Nationally, 17 such cases are pending in state courts. But, that number will climb to "a couple of hundred" within the next three years, said Joseph Kinney, executive director of the National Safe Workplace Institute, a non-profit research organization.

CHAPTER OVERVIEW

This chapter examines developments behind increasing state efforts to prosecute workplace death cases. It also highlights problems in federal investigations and prosecutions, and what many say are serious problems with the federal occupational safety and health laws themselves. The chapter originally appeared as a BNA White Paper in April 1989.

In addition, the chapter focuses on the cost—both financial and emotional—borne by the survivors of people killed on the job. State legislative initiatives to strengthen prosecutors' hands also are examined. Finally, the chapter looks at trench cave-ins, an all-too-common, yet easily prevented, workplace killer.

The second portion of the chapter contains a point-counterpoint discussion of federal preemption, a major legal issue surrounding workplace death prosecutions. The principal authors are attorneys George H. Cohen and Robert Gombar. (See also Appendices A through D.) In three months of reporting for this chapter BNA interviewed more than 125 prosecutors, private attorneys, labor and industry officials, and others to determine the magnitude of this workplace safety problem and what state and federal officials are doing to solve it. More than 1,000 pages of documents were reviewed.

Among BNA's findings:

- More than 100,000 workers may have died in job-related accidents since 1984 alone. While the numbers of workplace deaths vary and the number of negligent deaths is not known, no corporate officials have served jail sentences for federal violations resulting from an on-the-job fatality. Experts say just six have served time for state convictions.

- In all but four states, restrictive worker's compensation laws can prevent survivors from suing negligent employers, leaving families to get by on state and Social Security payments—which usually total substantially less than the dead worker's paycheck.

- Civil penalties under the federal Occupational Safety and Health Act, already limited, often are reduced drastically after negotiations; a similar situation exists in many states. As in the Studelska case, penalties of just a few hundred dollars are not uncommon.

- Criminal prosecutions of workplace deaths probably will remain a low priority under the Bush administration. Slashed under the Reagan administration, money for the Justice Department unit responsible for bringing charges in workplace deaths could be cut further under a Bush proposal to beef up funding for prosecution of obscenity cases.

- Many state prosecutors pursue job fatality cases aggressively, unlike their federal counterparts. Since 1980, Justice Department officials have chosen to prosecute only four of 30 criminal cases referred by the Occupational Safety and Health Administration, while local prosecutors have taken at least 50 similar cases in the past five years. A Justice decision not to pursue criminal charges also can stifle a state prosecution, sending it "down a black hole," one expert said.

- A lack of coordination between officials at OSHA and the Justice Department has delayed investigations and prosecutions. In some cases, officials have taken as long as three years to decide whether to prosecute or drop a case; local prosecutors, on the other hand, routinely bring charges within a few months.

- According to an internal OSHA memorandum, federal investigators have been discouraged from discussing cases with state officials. In some cases, local district attorneys contacted by BNA said they were unaware that federal officials had decided not to prosecute a case.

- Justice officials turn down many cases because OSHA fails to provide sufficient information to support criminal charges, while Labor

Department officials say they have no "leverage" to expedite Justice's review of cases.

- A perception exists, at least among some Justice Department officials, that most workplace death prosecutions do little to enhance a prosecutor's career. To win promotions, former officials told BNA, department lawyers feel they must take cases with a higher profile — such as drug crimes or white-collar prosecutions. "My instinct is that federal prosecutors want to get the biggest bang for the buck, so they want to pursue the highest visibility cases," said one expert.

'THE TOLL IS FAR TOO HIGH'

Figures vary on the number of people killed or injured in workplace accidents each year in the United States. The Bureau of Labor Statistics places the number of deaths at 3,500, while the National Safety Council estimates about 11,000. The council also estimates that more than 1 million workers are injured on the job annually.

On the basis of information from death certificates from all parts of the country, the National Institute for Occupational Safety and Health estimates about 7,000 workplace fatalities occur each year due to traumatic injuries. "But we know that's an underestimate," NIOSH spokeswoman Diane Porter told BNA.

The AFL-CIO says the rate may be even greater — more than 20,000 a year. "The toll is far too high," said Margaret Seminario, the AFL-CIO's associate director of safety, health, and social security. "Most of these fatalities could have been prevented."

One example is 61-year-old Stefan Golab, who died of exposure to cyanide used to extract silver from exposed X-ray film at a Chicago firm, Film Recovery Systems, Inc. Investigators said Golab and other workers, most of them illegal aliens who understood little or no English, worked routinely near unvented vats containing the potent poison.

In 1985, three Film Recovery executives were sentenced to 25 years in prison and fined $10,000 each on state convictions of murder and reckless conduct in the case. They have appealed.

Though rare, such convictions are cheered by survivors angered when employers avoid responsibility. Many corporations simply walk away, paying small fines and relying on insurance companies to deal with any "third-party" lawsuits or other litigation, they say.

Legal Barriers Fall, But Slowly

Since the middle of the 19th century, U.S. courts slowly have broken down legal barriers preventing corporations from being held criminally liable for their acts. Before then, the law held that corporations had no mind or soul and thus could not plan to commit a crime, according to William J. Maakestad, associate professor of business law at Western Illinois University.

In 1904, for example, the state circuit court for the southern district of New York let stand an indictment of manslaughter against a firm after almost 900 people drowned when a steamship owned by the company caught fire. Within the next few decades, courts further broadened corporate liability by allowing corporate officers to be charged with workplace crimes.

More recently, *Indiana v. Ford Motor Co.* was a watershed event for prosecutors seeking to hold companies accountable for their actions, Maakestad told BNA. In that 1979 case, Ford was charged by the state of Indiana with three counts of reckless homicide for allegedly designing and building an unsafe car, the Pinto. The case was brought after three teenagers died in a fiery explosion when the gasoline tank of a Pinto ruptured after the car was struck from behind by another automobile.

Although the automaker was acquitted after a long trial, the case "helped define for the public and prosecutors the kinds of crimes companies could be prosecuted for," Maakestad said.

"The case was a real turning point," he told BNA. "Here was the fourth largest company in the world being brought to a small court in southern Indiana to face charges. The case served a whole educational function" for the public and prosecutors.

No similar precedent has been set in federal court, however. Adding to the confusion is the unanswered question of whether OSHA criminal statutes preempt state prosecutions.

Besides delaying more than a dozen prosecutions in state courts, the question deters state prosecutors who otherwise would bring more new cases, experts said.

The preemption argument involves "a lot of competing interests" for the Bush administration, including the question of states' rights, Acting Solicitor of Labor Jerry Thorn told BNA.

"And you know how that plays politically in a Republican administration," Thorn said. On the other hand, "A lot of businesses are saying that they would rather answer to one agency than to 50 different state prosecutors."

Green Light for State Prosecutions

In recent months, however, new light has been shed on the preemption issue—the result of two legal developments that many experts told BNA will lead to more state prosecutions.

On Dec. 9, 1988, in a letter to the House Government Operations Committee, Justice said for the first time that it found "nothing in the OSH Act or its legislative history which indicates that Congress intended for the relatively limited criminal penalties provided by the act to deprive employees of the protection" of state laws.

(continued on page 13)

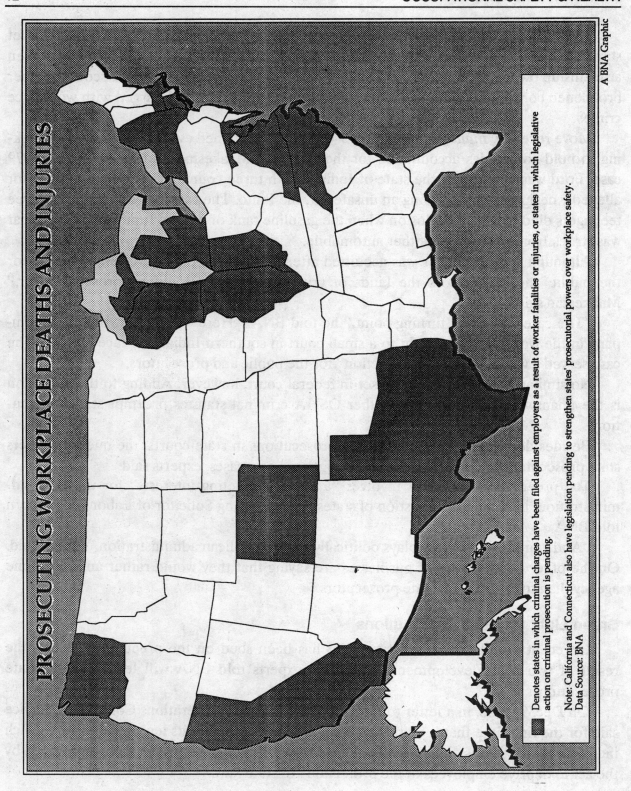

"If the Department of Justice says that federal law does not preempt state workplace prosecutions, that will undermine the argument that preemption exists," said W. William Ament, a consultant with Organization Resources Counselors Inc., a Washington, D.C., technical consulting firm whose clients include several *Fortune* 500 companies.

OSHA will not take a formal position on the preemption question until such a case goes to the U.S. Supreme Court, Thorn said. That case may be *Illinois v. Chicago Magnet Wire Corp.*, in which the Illinois Supreme Court ruled Feb. 2 that federal preemption does not preclude the prosecution of five executives alleged to have exposed workers to toxic substances.

Local prosecutors across the country applauded the state high court's decision. "A lot of effort and attention has focused on the *Magnet Wire* case. It would have been a serious blow" had the court ruled the preemption argument was valid, E. Michael McCann, district attorney for Milwaukee County, Wisc., told BNA. "You'll definitely see a lot more of these cases."

Industry Cites Potential for Abuse

Business representatives told BNA that prosecutions under state criminal laws that are tougher than the federal act may be warranted. "In some cases, we would agree that it was not the intent that the state criminal code would be preempted," said Susan Spangler, director of risk management for the National Association of Manufacturers. But business executives remain concerned that a reversal of the preemption argument could lead to "abuse" of prosecutorial discretion by local officials, she said.

State prosecutors should be able to bring charges only in cases involving "intentional conduct [by employers] that harms workers"—for example, "taking labels off containers of toxic chemicals, knowing employees were working with [those] substances," Spangler said.

Prosecutors discount such concerns. "The criminal justice system is organized so that defendants have enormous procedural protection against a willy-nilly exercise of authority," Kenneth Oden, prosecuting attorney for Travis County, Texas, told BNA.

MOST WORKER'S COMPENSATION LAWS BLOCK LAWSUITS

The families of workplace accident victim face further frustration—and often real financial hardship—because of the problem of civil redress.

Every state has laws under which employers, through worker's compensation insurance programs, pay to support spouses and minor children of workers killed on the job. In most cases, worker's compensation benefits amount to two-thirds of the worker's salary, supplemented with Social Security benefits if the surviving spouse has dependent children living at home.

(continued on page 15)

Pending Appellate Litigation

State	Defendant(s)	Trial Court	Appellate Court	Highest State Court	U.S. Supreme Court
Illinois	Chicago Magnet Wire Corp., Frank Asta, Gerald Colby, Anthony Jordan, Kevin Keane, Allan Simon	Cook County Court dismissed 12/13/85 aggravated battery, conspiracy, and reckless conduct charges, finding criminal prosecution preempted by OSH Act.	Illinois Appellate Court, First Judicial District, ruled 6/29/87 that OSH Act preempts criminal prosecution. State appealed.	Illinois Supreme Court ruled 2/2/89 that OSH Act does not preempt criminal prosecution. Defendants appealed.	State Supreme Court's decision stayed, pending filing of petition for certiorari with U.S. Supreme Court due by 5/3/89.
Illinois	Film Recovery Systems Inc., Metallic Marketing Systems, Steven O'Neill, Charles Kirschbaum, Daniel Rodriguez	Cook County Court 6/14/85 convicted officers of murder and reckless conduct and the company of involuntary manslaughter. Defendants appealed.	Appeal argued before Illinois Appellate Court, First Judicial District, 7/7/87. Scheduled for reargument 5/2/89.		
Michigan	Patrick Hegedus	Oakland County Circuit Court 11/4/86 granted defendant's motion to dismiss involuntary manslaughter charges. State appealed.	Michigan Court of Appeals ruled 6/6/88 that criminal prosecution preempted. State appealed.	Appeal argued before Michigan Supreme Court, 3/8/89. Under advisement.	
Michigan	General Dynamics Land Systems, Inc.	Michigan District Court for 37th District dismissed 10/23/84 charges of involuntary manslaughter, finding insufficient evidence. Michigan Circuit Court for McComb County affirmed the ruling. State appealed.	Michigan Court of Appeals reversed the dismissal 3/20/89, finding conflicting evidence warranting a trial. Defendant can either ask for a rehearing or appeal the ruling to the Michigan Supreme Court.		
New York	Pymm Thermometer Inc., Pak Glass Machiner Corp., William Pymm, Edward A. Pymm	Kings County Jury 11/13/87 convicted defendants of assault with mercury. Judge set verdict aside. State appealed.	New York Supreme Court, Appellate Division, heard state's appeal 2/27/89. Under advisement.		
Texas	Peabody Southwest, Inc.	Travis County Court 6/16/87 fined defendant $20,000 on defendant's no contest plea to negligent homicide charges. Defendant appealed.	Appeal pending before Texas Court of Appeals.		
Texas	Sabine Consolidated Inc., Joe Tantillo	Travis County Court 2/13/87 fined each defendant $10,000 on defendants' no contest plea to negligent homicide charges. Defendants appealed.	Texas Court of Appeals, Third District, ruled 8/31/88 that OSH Act preempts criminal prosecution. State appealed.	Decision appealed to Texas Court of Criminal Appeals. Case pending.	
Wisconsin	Larry Norman Cornellier, Sr.	Rock County Circuit Court 5/7/87 denied defendant's motion to dismiss charges of homicide by reckless conduct. Defendant appealed.	Court of Appeals, District IV, ruled 4/21/88 that OSH Act does not preempt criminal prosecution. Defendant bound over for trial.		

Data Source: BNA

A BNA Graphic

CRIMINAL PROSECUTIONS

However, in return for this coverage families are barred from suing an employer unless the death resulted from a deliberate act on the employer's part. In essence, worker's compensation programs are a "quid pro quo" system in that the families of workers killed are assured of income support while employers are assured of limited liability for workplace accidents, Eric J. Oxfeld, an attorney on the American Insurance Association's Worker's Compensation Committee, told BNA.

Only in Louisiana, South Dakota, Texas, and West Virginia have recent court rulings relaxed those constraints enough to allow survivors to file civil suits for deaths resulting from allegedly reckless or negligent disregard for safety, according to Deborah A. Ballam, assistant professor of business law at Ohio State University.

In the remaining 46 states, the rules are so strict that to win a suit the survivors "practically have to show that [the employer] put a gun to [the worker's] head and pulled the trigger," Donald Elisburg, administrator of the Occupational Health Legal Rights Foundation, a non-profit legal aid organization for workers, told BNA.

The situation dismays many prosecutors. "All the assumptions I'd had, that there was seldom criminal misconduct on the part of business organizations, and if there was, there were [civil sanctions to hold them accountable], turned out to be wrong," Oden said.

The 'Third-Party' Lawsuit Option

Although survivors at first may not do "too bad financially," the worker's compensation benefits in the majority of states are not adjusted for inflation, so benefits "become very inadequate in the long run," said Peter S. Barth, an economics professor at the University of Connecticut.

Survivors agree. "It's been devastating financially," said Barb Kohne, whose husband was killed in a Milwaukee sewer tunnel explosion in November 1988. Two other workers also died. "The income just stopped, period," Kohne, who has an 11-year-old daughter, told BNA.

An investigation into the accident is being conducted by the Milwaukee County District Attorney's Office. The office is awaiting the results of laboratory tests on equipment in the tunnel at the time of the explosion to determine if the case should be turned over to a board of inquest — Wisconsin's version of a grand jury, county Assistant District Attorney Kenneth Berg told BNA.

The board, usually comprising six people, will determine what course of action to take in the case, Berg said. The district attorney's office usually follows the inquest's recommendations, but it is not required to do so, he added.

In early 1989, Kohne began receiving regular Social Security survivor's benefits, and she has taken a part-time job as a supermarket cashier to support her daughter and herself. "My husband's gone and my daughter's father's gone, and everyone said, 'That's kind of too bad,'" she said. "I'd like to see some laws changed."

So would Studelska. "No one should be protected from the law when they do something wrong, especially where a life is concerned," she said. "I wish that by paying my car insurance that I would be protected from suits resulting from my negligent driving. It's the same thing, when you think about it."

Studelska has filed suit against the firm she said owned the building where her husband worked. In documents filed in the Hennepin County, Minn., District Court Oct. 14, 1988, she alleges that Oscar Roberts Co. leased space to Oscar Roberts Concrete Products Co.

Oscar Roberts Co. was negligent on two counts in the death of her husband, the suit states. The company did not "inspect, maintain and repair" the storage bin and failed to "warn and instruct" workers at the plant "of foreseeable dangers associated with the use of material storage bins," Studelska said in the lawsuit.

She is asking the court to award her at least $50,000 for her husband's death.

Oscar Roberts Co. denied the allegations in a Dec. 9, 1988, answer to the suit, saying it cannot be held responsible. John Studelska's death, the company said in court documents, was "caused by the conduct of others over whom the defendant has no control nor right to control."

Although such "third-party" tort suits can present another way for family members to try to recover punitive damages, survivors often are "unaware of [that] window of opportunity," Barth said. And even if the family is aware, they may "have a difficult time finding an attorney if the [potential] return is small," he added.

PROBLEMS IN THE OSHA SYSTEM

The OSHA law contains a fundamental flaw, critics charge: The 1970 act requires that the agency obtain a federal court order to halt an immediately dangerous practice or process occurring in a workplace.

Even when OSHA issues a citation, the act allows such operations to continue while a company challenges the citation before the Occupational Safety and Health Review Commission, a quasi-judicial tribunal independent from the agency. Decisions by the commission, in turn, may be appealed to the U.S. courts of appeals.

The AFL-CIO's Seminario and other observers compare these limitations to tougher provisions in the Federal Mine Safety and Health Act. The mine law allows inspectors, on their own authority, to shut down dangerous operations. It also requires mine operators to correct conditions cited by inspectors even if the owners challenge the citations.

OSHA inspectors should have the same authority when they find "hazards of similar magnitude," Seminario said. "Eighteen years of experience with OSHA shows that those actions are taken only in the most serious cases."

Reagan 'Self-Policing' Policy Criticized

OSHA critics also point to a 1981 Reagan administration decision to cut the agency's 1,700-member inspection force by one-third. At the same time, the administration changed its policy to exempt some companies from comprehensive inspections if company safety records show lower-than-average injury rates.

As a result, the agency has been forced to "rely on the good will of employers," Seminario said. In some cases, there have been "willful decisions by employers not to report injuries."

OSHA officials steadfastly defend the action, citing it as an efficient way of apportioning precious inspection time.

The change also enabled employers to escape inspections in which the agency might have uncovered dangerous working conditions, Kathleen Brickey, a Washington University business law professor, told BNA.

"The failure of that policy is nowhere more poignant than in the Film Recovery case," she said. The death of employee Stefan Golab occurred four months after an OSHA inspector visited the site, but limited his inspection to a review of company records.

Golab, a Polish immigrant who extracted silver from exposed X-ray films through a process involving a cyanide wash solution, died on Feb. 10, 1983, from acute cyanide poisoning, according to the medical examiner's report on the death. In finding the executives guilty of murder in the case, Cook County Circuit Judge Ronald J.P. Banks said Golab and other employees "were not properly warned of the hazards and dangers of working with cyanide."

OSHA, State Fines Seen as Too Small

The Film Recovery Systems case also highlights another problem with civil remedies used by OSHA and state agencies acting under OSHA's approval: Monetary fines levied by the federal and state agencies for safety and health violations usually are so small they do not deter employers from violating health and safety violations, critics say.

When OSHA subsequently conducted a full inspection of Film Recovery Systems after Golab's death, agency officials proposed a $4,000 penalty. "The company was able to bargain down to about $2,000," Richard Daley, now mayor of Chicago, told the House Government Operations Subcommittee on Employment and Housing in February 1988. Daley prosecuted the case as state's attorney for Cook County, Ill.

The misconduct alleged "surely merits more than a token $2,000 fine," Daley said.

Officials of OSHA's parent agency, the Department of Labor, defend the practice of negotiating with employers, saying the fines are reduced in return for companies quickly correcting hazards. Otherwise, they say, companies would be able to postpone action indefinitely while challenging an OSHA citation in court.

"We have had companies that litigated over a matter of years," Thorn said.

A similar situation exists at the state level. Terry Mueller, area supervisor for MOSHD, agreed the fine levied against John Studelska's employer was too low, but he told BNA the state agency is bound by the penalties set by legislators.

In its previous session the Minnesota legislature doubled to $20,000 the maximum fine for a willful workplace death and also raised the top fine for a serious violation from $1,000 to $2,000. Yet many workers and survivors think that is not sufficient, Mueller said.

"You can't put a price on a human life," he said. Penalties "could be $2 million and it still would not be enough to satisfy some people."

CONGRESS ALLEGES 'INSTITUTIONAL RELUCTANCE'

Many of the problems that hamper federal civil enforcement in workplace deaths also are present in the federal criminal system, according to safety experts and congressional critics of OSHA.

In its 1988 report, the House Government Operations Committee said the small number of cases brought by federal prosecutors is evidence of an "institutional reluctance" by federal agencies to pursue criminal charges against employers.

"OSHA's record in referring cases for criminal action is dismal," the committee said. "Part of the problem is that OSHA 'cannot' and part is it 'will not'."

"There is a need for OSHA to be more aggressive and timely in using available criminal sanctions," the committee said. "Unless the OSH statute is beefed up and vigorously enforced by OSHA to punish criminally those who show willful disregard for worker safety, some employers will continue 'to get away with murder'." (See Appendix A for full text of the committee report.)

'EVEN IF YOU GET A CONVICTION, SO WHAT?'

Former federal officials and others also told BNA that another reason the government has brought so few cases is a perception that prosecution of workplace deaths does little to enhance prosecutors' careers.

The maximum jail sentence under the act for a workplace death is six months. Thus, many Justice Department attorneys feel that "even if you get a conviction, so what?" a former federal official told BNA.

"Penalties are the same for criminal violations as for some civil violations," Kathleen Brickey, a professor of business law at Washington University in St. Louis, told BNA. "This provides very little incentive for prosecutors."

"It's such a minimal criminal sanction" that federal prosecutors often view it as "ridiculous," said Eula Bingham, a professor of environmental health at the University of Cincinnati, who headed OSHA from 1977 to 1981, when 21 cases were referred to the Justice Department.

Federal attorneys would pursue criminal charges more aggressively if they "think they can get some recognition out of [OSHA cases], if it's a priority for the administration," Bingham told BNA.

Justice Department officials deny that suggestion. "I like to think that prosecutors are not making a choice based on the publicity value," James Reynolds, deputy chief of the general litigation section in the department's Criminal Division, told BNA. The unit handles cases referred by the Labor Department.

Reynolds acknowledged, however, that "relatively low maximum fines" set under the Occupational Safety and Health Act may "deter some prosecutors" who have "more potential business than time.... It's a selection process."

Longer Jail Sentences Needed, Critics Say

Congress eliminated part of that problem in 1984 by raising the maximum criminal fines under the act to $500,000 for companies and $250,000 for individual corporate executives convicted of criminal violations, attorneys in the Justice and Labor departments say.

But even these are not enough to help ensure safe working conditions because they do not increase or expand prison sentences, according to the National Safe Workplace Institute's Kinney and other critics.

An increase in investigations occurred in cases involving alleged criminal violations of federal environmental laws after Congress added felony provisions to those laws in 1976, he said. In the past three years alone, according to Justice Department figures, the number of cases investigated by Justice that involve alleged criminal environmental violations jumped from 49 to 163 per year, and analysts at Justice expect the number to increase further to 215 by the end of 1989.

Jail sentences have "a greater impact" on white-collar offenders "than any amount of money," said Maryland Assistant Attorney General Elizabeth Volz, who has won such convictions for business executives in three cases involving environmental violations.

Justice Department officials agree that stricter jail sentences are needed. "The Department of Justice ... would be happy to see an increase in the period of imprisonment authorized for a criminal violation of OSHA safety standards," Assistant Attorney General Thomas Boyd said in a Dec. 9, 1988, letter to the House Government Operations Committee.

LEGISLATION PLANNED

Such help may be on the way. Aides to Rep. Tom Lantos (D-Calif), chairman of the committee's Employment and Housing Subcommittee, told BNA that Lantos planned to introduce legislation in 1989 to strengthen the criminal provisions of the act.

Lantos, who chaired 1988 hearings on the prosecution issue, may try to extend criminal penalties to any "outrageous safety practices that could cause a fatality or injury," and to workplace deaths that result from "reckless endangerment," according to Stuart Weisberg, subcommittee staff director. Criminal penalties now are limited to "willful" violations of the act.

Lantos also may try to extend penalties to situations in which workers are endangered by exposure to substances such as asbestos in which the health effects sometimes are not felt until "many years in the future."

Some judges also think the penalties should be increased. Cook County (Ill.) Circuit Judge Ronald J.P. Banks told BNA the maximum penalty for criminal misconduct that results in a workplace fatality should be 20 to 30 years, "a severe enough penalty so employers take better care of their employees. That's what it comes down to."

In 1985, Banks sentenced three officers of Film Recovery Systems Inc., the Chicago company that used a cyanide wash solution to extract silver from used X-ray film, to 25 years in jail after finding them guilty of murder in the death of an employee. Banks ruled that the officers failed to warn the worker about the dangers of exposure to hydrogen cyanide gas from the process.

Unless employers are held criminally liable for workplace deaths that result from negligent or reckless corporate behavior, they "can hide behind a corporate veil ... they can do that every day of the week. You have to have a way to punish them somehow," Banks said.

If Congress increases and extends criminal penalties in the act, the number of cases investigated by the Justice Department probably would increase tenfold in the next three years, to 100 or more, Kinney said.

'Beyond a Reasonable Doubt'

Reynolds also said one reason the department has prosecuted just four of 30 cases is because so few have come from Labor — an average of only three cases a year sent to Reynolds' office in Washington, D.C., or directly to the U.S. attorney for the area in which the death occurred.

The Justice Department in turn declined to prosecute many of those because the heavy burden of proof required by the act makes criminal cases difficult to win in court, Reynolds said.

"You have to prove beyond a reasonable doubt" that a fatal accident resulted from a violation of an OSHA standard, and that the standard was intentionally disregarded by the employer, Reynolds said. "You don't go indicting a marginal case. That's the principle under which the department operates."

Even in accidents that involve simple safety questions, it sometimes is impossible to show the necessary links between accident, cause, and intent, Reynolds added.

Factors in Trench Cases

Trench cave-ins are one such type of workplace accident where an intentional violation is difficult to prove, according to Reynolds.

More trench-collapse cases were referred by the U.S. Department of Labor to the Department of Justice for possible criminal prosecution from fiscal 1981 through fiscal 1987 than any other type of workplace accident case, according to OSHA. Seven of the 20 cases referred during that period involved trench cave-ins.

The Justice Department declined to pursue four of those seven trenching cases. U.S. attorneys still are considering two referrals, and a guilty plea was entered in another in 1982.

To bring a criminal case against an employer for a fatal trench collapse, federal prosecutors must show that the accident resulted from an employer's intentional violation of an Occupational Safety and Health Administration regulation, Reynolds told BNA.

Many factors go into a decision about when or how to slope or shore a trench to meet OSHA requirements. These include:

- The depth of the trench;
- Variations in the soil's water content;
- Expected changes in the soil from freezing, drying, or exposure to rain;
- Vibrations from traffic or blasting; and
- Pressure from nearby equipment, piles of dirt, and other objects.

For example, a collapse may result from an unintentional misjudgment about the stability of the soil, Reynolds said. "You're not always talking about someone taking a soil sample to a laboratory. These are judgments made by construction people on the nature of the soil."

Robert Peyton, director of safety and health for the Associated General Contractors of America, said "it's not necessarily a simple matter to determine when [trench safety] problems will occur."

Although "some classic signs" of soil instability exist, such as fracturing and heaving, "there are no hard and fast rules to say how trenches can be made safe all the time," Peyton told BNA.

"My experience is that, most of the time, fatalities occur in situations where somebody cut some corners—but that may not be the employer's fault," Peyton said. "The employee may have been told to wait for a period of time [before going into the trench], or to use some special equipment, and chose not to do so. Or there may have been a design problem" in the shoring.

STATE PROSECUTOR SEES LESS DIFFICULTY

However, others do not agree with Reynolds and Peyton that intentional violations of trenching standards are difficult to prove. Les Murphy, assistant director of health and safety for the AFL-CIO Building and Construction Trades Department, said there is "nothing complicated" about preventing trenches from caving in.

Given the known hazards of trenching work and the availability of information on safe trenching, construction companies "should have the knowledge of how to shore," Murphy said. "There's nothing exotic about it at all."

Texas prosecutor Oden told BNA that finding enough evidence of gross negligence to support homicide charges is not difficult in trenching cases, "simply because the risks posed by disregarding the safety regulations are so great, combined with the ease with which the employer can protect against that risk."

Four of the 17 cases pending in state courts in early 1989 involved charges of negligent homicide resulting from trench-collapse deaths. All four were brought by Oden, whose jurisdiction includes the Texas capital of Austin.

As of June 1989, three of the cases were in the Texas Court of Criminal Appeals, and the fourth was before the Travis County Court, pending determinations on whether local authorities are preempted by the federal OSH Act from bringing criminal charges.

Workers in trenches can face a "substantial risk of danger" if safety regulations are not followed, Oden said. "Earth is a heavy, unforgiving substance in a cave-in. It's a violent, gruesome death."

Ways to prevent cave-ins are well known and easy to implement—shoring (putting up barriers to prevent the walls of the trench from collapsing inward) and sloping (widening the top of the trench so that the walls are sloped back from its floor). "It's not a costly process to make trench work safe," Oden said.

CASES 'WASHED OUT' ON FIRST REVIEW

Of all the cases sent to Justice from OSHA, about half have been "washed out" on first review, Reynolds said.

He told BNA these cases were rejected for a variety of reasons. In some, Justice was "satisfied with [the amount of information in] the case file," but felt the case lacked "prosecutorial merit," Reynolds said. In other instances, cases were turned down because the information from OSHA was too "skimpy," he added.

"Different referrals come over with different degrees of thoroughness," he said.

OSHA officials say they do the best they can. "We do a very careful screening job and send only what we think are our best cases," said the Labor Department's Attwood. "Justice is a different agency with a different mission. They don't always take cases we think are our best."

Reynolds also defended the length of time it takes Justice officials to decide whether to prosecute a case. "Three years is on the long side—there's no question." He added that workplace death cases may involve complex technical questions that require detailed investigation.

"Any of these cases are difficult to bring," he said. "Most are very difficult."

Also, linking a fatal accident to a willful violation of an OSHA standard is a "painstaking" process, Reynolds said. So many factors may be involved in an accident that "if you ask experts, 'How did it happen?', they say, 'I don't know'."

Asking for technical help from another federal agency, such as the National Bureau of Standards, also may delay review of a case, Reynolds said. In one fatality investigation,

NBS "spent the better part of a year" responding to Justice's request for a technical assessment of evidence, he said. "These things are not quick."

Labor Department officials said they try to monitor cases after they have been referred to Justice, but have no way to expedite Justice's review. The department maintains no regular contact with Justice except to try to follow the progress of individual cases, they said.

"We don't have any independent litigation authority," Thorn told BNA. "When we call [Justice] on a matter, they simply tell us it's being looked at. They're not abrupt or anything. It's just that when [a case] is referred, it becomes their case. We don't have any leverage.

Added Thorn, "We can pick up a phone and call but their answer will be, 'We're still looking at it'."

Such arguments hold little sway for House subcommittee aide Weisberg. "Clearly, if I'm OSHA, and I've been waiting for three years for a decision, there's a communication problem," he told BNA. "That's putting it mildly."

1990 BUDGET WOULD CUT PROSECUTION FUNDS

Justice would receive no additional funding for criminal prosecutions under the Bush administration's budget proposal for fiscal 1990, and it remains to be seen whether OSHA will receive more money to bring cases, obervers said.

Indeed, in its fiscal 1990 budget request to congressional appropriations panels, the Bush administration asked for cuts of $200,000 from the $3.7 million previously appropriated for Justice's general litigation section. The administration also sought a reduction of authorized positions from 52 to 46.

The $200,000 would be used to create a new unit to prosecute cases brought under federal obscenity laws, Justice budget officers told BNA.

For OSHA, the administration requested $253 million, an increase of $8.4 million for the entire agency. But Seminario said spending would have to grow by 10 percent each year for the next five years, in real dollars, for OSHA to function effectively.

The proposed cuts anger some critics. Instead of trimming funds, they say, Justice should assign a higher priority to OSHA cases by establishing a special prosecution unit. "Back it up with some money," Bingham said.

"If OSHA made a priority out of fatality cases and devoted more resources, proportionately, to fatality cases, they probably could do some more [prosecutions]," said Frank White, a former OSHA deputy administrator and associate solicitor of labor.

But Reynolds said the proposed cuts, which would take money and positions from throughout the general litigation section, would not hurt the section's ability to prosecute workplace fatalities.

Kinney also said it would be possible for the section to absorb the loss by "playing musical chairs ... You're talking about a baseline next to zero, anyway."

STATE PROSECUTORS: 'WE GIVE THIS PRIORITY'

Many observers contrast the level of federal performance to the activities of state prosecutors who are bringing more charges against employers. These prosecutors openly scoff at Justice and OSHA arguments that the act makes it too difficult to prosecute. It would be better to make prosecutions a priority and collect as much evidence as quickly as possible, they say.

Middlesex County, Mass., for example, has requested that safety and health inspectors notify law enforcement offices of each workplace fatality they investigate, said former assistant district attorney Jim Howard.

And in Texas, prosecutor Oden keeps an investigator available 24 hours a day to visit the scene of any workplace fatality. "We give this priority because we're dealing with difficult cases, and we need to do our legwork," said Alia Moses, assistant district attorney to Oden.

Moses said it usually takes her office three to six months to bring an indictment. "Probably the longest we would go is a year, maximum ... With any criminal case, the longer it takes to prosecute, the worse the evidence gets, and the harder witnesses are to find."

Other prosecutors also challenge the notion that criminal cases involving worker fatalities need years to develop. "We would never leave anything open" for two or three years, Jan Chatten-Brown, special assistant to Los Angeles District Attorney Ira Reiner, said. "In most cases, we have a fairly good idea in a couple of months."

Chatten-Brown's boss, Los Angeles County District Attorney Ira Reiner, has pioneered a special "roll-out" unit to investigate workplace fatalities. Reiner has prosecuted 23 such cases since 1985, and has referred more than a dozen to city prosecutors in his region, according to Chatten-Brown. Nationally, five of the six jail sentences served for workplace violations resulted from cases brought by Reiner.

Some former OSHA officials also say the federal process could move faster. One way would be for the Labor Department to take a referral directly to the U.S. attorney for the region in which the case arose, former OSHA official White told BNA. "They have a sense of what's appealing to the U.S. attorney."

'When the Feds Drop the Ball'

In a memorandum last summer, then-OSHA Administrator John A. Pendergrass told agency personnel not to turn case files over to state or local authorities without first consulting with regional Labor Department attorneys about the possible effects of a state or local prosecution on a pending federal criminal or civil case.

State prosecutors told BNA such actions put them at a disadvantage and illustrate the lack of coordination and communication between state and federal agencies.

"When the feds drop the ball, it really screws up a case," said Michael Modelski, assistant prosecuting attorney for Oakland County, Mich. "If we have to find witnesses, and do testing, and things of that sort, the trail would be cold."

BNA examined nine cases that U.S. attorneys declined to prosecute between 1980 and 1988 and found that:

- Only in one instance did local prosecutors bring charges against a defendant after federal authorities declined to prosecute. But that case, against a Wisconsin fireworks maker, has been delayed by preemption arguments.

- In a second instance, involving an explosion that killed 21 workers at a fireworks plant in Oklahoma, Pawnee County District Attorney Lawrence Martin declined to file charges, saying he agreed with the Justice Department's decision. The explosion "was an accident," apparently caused by employee error, he told BNA.

- In the remaining seven cases, local prosecutors said they were unaware of the federal government's actions. Several were surprised to learn from BNA that an OSHA investigation had been referred to Justice. "When federal prosecutors [turn] down a case, it [goes] down a black hole," Western Illinois University professor Maakestad said.

Reynolds of Justice said department officials try to stay attuned to the needs of state and local prosecutors by meeting with representatives of the National Association of Attorneys General and the National District Attorneys Association in an ongoing working group.

"To my knowledge, [lack of coordination on OSHA prosecutions] has not been raised as a problem" in those sessions, he said.

Stronger State Laws Sought

At the local level, prosecutors have chosen to take action on their own. For them, the ideal situation would be one in which health and safety agencies, employers, and the legal system work together effectively to safeguard workers.

"We don't want to get involved in being another regulatory agency. We want what industry itself generally wants: To take the bad actors out of business," Howard of Massachusetts said. "You can [either] regulate beforehand and try to prevent injuries or you can let things go and prosecute them when injuries occur."

Until such a time, Howard and others say, aggressive pursuit of cases is the most effective way to stop corporate officials from skirting safety and health regulations.

In several states, district attorneys and labor leaders are asking their legislatures to increase criminal penalties for violations of state workplace health and safety laws.

In other states, prosecutors and regulatory agencies are beginning to pursue alleged offenders under existing criminal provisions that were seldom used in the past.

"The feeling is, if it is manslaughter, then call it that," said Virginia Labor Commissioner Carol Amato. Recently, Amato announced that the state will work with local prosecutors to bring involuntary manslaughter charges against employers whose "callous or reckless disregard for human life" leads to a workplace death.

In California, critics charge that prosecutors' efforts to get tough with employers whose policies and practices cause worker injuries and deaths often are politically motivated. For example, some business groups claim that workplace safety legislation proposed by the Los Angeles County District Attorney's Office (LACDAO) is unnecessary and merely a tool to further the careers of the district attorney and his assistants.

"Of course, it's just politics," said Robert C. Kline, a lobbyist for several California industry groups that have fought legislation proposed by LACDAO. "Everyone wants to womp on the companies."

While the large majority of businesses are safety-conscious, Kline admitted "there are bad apples in every barrel. [But] we don't need over-inclusive penalty legislation ... when only a few are troublesome."

Bills Proposed by District Attorney

LACDAO, which has prosecuted more job-related death and injury cases than any other state or federal agency, has proposed workplace safety bills over the past several years, including a 1988 measure that would have allowed prosecutors to bring misdemeanor and felony charges simultaneously against employers who have caused injury or death among their employees.

The General Assembly passed the measure, but Gov. George Deukmejian (R) vetoed it after at least 12 business groups voiced opposition. Despite the veto, the bill and other similar measures help educate the public and politicians about workplace safety issues, Jan Chatten-Brown, Los Angeles County deputy district attorney, told BNA.

LACDAO plans to have the same bill introduced in the 1989 General Assembly, plus several others, including a measure that would allow employers to be convicted of a misdemeanor for workplace injuries even where the prosecution cannot prove that they willfully caused the accident, Chatten-Brown said.

The office also is supporting the California Corporate Criminal Liability Act, which would subject employers to fines as great as $250,000 and up to three years in jail for failing to disclose dangerous workplace conditions to state officials.

The deputy district attorney said she is not surprised by employers' opposition to legislative initiatives. But Chatten-Brown denied she and other district attorneys are seeking personal gain from pursuing tougher worker safety laws.

"These are meritorious prosecutions," she told BNA. "We don't expect them to be popular with business."

Push from Attorney General in Minnesota

In Minnesota, it is the state's attorney general who is pushing local prosecutors to bring more cases to court, said Nancy Leppink, special assistant attorney general.

Many Minnesota district attorneys, especially those in smaller and rural districts, often refuse to investigate or prosecute such offenses because the penalties are so slight, and because many do not believe that circumstances which cause or contribute to worker deaths constitute a crime, Leppink explained.

"The perception by law enforcement and prosecutors is that these cases are accidents ... that don't have a criminal component," she said.

Under the attorney general's proposal, employers who knowingly create a threat of serious harm to employees could be fined up to $30,000 and sentenced to a maximum of six months in jail. The maximum penalty would increase to $40,000 and a year in jail if anyone is killed or injured as a result of employer negligence. Employers who willfully cause the death of an employee could be fined up to $50,000 and be sent to jail for up to seven years.

Current state law provides fines of up to $20,000 and/or up to six months in jail for first-time offenders and $35,000 and a year in jail for repeat offenders, Leppink told BNA.

Like most states, Minnesota does not allow the attorney general to bring a case unless it first is referred by a local prosecutor. No local prosecutor has done this in a workplace death or injury case, and none has prosecuted such a case on his or her own, Leppink said. Several local prosecutors told BNA they would prosecute such cases if the state asked them.

Leppink said she will find out from local prosecutors how they handle reports of workplace fatalities, an effort she hopes will lead to a special line item in the attorney general's office budget for a program to provide local officials with the information and expertise they need to take these cases to court.

"We have to have more than just an ad hoc strategy."

The 'Barking Dog' of Deterrence

State and local prosecutors' push for deterrence is spreading. For example, criminal charges were filed in a workplace fatality case for the first time in Indiana after a grand jury brought an indictment of criminal recklessness against an Indianapolis company for the Nov. 2, 1988, death of a worker in an explosion.

(continued on page 29)

Prosecutorial Discretion

Under what circumstances are district attorneys or state's attorneys most likely to file criminal charges against an employer in connection with a workplace accident?

BNA found on the basis of its interviews with prosecutors and an examination of criminal cases nationwide that 10 elements, in various combinations, were present in each instance. The survey indicates that companies or business executives are most vulnerable to criminal prosecution when these elements are present:

- An accident occurs in which an employee is killed or seriously injured.

- An accident can be attributed to misconduct by the company or one or more of its officers.

- One or more persons witnesses an accident that kills or seriously injures an employee.

- Evidence is available pertaining to the facts of an accident that injures or kills an employee.

- A local prosecutor is notified quickly about an accident that injures or kills an employee.

- A local prosecutor is under pressure to investigate workplace accidents for possible criminal prosecutions.

- A prosecutor believes that a workplace accident involves criminal misconduct by employers.

- Employees perform tasks so inherently dangerous, such as working in a trench or manufacturing fireworks, that any accident could seriously injure or kill them.

- Work is regulated under a government safety standard or is covered by industry safety guidelines.

- Government standards or industry guidelines are disregarded.

A BNA Graphic

Copyright © 1989 by The Bureau of National Affairs, Inc.

In Tennessee, Knox County District Attorney William E. Dossett is investigating the deaths of two employees in a trench cave-in. If criminal charges are filed, it would mark the first time an employer was prosecuted in the state for a workplace death.

In Maryland, prosecutors are "dusting off" a 20-year-old electrical safety law to bring criminal charges against employers in accidental workplace electrocutions, according to Craig Lowry, director of compliance for the Maryland Occupational Safety and Health Administration.

Lowry said state prosecutors have charged four employers with criminal misconduct under that law for allegedly failing to protect workers from contact with overhead power lines. The cases are pending.

And in Connecticut, where federal prosecutors last year declined to file charges against companies involved in the 1987 Bridgeport building collapse that killed 28 workers, a bill pending in the state legislature would create a special workplace accident investigation unit in the state Department of Criminal Justice. The measure has been approved by one legislative committee and is pending before another.

"The bottom line is not how many people you've prosecuted, but how many lives you've saved," said McCann of Wisconsin.

Prosecutions, he explained, are "like a barking dog. How many burglaries do barking dogs prevent annually? Probably tens of thousands."

• • •

A LEGAL POINT-COUNTERPOINT

The first section of this chapter discussed the lack of federal enforcement of workplace deaths. According to many experts, this problem is compounded by the issue of whether the Occupational Safety and Health Act of 1970 preempts state prosecution under general crimnal statutes of employers who expose employees to harmful workplace conditions.

That issue, recently addressed by the Illinois Supreme Court in Illinois v. Chicago Magnet Wire Corp., ultimately may be decided by the U.S. Supreme Court. The question is also the subject of the following legal point-counterpoint. The first view, by George Cohen, senior partner with the Washington, D.C. law firm of Bredhoff & Kaiser, argues that states are not preempted from prosecuting employers for negligent workplace deaths.

The second article, by Robert Gombar, Willis Goldsmith, Frank White, and Arthur Sapper of the Washington, D.C., office of the law firm of Jones, Day, Reavis & Pogue, asserts that it is up to Congress, not the courts, to determine the preemption issue.

Preemption: A Union Lawyer's View

By George H. Cohen*

In 1970 Congress enacted the Occupational Safety and Health Act, 29 USC 651 *et seq*. and "declare[d] it to be its purpose and policy ... to assure so far as possible every working man and woman in the Nation safe and healthful working conditions ...." (*id.* at 651(b).) After almost 20 years of observing the manner in which the act has been implemented from the perspective of the interests of labor unions and the workers they represent, this much seems clear beyond any doubt: the lofty goal reflected by that congressional declaration has not been achieved. A critical ingredient is still missing. Top-level officials within a multitude of business enterprises throughout the country have yet to demonstrate a commitment to that goal by directing their managers to take all necessary steps to make their workplaces safe and healthful.

Consequently, today we now know that state-of-the-art technology can eliminate or substantially reduce workplace hazards. Some employers have shown an exemplary capacity to utilize that technology.[1] However, the technology—whether it is engineering controls, such as modernized ventilation systems, protective clothing and equipment, or comprehensive work practices, such as increased maintenance—can be implemented only when employers are prepared to spend enough money and employ the manpower necessary to get the job done.

Therefore, in assessing the success of the act, the fundamental question is whether the administration of that law has provided employers with a sufficiently powerful inducement to "pay the price," that is, to earmark and devote their resources so that safe and healthful workplaces become a reality.[2]

Regrettably, the lesson of the past two decades is that the programs and actions undertaken by the Occupational Safety and Health Administration (OSHA) have been far too little and much too late to provide any such meaningful inducement. As envisioned by Congress, the secretary of labor was to promulgate expeditiously a series of stringent worker-protective safety and health standards that would then effectively be enforced by a cadre of federal inspectors armed with authority to issue citations and abatement orders,[3] post workplace notices, and propose civil penalties against employers that did not comply with those standards.

---

*Mr. Cohen is a senior partner with the Washington, D.C., law firm of Bredhoff & Kaiser. The author was assisted by his colleague, Virginia Seitz, in the preparation of this article. Mr. Cohen assumes full responsibility for the content of this article.

Copyright © 1989 by The Bureau of National Affairs, Inc.

The plain fact is that the all-important standard-setting function, by any objective yardstick, has moved at a snail's pace. OSHA has issued only 24 substance-specific health regulations since its creation,[4] leaving many hundreds of known toxic chemicals unregulated. Furthermore, the secretary of labor's enforcement program has been woefully inadequate whether measured in terms of the number of inspectors, the number of workplaces inspected, the number of hazards eliminated and/or the civil penalties typically imposed when violations were found.[5]

All of these factors, coupled with the Reagan administration's abiding commitment to a policy of "deregulation" with respect to federal environmental and occupational health and safety programs, inevitably conveyed the wrong message to corporate America. Many employers were lulled into believing that continuing to do precious little to eliminate workplace hazards was an acceptable course of conduct. At bottom, the act to date has provided only a minimal inducement for employers to clean up their own acts.

Specter in the Boardroom

Precisely because the administrative scheme of the act has not served its intended purpose, labor unions and public interest groups increasingly have turned their attention to other initiatives that might cause employers to sit up and take notice of the risks they confront if they maintain the status quo. One of those initiatives is the subject of this article: the increased use by state law enforcement agencies of criminal prosecutions against individual corporate officers and/or their corporations in appropriate workplace-related situations.

It bears emphasizing at the outset that the purpose of such prosecutions is essentially to punish the charged party and thereby to deter others from engaging in conduct that constitutes a "crime" either under common law or by virtue of a state law. This purpose stands in marked contrast with the purpose of the administrative regulatory scheme established by the act.[6]

The salutary deterrent effect that the threat of possible criminal prosecution could have on corporate officials is illustrated by this scenario: the chief executive officer of Company A is presiding over a board of directors meeting in which that company is considering the advantages and disadvantages of spending several million dollars to purchase and implement a new ventilation system to reduce worker exposure to asbestos in several of its older plants. The company's key finance officers have made a presentation emphasizing that the company's projected small profit margin for that year would be jeopardized by this proposed major capital expenditure and that, therefore, less expensive "patchwork" steps should be taken to deal with the problem.

Even assuming that OSHA would inspect those facilities, if a modest civil penalty and an abatement order were the maximum legal liabilities that the company would face were it to decide not to purchase a new ventilation system, there would be little incentive to appropriate the funds for that system. Let us assume, however, that another consideration entered into the decisionmaking calculus.

Suppose that the CEO knew that he and his fellow board members could be subjecting themselves, individually, to possible criminal prosecution because the vice president for occupational and environmental affairs already had advised the board that workers were being exposed to excessive levels of asbestos. Assume, further, the board was on notice that, unless a new control system was put in place, there was a substantial certainty that a significant number of workers would contract asbestosis and/or various types of cancer as a direct result of continued exposure to asbestos. In these circumstances, the CEO would understand that a decision not to appropriate the funds for the ventilation system could hardly be explained away as a "legitimate business judgment"; instead, it might well be tantamount to committing a crime.

Perhaps this scenario may help to explain why it appears that the specter of state criminal prosecutions may be serving as a more powerful incentive for corporate decision makers to take actions that comport with the underlying purposes and policies of the act than have the act's core administrative enforcement procedures. The validity of that thesis is bolstered by the degree of anguish that has been generated within the corporate community by these state criminal prosecutions and the extensive counteroffensive it has launched, through counsel, in an attempt to preclude such prosecutions on federal preemption grounds.

Statutory Shortcomings

What has caused this unlikely trend whereby state criminal prosecutions have increasingly served as a deterrent to employer misconduct? In part, this trend is explained by the fact that the act itself contains only one criminal provision of any significance; it does not suffice to inhibit employer misconduct. Section 17(e) (29 USC 666(e)) states that:

> Any employer who willfully violates any standard, rule, or order promulgated pursuant to section 6 of this title, or of any regulations prescribed pursuant to this chapter, and that violation caused death to any employee, shall, upon conviction, be punished by a fine of not more than $10,000 or by imprisonment for not more than six months, or by both; except that if the conviction is for a violation committed after a first conviction of such person, punishment shall be by a fine of not more than $20,000 or by imprisonment for not more than one year, or by both.

That provision has been administered in a particular manner: In carrying out its inspection function where workplace fatalities have taken place, OSHA determines whether the employer's conduct in a given case warrants criminal prosecution. If OSHA makes an affirmative determination, it can refer the case to the Department of Justice. However, a federal criminal prosecution will be initiated only if the Justice Department recommends that action and the local U.S. attorney's office responsible for prosecuting the case concurs with that recommendation.

Over the first 18 years of the act, in excess of 125,000 workers have died at their workplaces. Only 42 cases have been referred by OSHA for criminal prosecution and, of that number, only 14 cases were actually prosecuted, resulting in 10 convictions.[7] But no defendant has ever spent one day in jail for a criminal violation of the act.[8] (*Getting Away With Murder in the Workplace: OSHA's Nonuse of Criminal Penalties for Safety Violations*,

Rep. of the Comm. on Government Operations, HRep No. 100-1051, 100th Cong. 2d Sess., at 4 (Oct. 4, 1988).)

High Visibility Cases

In the face of the federal government's paralysis, a growing number of state and local prosecutors vigorously began to investigate the facts surrounding workplace fatalities and serious injuries and illnesses with a view toward determining whether the conduct of individual corporate officials or the corporation itself constituted criminal behavior under the laws they enforce.[9] Without question, the national attention that has been focused on certain particularly egregious, high visibility cases has spurred this development. The earliest such case involved Film Recovery Inc., a company that operated a silver reclamation facility located in the Chicago suburbs. In the wake of an investigation into the death of a 45-year-old Polish immigrant from exposure to lethal levels of cyanide fumes in May 1983, the Illinois state's attorney indicted several company officials and charged them with murder and lesser offenses. Among other things, the indictment alleged that the defendants intentionally painted over manufacturer's warning labels on 55-gallon drums of toxic chemicals so that the workers would not be able to understand the serious risks they confronted.

After trial, in July 1985 three officers were convicted of murder and sentenced to 25 years in prison. Originally, the defendants appealed the conviction but not on the grounds of federal preemption. But in short order that changed—in the past few years, as the number of state criminal prosecutions increased significantly, the uniform battle cry of indicted corporate officials is that their state prosecutions are preempted by the federal act.[10]

STAGE SET FOR LEGAL SHOWDOWN

Two recent cases provide an appropriate point of departure for this part of our discussion. In one, the state of New York brought criminal charges of assault in the first degree, assault in the second degree, reckless endangerment in the second degree, conspiracy to falsify business records in the fifth degree, and falsifying business records in the first degree against two corporate officers—William Pymm and Edward Pymm, Jr.—and against the Pymm Thermometer Corp., and the Pak Glass Machinery Corp. (See New York Penal Law 105.51(1), 175.10, 120.10(4), 120.05(4), and 120.20.)

Specifically, the state charged that these individual and corporate defendants knowingly and intentionally subjected their employees to continuous exposure to a toxic substance—mercury—and thereby recklessly endangered the health of those employees. The defendants also were charged with conspiring to hide the existence and conditions giving rise to this toxic exposure from OSHA inspectors. One employee was crippled and suffered brain damage as a result of defendants' conduct.

After a trial in New York Supreme Court in Brooklyn, the jury found the defendants guilty on all counts for causing serious physical injury to that employee, by knowingly and continually exposing him to mercury, and for creating a substantial risk of serious injury to other employees by the same conduct.[11]

The defendants filed a motion to set aside the jury verdict, asserting that, as a matter of law, even if they were guilty of the foregoing crimes as charged, their prosecution and conviction is preempted by the federal act because the toxic substance to which the employees were exposed is regulated by OSHA. On Nov. 18, 1987, the trial judge set aside the verdict on this ground.[12] The state of New York filed a timely appeal seeking the reinstatement of the jury verdict against defendants. That appeal was *sub judice* in early 1989. (See *People of the State of New York v. William Pymm, et al.*, Docket No. 1971E/88.)

In the second case, the state of Illinois brought criminal charges of aggravated battery, reckless conduct, and conspiracy to commit aggravated battery against Chicago Magnet Wire Corp. and five of its corporate officials. (See Ill. Rev. Stat. 1985, Ch. 38, 12-4(a), 12-4(c), and 8-12(a).) The state alleged that the defendants, acting in their official capacities, committed these common law crimes by knowingly and recklessly exposing 41 employees to "poisonous and stupefying substances," causing them serious injury.

Unlike the defendants in *Pymm*, these defendants raised federal preemption in a pre-trial motion to dismiss. The trial judge in the Circuit Court of Cook County dismissed the charges against the defendants on this ground, and an appellate court affirmed this decision. On Feb. 1, 1989, the Supreme Court of Illinois issued its decision unanimously reversing the lower court's dismissal and remanding to the circuit court for a trial.[13] Shortly before that decision was issued, the U.S. Department of Justice publicly declared that its position is that the criminal penalty provisions of the act were not intended to preempt state prosecutions of employers for workplace-related crime:

> As a general matter, we see nothing in the act or its legislative history which indicates that Congress intended for the relatively limited criminal penalties provided by the Act to deprive employees of the protection provided by the state criminal laws of general applicability.[14]

The Department of Labor, on the other hand, has consciously chosen to remain silent on this important issue. The department's solicitor did not submit an amicus brief in either the Illinois or New York state court proceedings.[15]

Illinois is presently the only state whose highest court has decided the preemption issue. Given the importance of the issue, it is highly likely that the U.S. Supreme Court will choose to have the last word on this subject. It remains to be seen, however, whether the Supreme Court will grant certiorari in *Chicago Magnet Wire* at this preliminary stage before the defendants have had to stand trial or, at the very least, choose to wait to see if the defendants are convicted. Alternatively, the Supreme Court may prefer to wait to see if the highest court in another state reaches a contrary decision to *Chicago Magnet Wire* before granting review.

STATES NOT PREEMPTED[16]

The inquiry whether the federal act preempts state criminal prosecutions resulting from workplace fatalities, injuries, or illnesses must begin with a review of the relevant provisions of that legislation.[17]

An Overview and Summary

The discussion that follows first shows that the state's authority to establish and enforce its criminal laws is at the core of its historic police power, and that such authority is not preempted by federal statutory law unless Congress unmistakably manifests an intention to preempt. There is no evidence of a congressional intent to preempt state criminal law on the face of the act or in its legislative history; to the contrary, 4(b)(4) of the act clearly reveals Congress' intent that the state's authority in this respect be preserved.

The discussion then shows that any reliance on Section 18 of the act for the contention that the act preempts state criminal prosecutions is entirely misplaced. There is a fundamental conceptual distinction between two different types of laws concerning worker safety and health: *viz.*, those laws that compensate victims of industrial injury and punish employers who wrongfully inflict such injury, and those laws that set preventive, regulatory standards. The difference between these types of laws cuts the line of demarcation between Section 4(b)(4)'s broad preservation of state authority (see discussion below) and the limited preemption of state authority worked by Section 18 of the act (see discussion below). The latter provision reflects Congress' determination that a unitary federal system of mandatory, minimum, preventive occupational safety and health regulations is necessary to supplement the historic tort and criminal law requirements by which the states have redressed employer actions injurious to their employees. This determination was not intended to, and does not in any way, limit state authority to compensate employees injured by employer torts or to punish employers for criminal wrongs.

State's Right Expressly Preserved

To begin with, as noted, there is a strong presumption against preempting a state's exercise of its historic police power to establish and enforce criminal prohibitions to protect its citizens against physical harm. The state plainly has a special interest in controlling conduct that is motivated by a malicious or reckless mental state and that endangers the lives of its citizens, whether the conduct occurs in the context of operating a car or operating a workplace. (See *e.g., Hoag v. New Jersey*, 356 U.S. 464, 468 (1958) ["it has long been recognized as the very essence of our federalism that the States should have the widest latitude in the administration of their own systems of criminal justice"]).[18]

Consistent with that principle, the U.S. Supreme Court has repeatedly emphasized that preemption analysis must "start with the assumption that the historic police powers of the States were not to be superseded by [federal law] unless that was the clear and manifest purpose of Congress."

(*Hillsborough County, Fla. v. Automated Medical Laboratories, Inc.*, 471 U.S. 707, 710 (1985), and cases cited.)

Indeed, in *San Diego Unions v. Garmon*, 359 U.S. 236, 243-44 (1959), the U.S. Supreme Court made it plain that, even when Congress generally preempts state law tort remedies in favor of a uniform federal scheme, as, for example, under the National Labor

Relations Act, the federal law does not preempt state laws "deeply rooted in local feeling and responsibility," most particularly those that prohibit the intentional, knowing, or reckless infliction of physical harm upon individuals. What this means is that, unless Congress expressly declares otherwise, "the state still may exercise 'its historic powers over such traditionally local matters as public safety and order'." (*United Construction Workers v. Laburnum Construction Corp.*, 347 U.S. 656, 664 (1954) (quoting *Garner v. Teamsters Union*, 346 U.S. 485 (1953) and *Allen-Bradley Local v. Wisconsin Board*, 315 U.S. 740, 749 (1946) (NLRA does not preempt state law tort action against union for intimidation and threats of violence).) (See also *United Automobile Workers v. Russell*, 356 U.S. 634, 646 (1958) (same).)

In recognition of this principle, courts have not applied the preemption doctrine to foreclose states from enforcing their "general criminal statutes governing reckless and intentional conduct that threatens public safety." (Note "Getting Away With Murder: Federal OSHA Preemption of State Criminal Prosecutions for Industrial Accidents," 101 *Harv. L. Rev.* 535, at 546 & n. 64 (1987), and cases cited). As one recent decision aptly observed: "the fact that [criminal] conduct may in some respects violate OSHA safety regulations does not abridge the state's historic power to prosecute crimes." (*State ex rel. Cornellied v. Black*, 425 NW2d 21, 25; 144 Wis. 2d 745, 755 (Wis. App.) (review denied, 430 NW2d 351 (WI 1988).)

Furthermore, in enacting the Occupational Safety and Health Act Congress was not silent on this issue—it expressly provided that a state shall not be preempted from enforcing its criminal laws. All states make it a crime for "A" willfully and wrongfully to poison "B," thereby inflicting grievous bodily injury. So far as we are aware, the fact that "A" is an employer and "B" is his or her employee has never constituted a defense to an otherwise valid indictment for that crime. The federal act, moreover, provides no criminal penalties for such an act by an employer or by anyone else.

Heavy Burden for Proponents

The proponents of the preemption argument must shoulder a very heavy burden. They must establish that in passing the act—whose overall purpose is to promote worker safety and health, 29 USC 651(b)—Congress was animated in part by an intent to deprive the states of their longstanding authority to enforce their criminal laws against employers who knowingly and willfully poison their employees, thereby insulating such employers against any criminal penalty for that kind of intentional wrongdoing. Nothing in the act or its legislative history provides any support for attributing such an unlikely intention to Congress; in fact, Section 4(b)(4) of the act is an express anti-preemption provision:

> Nothing in this chapter shall be construed to supersede or in any manner affect any workmen's compensation law or to enlarge or diminish or affect in any other manner the common law or statutory rights, duties, or liabilities of employers and employees under any law with respect to injuries, diseases, or death of employees arising out of, or in the course of, employment.

[29 USC 653(b)(4)]

Thus, by its terms, the act does not "diminish or affect in any other manner the common law or statutory ... duties, or liabilities of employers ... under any law with respect to injuries, diseases, or death of employees arising out of, or in the course of, employment."

A fair reading of the quoted words leads directly to the conclusion that state criminal charges alleging such employer conduct are not preempted. Section 4(b)(4) speaks in broad terms: "any law" is a clear and sweeping phrase, obviously including criminal laws.

The stricture against preemption is not limited to state civil laws generally or even to state tort laws; by its terms Section 4(b)(4) applies, as well, to state criminal laws, both "common law" and "statutory." Employers, moreover, obviously have a "dut[y]" under the criminal "law" of their respective states not to commit assault or to engage in reckless conduct that "injur[es]" individuals "in the course of employment." Far from disclosing a clear and explicit congressional intent to preempt state law criminal actions, Section 4(b)(4), instead, makes it manifest that such actions are not preempted.

Strained Interpretations

In *Chicago Magnet Wire* and *Pymm Thermometer,* supra, as in numerous other cases, the defendants offered two strained interpretations of Section 4(b)(4) to support their claim that this provision does not demonstrate a congressional decision against preempting state criminal laws.

First, the defendants noted that the provision refers only to the "rights, duties, or liabilities" of employers and employees, and not to those of the states. The defendants maintain that this omission is significant because only the states have a "right" to enforce the criminal laws. The simple answer is that Section 4(b)(4) speaks not only of rights but also of the "common law or statutory ... duties or liabilities of employers." Criminal laws of the type at issue here place "duties" upon employers. Such provisions thus fall precisely into one of the categories covered by Section 4(b)(4). This argument is therefore nothing more than sophisticated word play.

Second, the defendants argued that the phrase "any law" in Section 4(b)(4) modifies the phrase "with respect to injuries, diseases, or death of employees," and therefore that this provision saves only those state laws that pertain specifically to occupational health and safety, while it preempts all other laws. This interpretation is entirely counter-intuitive, as it would mean that Section 4(b)(4) would save from preemption only those state laws that touch most closely on the area of federal regulation, while it would preempt all others. Moreover, this argument would nullify the language of Section 4(b)(4) that preserves the duties and liabilities of employers under the "common law," because, so far as can be determined, there is no common law specifically pertaining to occupational safety and health.

Consistent with Section 4(b)(4)'s plain meaning, it is well settled that the act does not preempt state law actions seeking money damages for injuries caused by occupational safety and health hazards. (See *Atlas Roofing Co. v. OSHRC,* 430 U.S. 442, 445 (1977) (act leaves "existing state statutory and common law remedies for actual injury and death ...

unaffected"); *United Steelworkers v. Marshall*, 647 F.2d 1189, 1235-36 (D.C. Cir. 1980), *cert. denied*, 453 U.S. 913 (1981); *Frolich Crane Service v. OSHRC*, 521 F.2d 628, 631 (10th Cir. 1975); *Smith v. Western Electric Co.*, 643 S.W.2d 10, 14 (Mo. App. 1982); *Berardi v. Getty Refining and Marketing*, 107 Misc. 2d 451, 435 N.Y.S.2d 212 (1980).)[19]

These principles have been established even though it is axiomatic that the power to impose compensatory damages for such injuries carries with it the ability to regulate primary conduct:

> [R]egulation can be as effectively exerted through an award of damages as through some form of preventive relief. The obligation to pay compensation can be, indeed is designed to be, a potent method of governing conduct and controlling policy. [*San Diego Unions v. Garmon, supra*, 359 U.S. at 247. See also *Smith v. Wade*, 461 U.S. 30, 54-55 (1983).]

Integral Part of Tort Remedies

This point is even more compelling when we pause to consider that punitive damages are an integral part of traditional tort remedies. Where, as in the act, Congress expressly precludes preemption of state law tort actions, it is presumed that all traditional tort law remedies, including punitive damages, are available, unless Congress directs otherwise. (See, *e.g.*, *Silkwood v. Kerr-McGee Corp.*, 464 U.S. 238, 256 (1984) ("Congress assumed that State law remedies, in whatever form they might take, were available to those injured").)

There is little, if any, difference in kind between punitive damages available in tort and criminal penalties. (See *Smith v. Wade, supra*, 461 U.S. at 49 (punitive damages are designed to punish and deter outrageous behavior, that is, behavior linked with a criminally culpable mental state); *Restatement of the Law of Torts, Second* 908, comment b (1977) ("[punitive] damages can be awarded only for conduct for which this remedy is appropriate—which is to say, conduct involving some element of outrage similar to that usually found in crime"); *Prosser on Torts*, 2, at 9 (5th ed. 1984) (punitive damages are the civil analogue of criminal penalties).) In *Silkwood v. Kerr-McGee Corp.*, 464 U.S. 238 (1984), for example, the U.S. Supreme Court rejected the argument that because a "State-authorized award of punitive damages ... punishes and deters conduct relating to radiation hazards," id. at 249, it is preempted by the Atomic Energy Act:

> In sum, it is clear that in enacting and amending the Price-Anderson Act, Congress assumed that state-law remedies, in whatever form they might take, were available to those injured by nuclear incidents. This was so even though it was well aware of the NRC's exclusive authority to regulate safety matters. No doubt there is tension between the conclusion that safety regulation is the exclusive concern of the federal law and the conclusion that a State may nevertheless award damages based on its own law of liability. But as we understand what was done over the years in the legislation concerning nuclear energy, Congress intended to stand by both concepts and to tolerate whatever tension there was between them. We can do no less. It may be that the award of damages based on the state law of negligence or strict liability is regulatory in the sense that a nuclear plant will be threatened with damages liability if it does not conform to state standards, but that regulatory consequence was something that Congress was quite willing to accept.

Id. at 256.

In short, since Section 4(b)(4) of the act preserves state law tort actions for compensatory and punitive damages from preemption, and since punitive damages and criminal penalties serve virtually identical law enforcement purposes, there is no statutory or logical reason to preserve the one and preempt the other. Given that Section 4(b)(4) preserves state law tort actions—which assuredly regulate conduct by imposing a generally applicable state law standard of care—and given that the act's anti-preemption provision simply does not expressly distinguish between criminal actions and tort actions, there is no basis for act preemption of criminal actions. (See *Chicago Magnet Wire, supra*, Op. at 10 ("if Congress, in OSHA, explicitly declared it was willing to accept the incidental regulation imposed by compensatory damages awards under State tort law, it cannot plausibly be argued that it also intended to preempt State criminal law because of its incidental regulatory effect on workplace safety").)

One final point in this regard. The fact that the criminal indictments in *Pymm Thermometer* and *Chicago Magnet Wire* specifically referred to OSHA standards governing employee exposure to toxic chemicals is of no moment. Even if it became necessary for the state to implicate a federal safety or health standard in order to prove that a crime has been committed, that consideration would not counsel a result in favor of preemption. The weight of authority is that state law tort actions in which private plaintiffs rely on OSHA standards to establish a per se negligence claim or as evidence that an employer did not act reasonably are not preempted. As the U.S. Court of Appeals for the First Circuit explained:

> [W]hile the Fifth Circuit has found, along with every other court that has considered this issue, that section 653(b)(4) was designed to ensure that OSHA did not permit injured employees to bypass applicable state worker's compensation schemes through a private action in federal court, it has distinguished this limitation on the uses of OSHA from a limitation that would prevent violations of OSHA regulations from having the same consequences under existing common law and statutory schemes as violations of any other regulatory statute. [*Practico v. Portland Terminal Co., supra*, 783 F.2d, at 265 (footnotes omitted).]

(See also, *e.g., Rabon v. Automatic Fasteners, Inc.*, 672 F.2d 1231, 1238 (5th Cir. 1982); *Knight v. Burns, Kirkley & Williams Constr. Co.*, 331 So. 2d 651, 654 (Ala. 1976). But see *Wendland v. Ridgefield Const. Services*, 184 Conn. 173, 439 A.2d 954, 956-57 (Conn. 1981).) That reasoning applies as well in a criminal case.

Narrow Criminal Provisions

The proponents of preemption are reduced to pointing to the narrowly circumscribed criminal provisions in the act and then claiming that it is proper to imply a congressional intent to preempt all state criminal laws touching on employee health and safety from these sections. That claim, as the Department of Justice has concluded,[20] is incorrect. The enactment of federal criminal provisions whose coverage overlaps in one minimal respect with that of state criminal laws does not, by itself, manifest a congressional intent to occupy the field and preempt the states. In *California v. Zook*, 336 U.S. 725 (1949), the U. S. Supreme Court, in holding that the Federal Motor Carrier Act does not preempt a California statute prohibiting the sale of transportation over the state's public highways without a permit from the Interstate Commerce Commission, stated:

> [The] automatic 'coincidence means invalidity' theory, applied in an area as imbued with the State's interest as is this one, ... would lead us to the conclusion that a state may not make a dealer in perishable agricultural commodities respect its laws on the fraudulent nonpayment of an obligation, if that fraud occurred after an interstate shipment, 7 USC 499(b)(4), for Congress has not explicitly saved such prosecutions. We would hold, too, that extortion or robbery from interstate commerce under 18 USC 1951 or 18 USC 2117 is immune from state action; that the wrecking of a bridge over an interstate railroad is an "exclusively federal" offense, 18 USC 1992; that the transmittal of a ransom note in interstate commerce cannot be punished by local authorities, 18 USC 875.... *In short, we would be setting aside great numbers of state statutes to satisfy a congressional purpose which would be only the product of this Court's imagination.*

Id., at 732-733 (emphasis supplied).

In this connection, it should be kept in mind that general criminal laws protect different community concerns than do regulatory standards, even when they have overlapping governance of the same conduct. Most significantly, criminal law deters, punishes, and stigmatizes blameworthy behavior.

The fact that a state administrative scheme regulating occupational safety and health might be preempted by the act does not mean that state criminal laws, which serve different concerns, should be similarly treated. (See *Chicago Magnet Wire, supra*, Op. at 6 (unlike administrative enforcement of act standards, criminal law "serves to punish as a matter of retributive justice," "reaches out to regulate conduct in society in general," and punishes conduct where the defendant possesses "the charged mental state"); *New Jersey State Chamber of Commerce v. Hughey*, 774 F.2d 587, 593 (3d Cir. 1985) (OSHA disclosure standard preempts similar state standard only insofar as the latter deals with employee protection and does not preempt provisions of state law designed to protect against broader environmental and community hazards). See also Note, 101 *Harv. L. Rev., supra*, at 546.)

NO INTENT TO NULLIFY PROTECTION

In sum, as the Illinois Supreme Court concluded in *Chicago Magnet Wire, supra*, Op. at 7: "[i]t cannot be said that it was the clear and manifest purpose of Congress to preempt the application of State criminal laws for culpable conduct of employers simply because the same conduct is also governed by OSHA occupational safety and health standards."

Certainly the act's criminal provisions are not the kind that can be said to raise a rational inference that Congress intended to occupy the field of criminal law relating to employee health and safety. The act makes it criminal (a) to give unauthorized advance notification of an OSHA inspection (17(f)); (b) to knowingly make false statements on any filing with OSHA (17(g)); and (c) to willfully violate an OSHA standard, rule, or order if "that violation caused death to any employee" (17(e)). The penalty for the intentional violation of an OSHA standard that causes the death of an employee is "a fine of not more than $10,000 or ... imprisonment for not more than six months, or ... both." *Id.*

The first two criminal penalties were intended only to ensure the integrity of the act's administrative scheme. The third—the penalty for murder—is a narrowly circumscribed criminal penalty. It is simply implausible to contend that Congress, in enacting a statute to ensure worker health and safety, would nullify all criminal law protection of employees by penalizing employers only if their willful conduct kills an employee, ignoring every other circumstance in which an employee is intentionally injured.

Moreover, had Congress intended the act's criminal provisions to serve as a complete substitute for state criminal law in this regard, common sense dictates that the penalty provision it enacted surely would have exceeded the mere "slap on the wrist" provided by Section 17(e).[21]

In these circumstances, the *Chicago Magnet Wire* court correctly concluded:

> OSHA provides principally civil sanctions and only a few minor criminal sanctions for violations of its standards. Even for willful violations of OSHA standards which result in an employee's death an employer can be sentenced only to a maximum of six months' imprisonment. There is no penalty provided for conduct which causes serious injury to workers. It seems clear that providing for appropriate criminal sanctions in cases of egregious conduct causing serious or fatal injuries to employees was not considered. Under these circumstances, it is totally unreasonable to conclude that Congress intended that OSHA's penalties would be the only sanctions available for wrongful conduct which threatens or results in serious physical injury or death to workers.

Op. at 8.

Section 18 Not Relevant to Preemption Issue

The proponents of preemption seek to evade the dispositive force of Section 4(b)(4) by claiming that Sections 18(a) and (b) of the act expressly forbid the application of any state law that in any manner regulates conduct already regulated by OSHA. This contention is, we submit, wrong. Section 18 is of no relevance to the issue under consideration because it has no application to state law criminal actions. Rather, the singular focus of Section 18 is directed to an entirely different federal concern and preempts only one, narrow area of state action—that is, state health and safety standard setting.

There is a fundamental conceptual distinction between two different types of laws that concern worker health and safety. On the one hand, battery, negligence, and worker compensation laws—such as those referenced in Section 4(b)(4)—make conduct by an employer that in fact injures a particular employee actionable after the injury. The theory of these laws is that conduct which in fact causes harm should result in compensation to the victim or punishment of the person who causes the harm, or both.

So far as we are aware, as of 1970 when the act became law, every state made employer conduct that in fact injured an employee actionable—in worker's compensation in almost all circumstances, in tort in some circumstances, and, if the action was the product of a criminal intent, in the criminal law as well. And, as far as we have been able to ascertain, the employer-employee relationship has never been held to privilege employer action against an employee that constitutes criminal assault or battery under the state's common law or statutory definitions of those crimes.

In contrast, employee health and safety may also be protected by prophylactic regulation of employer conduct enforceable by injunction and/or fines for breach of the protective rules that are promulgated without regard to whether a particular employee of the employer against whom enforcement action is taken has yet been harmed. The federal administrative scheme implemented by the act is an example of such a prophylactic approach.

While such preventive regulation has the same general aim as laws that make harmful conduct actionable after the harm occurs—*viz.*, to ensure a higher level of safe employer conduct—the means employed to reach that end are, as we have noted, qualitatively different. The preventive regulatory regime requires the promulgation of quite specific standards covering workplace conduct, a system of inspection, citations for failure to meet the standards, and fines and/or abatement orders to correct any failure to comply with the regulations.

This regulatory approach to protecting employees was not widespread in 1970. Only a small number of states had an extensive system for preventing injuries, and the act was enacted to fill this need. (See *Whirlpool Corp. v. Marshall*, 445 U.S. 1, 12 (1980) ("The [OSH] Act does not wait for an employee to die or become injured. It authorizes the promulgation of health and safety standards and the issuance of citations in the hope that these will act to prevent deaths or injuries from ever occurring").) Congress in enacting OSHA had concluded that there would be no further progress in this regard unless a federal system was put in place that empowered the secretary of labor to set standards establishing meaningful workplace protections:

> No one has seriously disputed that only a relatively few states have modern laws relating to occupational safety and health and have devoted adequate resources to their administration and enforcement. Moreover, in a state-by-state approach, the efforts of the more vigorous states are inevitably undermined by the shortsightedness of others ...
>
> In sum, the chemical and physical hazards which characterize modern industry are not the problem of a single employer, a single industry, nor a single state jurisdiction. The spread of industry and the mobility of the workforce combine to make the health and safety of the worker truly a national concern. [S. Rep. 91-1282, 91st Cong., 1st Sess., p. 144 (1970).]

As the *Chicago Magnet Wire* court explained:

> We judge that the purpose underlying section 18 was to ensure that OSHA would create a nationwide floor of effective safety and health standards and provide for the enforcement of those standards. (See *United Airlines, Inc. v. Occupational Safety & Health Appeals Board* (1982), 32 Cal. 3d 762, 654 P.2d 157, 187 Cal. Rptr. 387.) It was not fear that the States would apply more stringent standards or penalties than OSHA that concerned Congress but that the States would apply lesser ones which would not provide the necessary level of safety.

Op. at 8.

COMPREHENSIVE REGULATORY SCHEME

Thus, the 1970 Congress recognized the need for a comprehensive preventive regulatory scheme, but that Congress also ensured that the widespread system of after-the-fact compensation of victims of industrial injury and of punishment of employers who wrongfully inflict such injury would continue as it existed at the time. Congress, as we have seen, provided that assurance in Section 4(b)(4)'s anti-preemption language. Sections 18(a) and (b) of the act, upon which the proponents of preemption rely, provide:

(a) Nothing in this chapter shall prevent any State agency or court from asserting jurisdiction under State law over any occupational safety or health issue with respect to which no standard is in effect under Section 6 of this title.

(b) Any State which, at any time, desires to assume responsibility for development and enforcement therein of occupational safety and health standards relating to any occupational safety or health issue with respect to which a Federal standard has been promulgated under Section 6 of this title shall submit a State plan for the development of such standards and their enforcement.

This section preempts only the states' jurisdiction over an "occupational health and safety issue" covered by a "federal standard" absent the state's affirmative "assum[ption of] responsibility for development and enforcement of an occupational safety and health standard" on the same issue. Sections 18(a) and (b), in other words, forbid states to assert jurisdiction over, develop, or enforce occupational safety and health standards if a federal safety or health standard is already in effect, unless the state operates under a plan approved by the secretary of labor. (See *United Steelworkers of America v. Auchter*, 763 F.2d 728, 738 (3d Cir. 1985).)[22]

Section 18's language shows a congressional intent, then, to preempt state authority only with respect to prophylactic regulation of employer conduct through a regulatory regime. Section 18 simply does not reach state law actions designed to compensate an injured employee or to punish an employer who injures an employee while acting with criminal intent.[23]

Consistent with the words of Section 18(a) and (b), the structure of Section 18 taken as a whole reveals that Congress' focus in this provision was on state standard setting.

In Section 18(c) — the subsection immediately following the preemption provision — Congress sets out the prerequisites for state "development and enforcement of safety and health standards": the creation of an administrative agency, the formulation of standards that will be "at least as effective" as those promulgated by the federal agency, and the creation of an administrative inspection and enforcement mechanism that will be "at least as effective" as that of the federal agency. The remainder of Section 18 sets out further administrative requirements governing the submission, rejection, supervision, and termination of state plans. (See 18(d), (e), (f), (h).) None of those requirements has anything to do with the passing and enforcing of a general state law provision prohibiting, for example, criminal assault.[24]

In sum, it was only the administrative regulation of occupational safety and health standards by states lacking an approved plan that Congress decided to forbid.

CONCLUSION

We believe that it would be paradoxical in the extreme for any court to hold that a Congress whose primary goal was "to assure so far as possible every working man and woman in the Nation safe and healthful working conditions" (29 USC 651(b)) nonetheless enacted a statute that deprived employees of the longstanding protection provided by state criminal laws.[25]

ENDNOTES

[1] Experience has shown that employers can derive substantial economic benefits by virtue of increased efficiency and productivity resulting from implementing state-of-the-art technology. This was precisely what happened when cotton textile manufacturers implemented the necessary engineering controls to comply with the cotton dust standard.

[2] Experience also has shown that employers have a penchant for exaggerating the costs of achieving compliance with health standards. Achieving compliance with the vinyl chloride standard, for example, cost only a small percent of the estimates tendered during rulemaking.

[3] One serious deficiency in the act itself is that any employer can obtain a stay of any obligation to abate a hazard merely by filing a notice of contest (29 USC 659).

[4] The refusal of the Reagan administration to promulgate more health standards is especially disquieting given that many previously identified toxic chemicals had yet to be regulated and that the U.S. Supreme Court at the beginning of that administration made clear that the secretary of labor has the authority—indeed, the responsibility—to promulgate the most worker-protective health standards, subject to only one real limitation: technological feasibility. *American Textile Manufacturers Institute v. Donovan, 452 U.S. 490 (1981).*

[5] Recently and for the first time, OSHA has, in several cases imposed substantial civil penalties on companies that have engaged in a pattern of illegal conduct.

[6] See discussion beginning on page 35.

[7] Ironically, one explanation offered for the reluctance of local U.S. attorneys to initiate criminal prosecutions is that because of the question "whether a sentence sufficient to the seriousness of the offense and to the necessary investment of prosecutorial resources could be obtained in the event of a conviction. In this regard, we think it likely that the increased fines available will increase the prosecutive appeal of OSHA cases." Letter from Thomas M. Boyd, assistant attorney general, to Rep. Tom Lantos, at 2-3.

[8] Apart from its dismal record in enforcing 17(e), OSHA also has shown a similar reluctance to file civil suits to obtain court-ordered injunctions to enjoin employers from maintaining workplace conditions and practices that pose "imminent danger" to workers. (See 13(a)-(c) of the act, 29 USC 662(a)-(c).)

[9] For example, the state of California alone has successfully prosecuted 112 cases in the past eight years (Rep. of the Comm. on Government Operations, supra at 4).

[10] See "Prosecutions for Workplace Hazards," 121 LRR 280 (April 21, 1986), an article that summarizes a variety of other prosecutions initiated by state and local law enforcement authorities.

[11] The jury also found defendants guilty of falsifying business records in an attempt to conceal their acts.

[12] Although it is of no consequence to our discussion, the trial court also set aside the defendants' conviction on two specific counts of the indictment, charging conspiracy in the fifth degree and reckless endangerment in the second degree, on the ground that the evidence produced was not legally sufficient to support a finding of guilt on these two counts.

[13] On Feb. 21, 1989, the Supreme Court of Illinois issued a stay of its mandate pending the defendants' application to the U.S. Supreme Court for a writ of certiorari.

[14] See letter from Thomas M. Boyd, assistant attorney general, Office of Legislative and Intergovernmental Affairs, to Rep. Tom Lantos, chairman, House Subcommittee on Employment and Hiring, Committee on Government Operations (dated Dec. 9, 1988 at 3). The entire letter is reprinted as Appendix B of this report.

[15] As the solicitor has explained, the Department of Labor has been watching developments "percolate" and intends to stay its hand until a case is presented to the U.S. Supreme Court.

[16] This portion of the article is a synopsis of several amicus briefs filed on behalf of the AFL-CIO with respect to the preemption issue. The author participated in the preparation of those briefs in conjunction with Laurence Gold, AFL-CIO general counsel, and two of Mr. Cohen's colleagues, Robert M. Weinberg and Virginia Seitz.

[17] 4(b)(4) provides:

Nothing in this chapter shall be construed to supersede or in any manner affect any workmen's compensation law or to enlarge or diminish or affect in any other manner the common law or statutory rights, duties, or liabilities of employers and employees under any law with respect to injuries, diseases, or death of employees arising out of, or in the course of, employment.

17(e), (f) and (g) provide:

(e) Willful violation causing death to employee

Any employer who willfully violates any standard, rule, or order promulgated pursuant to section 6 of this title, or of any regulations prescribed pursuant to this chapter, and that violation caused death to any employee, shall, upon conviction, be punished by a fine of not more than $10,000 or by imprisonment for not more than six months, or by both; except that if the conviction is for a violation committed after a first conviction of such person, punishment shall be by a fine of not more than $20,000 or by imprisonment for not more than one year, or by both.

(f) Giving advance notice of inspection

Any person who gives advance notice of any inspection to be conducted under this chapter, without authority from the Secretary or his designees, shall, upon conviction, be punished by a fine of not more than $1,000 or by imprisonment for not more than six months, or by both.

(g) False statements, representations, or certification

Whoever knowingly makes any false statement representation, or certification in any application, record, report, plan, or other document filed or required to be maintained pursuant to this chapter shall, upon conviction, be punished by a fine of not more than $10,000, or by imprisonment for not more than six months, or by both.

18(a) & (b) provide:

(a) Assertion of State standards in absence of applicable Federal standards

Nothing in this chapter shall prevent any State agency or court from asserting jurisdiction under State law over any occupational safety or health issue with respect to which no standard is in effect under section 6 of this title.

(b) Submission of State plan for development and enforcement of State standards to preempt applicable Federal standards

Any State which, at any time, desires to assume responsibility for development and enforcement therein of occupational safety and health standards relating to any occupational safety or health issue with respect to which a Federal standard has been promulgated under section 6 of this title shall submit a State plan for the development of such standards and their enforcement.

[18] See also *Patterson v. New York*, 432 U.S. 197, 201 (1977) ("preventing and dealing with crime is much more the business of the States than it is of the Federal Government"); *Abbate v. United States*, 359 U.S. 187, 195 (1959) ("the States under our federal system have the principal responsibility for defining and prosecuting crimes"); *Kelly v. Washington ex rel. Foss Co.*, 302 U.S. 1, 13 (1937) (the principle that states may exercise their

police power unless it directly conflicts with federal action "is strongly fortified where the State exercises its power to protect the lives and health of its people").

[19] It is noteworthy that the courts have consistently found that in 4(b)(4) Congress sought to ensure that the enactment of the act neither expanded nor contracted the pre-existing state law rights and responsibilities of employers and employees for injuries arising out of employment. See, e.g., *Practico v. Portland Terminal Co.,* 783 F.2d 255, 266 (1st Cir. 1985) the reading " ... that observes both [4(b)(4)'s] language and ... spirit is that no new duty is to be exacted [by reason of the act] where none existed before and, equally, no existing duty is to be excused by reason of [act] regulation") (quoting *Provenza v. American Export Lines, Inc.,* 324 F.2d 660, 665 (4th Cir. 1963), cert. denied, 376 U.S. 952 (1964).)

Although not directly relevant here, *Practico* makes clear that one of Congress' purposes in 4(b)(4) was to ensure that the act did not create a new private cause of action for employees injured on the job. See, e.g., *United Steelworkers v. Marshall,* 647 F.2d 1189, 1235-36 (D.C. Cir. 1980), cert. denied, 453 U.S. 913 (1981). In other words, the case law recognizes that while the act does establish an administrative scheme to enforce promulgated occupational safety and health standards, the act does not create a federal private cause of action in tort for damages in addition to those already existing in state law.

[20] See discussion note 14.

[21] In point of fact, the penalty provision was incorporated into the act at the last minute in a context which further demonstrates that Congress did not make a considered judgment to nullify application of state criminal laws by enactment of this lone provision. See Conference Report on S.2193 at 41, reprinted in Legislative History of the Occupational Safety and Health Act of 1970, at 1194, 1210 (statement of Rep. Daniels).

[22] It should be noted that our discussion focuses on 18(b), not 18(a), because the former section states the applicable rule when a federal standard has been adopted. Subsection (b) extends solely to the development and enforcement of state standards relating to occupational safety and health issues. It does not refer to state jurisdiction over occupational safety and health issues that is, in any event, expressly preserved over those issues as to which no federal standard has been adopted in subsection (a).

[23] See also *Chicago Magnet Wire,* supra, Op. at 5 ("section 18 refers only to a State's development and enforcement of 'occupational health and safety standards.' (29 USC 667(a) (1982).) Nowhere in section 18 is there a statement or suggestion that the enforcement of state criminal law as to federally regulated workplace matters is preempted unless approval is obtained from OSHA officials").

[24] This construction of section 18 is further supported by the narrow definition of the term "occupational safety and health standard" which appears in another section of the act:

The term 'occupational safety and health standard' means a standard which requires conditions, or the adoption or use of one or more practices, reasonably necessary or appropriate to provide safe or healthful employment and places of employment. [29 USC 652(8).]

Section 18 forbids states without an approved plan to set or enforce "standards" within the meaning of the act. As noted, occupational safety and health standards by their very nature are forward-looking and preventive; their point is to establish norms that are enforced through inspections, injunctive orders and civil fines without regard to whether an injury has yet occurred. "There is no indication that Congress intended the term 'standards' to include general criminal laws or for section 18 to preempt or require federal approval of state enforcement of general criminal laws." (Note, 101 Harv. L. Rev., supra, at 542.) OSHA's administrative scheme bears no relationship to statutes compensating injured workers after the fact or punishing criminal wrongdoing that leads to occupational injuries. *See id.,* at 543 ("[criminal laws] do not prescribe specific practices, means, methods, operations, or processes' as such, but rather focus after the fact on whether conduct causing an injury in particular circumstances was blameworthy, as measured by general societal norms"). Cf. *P&Z Co. v. District of Columbia,* 408 A.2d 1249, 1250 (D.C. App. 1979) (act does not preempt D.C. accident reporting requirements); *West Va. Mfgrs. Assn. v. West Virginia,* 714 F.2d 308 (4th Cir. 1983) (OSHA's toxic substance standards do not preempt state regulations of notice and posting requirements for toxics).

CRIMINAL PROSECUTIONS

[25] Were the U.S. Supreme Court to decide otherwise, this would necessarily trigger the need for a major overhaul of section 17 of the act to incorporate much more comprehensive criminal provisions and much more severe criminal penalties. Even were those statutory changes in place, there would remain the nagging question whether federal prosecutors thereafter would devote the manpower and resources to effectively enforce the amended statutory scheme.

Use of General Criminal Laws

By Robert C. Gombar, Willis J. Goldsmith, Frank A. White, and Arthur G. Sapper*

The recent decision of the Illinois Supreme Court in *Illinois v. Chicago Magnet Wire Corp.*[1] is the latest decision on the legality of state and local criminal prosecutions against employers for workplace injuries to employees. The legal question presented by *Chicago Magnet Wire* and similar cases can be summarized as follows: Does Section 18 of the Occupational Safety and Health Act of 1970, 29 USC 651-678, allow a state without a federally approved state plan to prosecute an employer under general criminal statutes for exposing employees to harmful workplace conditions governed by federal occupational safety and health standards?

Section 18 of the act preempts states from "asserting jurisdiction under State law over any occupational safety and health issue" governed by a federal Occupational Safety and Health Administration (OSHA) standard. OSHA must approve a state's occupational safety and health plan before the state can take action. Twenty-two states have federally approved plans covering private-sector employment. Illinois is one of the 28 states that have chosen not to have such a plan. When a state without a federally approved plan attempts to apply its general criminal laws to workplace conditions governed by OSHA standards, it is "asserting jurisdiction under state law" in a way that is prohibited by the broad language of Section 18.

In *Chicago Magnet Wire*, however, the Illinois Supreme Court found that the federal act does not preempt state or local criminal prosecutions. The court thus widened a split among the state courts that have considered the preemption issue.[2] Only a ruling by the U.S. Supreme Court or clarifying legislation by Congress is likely to end the controversy.

A Red Herring: Getting Away With Murder

In *Chicago Magnet Wire*, the Illinois attorney general argued that preemption would allow employers to kill their employees purposefully. Preemption would allow an employer who lusted after an employee's wife to place a fatal dose of an OSHA-regulated substance purposefully in the employee's coffee, the state argued.[3]

Such hypothetical situations are irrelevant and are designed to mislead. Purposeful killings raise no "occupational safety and health issues" within the meaning of Section 18 of the act.[4] They do not concern an employee's working conditions. Purposeful killings are not related to the victim's status as an employee or to the employer's efforts to manufac-

---

The authors practice occupational health and safety law in the Washington, D.C., office of the law firm of Jones, Day, Reavis & Pogue.

ture a product or render a service. Occupational safety and health issues are issues raised by workplace conditions—however severe—to which employees may be exposed as a consequence of their employment. No employer bent on poisoning an employee by adulterating the employee's coffee can escape criminal prosecution by relying on the preemption provisions of Section 18 of the act.

Section 18 Means What It Says

Unlike many federal laws, the Occupational Safety and Health Act declares when state law will, and will not, be preempted.[5] Section 18 of the act, 29 USC 667, states in part:

> (a) Nothing in this Act shall prevent any State agency or court from asserting jurisdiction under State law over any occupational safety or health issue with respect to which no standard is in effect under section 6.
> (b) Any State which, at any time, desires to assume responsibility for development and enforcement therein of occupational safety and health standards relating to any occupational safety or health issue with respect to which a Federal standard has been promulgated under section 6 shall submit a State plan for the development of such standards and their enforcement.

The federal courts uniformly have held that Section 18 "expressly preempts state law pertaining to issues that are addressed by OSHA standards unless a state plan has been approved by the Secretary."[6] Moreover, an OSHA interpretive regulation, 29 CFR 1901.2, adopted soon after the passage of the act, construes Section 18 as: "preventing any State agency or court from asserting jurisdiction under state law over any occupational safety or health issue with respect to which a Federal standard has been issued under section 6 of the Act."

Thus, federal case law and OSHA's regulation make clear that no state agency or court may assert jurisdiction under state law over any occupational safety or health issue addressed by a federal standard. It seems crystal clear that a criminal prosecution under a general criminal statute for a workplace condition covered by an OSHA standard constitutes an assertion of "jurisdiction" by a "state agency or court" over a "safety or health issue" addressed by a federal standard. That conclusion is buttressed further when one looks at the charges that were brought in *Chicago Magnet Wire*.

OSHA Regulation In Disguise

After a full investigation by OSHA, resulting in the issuance of citations for violations of OSHA standards in December 1983, the grand jury for Cook County, Ill., indicted Chicago Magnet Wire Corp. and five of its officers for the crimes of "reckless conduct" and "aggravated battery." The essence of the criminal charges was that the defendants exposed employees to various toxic substances such as xylene and naptha, failed to provide safety instructions and safety equipment such as respirators and impervious clothing, and failed to provide adequate ventilation. The prosecutors alleged that, as a result of these failures, workers suffered harmful exposures to toxic substances. In addition, it was alleged that the defendants violated a "duty to provide ... a safe workplace." No one disputed that the issues addressed by the indictments are governed by OSHA standards.[7]

The indictments in *Chicago Magnet Wire* do not expressly allege that the defendants violated any OSHA standards. They were couched in terms of crimes such as "aggravated battery." But a criminal prosecution focusing on workplace conditions is an OSHA case in disguise.

First, any such criminal prosecution will inevitably mimic an OSHA enforcement action. If the defendant violated an OSHA standard, the prosecutor probably would introduce evidence of the violation to show that the defendant acted unreasonably.[8] A defendant who complied with the OSHA standard, on the other hand, will point to its compliance as strong evidence that it acted reasonably. In either case, the parties will be litigating an OSHA violation. As a result, inexpert state and county prosecutors, judges, and juries will interpret and apply OSHA standards, determine the reliability of sampling methods and laboratory analysis, weigh expert testimony about the existence or harmfulness of the employees' exposure, judge whether engineering controls are feasible, and so on — the very heart of a typical OSHA case.

Second, if neither party mentions the OSHA standard, or if the prosecutor argues that compliance with it is not enough, the prosecution will implicitly attempt to fashion a standard of conduct *ad hoc*. The trial will become an OSHA mini-rulemaking proceeding. The prosecutor will introduce scientific literature and expert testimony establishing, for example, the concentration and exposure duration at which an airborne chemical is harmful, whether the employee received a harmful dose, and whether the employer failed to use technologically and economically feasible measures to control exposures. The defense will attempt to rebut this evidence, and the jury will weigh it. All of these issues are, of course, central to the fashioning of an OSHA standard.[9]

Third, and most important, through a jury verdict and appellate case law, the prosecutor will establish duties of care identical to, similar to, or stricter than those imposed by OSHA standards. A criminal conviction will establish not only that the defendant had a wrongful state of mind, but that a condition in the workplace was unreasonably dangerous. Local juries and prosecutors will determine, for example, how much ventilation is necessary or what sort of protective gear must be worn. Leadership in the occupational safety and health field will pass from OSHA and be shared among state and local prosecutors. Confusion and uncertainty will replace uniformity. Candid state and local prosecutors have said they bring criminal prosecutions to regulate workplace safety. Some even have established their own administrative programs to manage their regulatory efforts. For example, Brooklyn District Attorney Elizabeth Holtzman, who has sharply criticized what she describes as "OSHA's do-nothing approach," told a congressional committee:

> [T]he problem of danger in the workplace is one ... that is serious and national in scope .... As a district attorney, I have tried to respond to the problem by setting up in my own office a special Environmental Crimes and Work Site Safety Unit which seeks to uncover and prosecute violations of State law that occur when employers do not protect employees from dangerous substances.[10]

During the same hearing, Kenneth Oden, Travis County, Texas, district attorney, discussed OSHA's shortcomings in explaining his efforts to prosecute employers for workplace conditions:

> In 1986, the State of Texas, a pretty big place, had only 87 OSHA inspectors for a work force of between six and seven million people strung out over a large geographic area. ... These are people who are asked to do an impossible job and ... [t]hey are not getting the job done.[11]

The Preemption Decisions

Contrary to the Illinois Supreme Court, other state courts have recognized that the use of general criminal laws to regulate occupational safety and health is inconsistent with Section 18 of the act. In *Colorado v. Kelran Construction Inc.*, 13 OSHC 1898, 1900 (Colo.Dist.Ct. 1988), the court rejected the argument that prosecutions under general criminal laws are not preempted:

> The determinative factor is not whether the State is seeking to enforce a criminal law of general application rather than specific workplace health and safety regulations but rather what conduct the State is seeking to regulate .... [T]he conduct the People here seek to regulate is the conduct related to working conditions which is exactly the same as that regulated by OSHA.

Similarly, in *Accord, Sabine Consolidated Inc. v. Texas*, 756 S.W.2d 865, 13 OSHC 1881, 1883 (Tex.Ct.App. 1988) the court ruled that the "practical effect of such charges is to set up a body of state law affecting workplace safety issues already governed by federal standards ...." These decisions correctly analyze the issue. As the U.S. Supreme Court has stated, "It is the conduct being regulated, not the formal description of governing legal standards, that is the proper focus of concern [in a preemption case]."[12]

The Illinois Supreme Court, however, did not accept this view. It found nothing in the "structure of OSHA" or its legislative history indicating that "Congress intended to preempt the enforcement of State criminal law prohibiting conduct of employers that is also governed by OSHA safety standards." The court also stated that "the purpose underlying section 18 was to ensure that OSHA would create a nationwide floor of effective safety and health standards .... It was not fear that the States would provide more stringent standards or penalties ... that concerned Congress but that the States would apply lesser ones ...."[13]

Despite its references to the "structure of OSHA" and its legislative history, the court hardly discussed either. It also did not examine closely the language of Section 18, which draws no distinction between criminal and civil laws. A review of the structure and legislative history of the act shows that Section 18's sweeping language was chosen deliberately. It reveals that Congress was well aware of state efforts to regulate workplace safety using criminal laws and that Congress intended to preempt these efforts with an enforcement scheme heavily weighted in favor of civil sanctions.

CONGRESS MEANT TO PREEMPT STATE CRIMINAL LAWS

The legislative history of the act establishes that Congress had full knowledge of how the states had gone about the regulation of occupational safety and health. The departments of Labor and Health, Education, and Welfare had submitted to Senate and House committees considering workplace safety legislation several comprehensive reports describing the states' programs.[14]

From these documents, Congress knew that the states had passed hundreds of occupational safety and health regulations and that they were commonly enforced by criminal sanctions.[15] Some documents submitted to the Senate committee even noted the possibility of state prosecutions of employers for criminal negligence.[16] Significantly, Congress nevertheless found that the states' frequent use of criminal sanctions had proven counterproductive.[17]

Congress also concluded that, although some states had excellent state enforcement programs,[18] state laws generally were so inadequate that oversight of state efforts by the U.S. secretary of labor was needed to ensure "a comprehensive, nationwide approach."[19] Congress found that even states with superior programs had very different standards and that the imposition of uniformity over "a jumble of differing State laws"[20] was desirable.

In crafting the Occupational Safety and Health Act, Congress could have preserved the better state programs, just as it had only a year before in the Federal Coal Mine Health and Safety Act of 1969.[21] That statute, drafted by the same Senate and House committees that drafted the Occupational Safety and Health Act, stated expressly that more stringent state laws would not be superseded by federal regulations.[22] The coal mine act thus allowed existing state mining laws to supplement federal mining standards without going through a federal approval procedure.

But Congress allowed no such supplementation of federal enforcement under the Occupational Safety and Health Act. Indeed, it rejected proposals to follow the coal mine act's supplementary enforcement model.[23] Congress decided instead to displace state laws and establish a system for the "development and administration, by the Secretary of Labor, of uniformly applied occupational safety and health standards."[24] In short, Congress intended to have Section 18 strip all states of their workplace health and safety powers until each state, on a state-by-state basis, obtained federal approval to recover such powers.[25]

The legislative history also shows that Congress decided, after considerable thought, that regulation of occupational safety and health was best accomplished through an enforcement scheme heavily weighted in favor of civil sanctions. For example, HR 4294, introduced by Rep. Perkins, chairman of the House Committee on Education and Labor, provided:

> (c) Any person who willfully violates or fails or refuses to comply with [the Secretary's standards] shall be guilty of a misdemeanor, and upon conviction shall be punished by a fine of not more than $5000 or by imprisonment for not more than six months, or by both ....[26]

The bills that ultimately were reported and passed by the House eliminated these broad criminal sanctions for violations of standards and substituted civil penalties for willful as well as other violations.[27]

The first administration-backed bill in the Senate, S 2788, and early substitute bills, such as S 4404, introduced by Sen. Dominick, provided only civil penalties for willful violations of standards and no penalties for other violations.[28] The bill ultimately passed by the Senate, S 2193, included civil penalties for all violations of standards except willful violations, for which criminal sanctions identical to those in the Perkins bill were included.

What finally emerged from the House-Senate conference committee was an enforcement scheme that adopted the civil penalty structure of the House-passed bill and added a scaled-down version of the Senate's criminal provision, limiting criminal penalties to willful violations of standards associated with the death of an employee.[29]

Conscious Rejection

The legislative history provides insight into why Congress favored civil rather than criminal sanctions. Congress was aware that, of the states with the poorest performance in assuring safe and healthful workplaces, many used either general or specific criminal sanctions to enforce various provisions of state safety and health law. In most cases, violations of state law were considered misdemeanors, punishable by fines or imprisonment or both.[30] Sen. Dominick, in remarks supporting his preference for civil sanctions over criminal penalties for willful violations in S 4404, explained:

> We did it this way because ... most of us know how difficult it is to get an enforceable criminal penalty in these types of cases. Over and over again, the burden of proof under a criminal type allegation is so strong that you simply cannot get there, so you might as well have a civil penalty instead of the criminal penalty and get the employer by the pocketbook if you cannot get him anywhere else.[31]

Thus, Congress consciously considered and rejected enforcement schemes that relied heavily on criminal sanctions because it believed that criminal remedies were not an effective method for assuring a safe and healthful workplace. It, therefore, would fly in the face of the act's legislative history to argue that Congress did not intend state criminal prosecutions to be among the assertions of jurisdiction by state agencies or courts that Section 18 would preempt.

It is important to bear in mind that if a state desires to establish more rigorous standards or enforcement tools, it is allowed to do so under Section 18(c) if it receives OSHA approval for such enhanced provisions. Thus, a state is not prevented from enforcing a criminal statute like Arizona's, which makes any employer whose knowing violation of the law results in an employee's death guilty of a felony punishable by a maximum penalty of a $150,000 fine and imprisonment for up to 18 months.[32] In California, any employer who knowingly or negligently commits any serious violation or who repeatedly violates any standard may be convicted of a misdemeanor.[33]

Like these states with OSHA-approved plans, several other states with approved plans have criminal provisions that are considerably more stringent than the provisions of the act.[34] In some cases, states with state plans may prosecute under their general criminal laws.[35]

Section 18(c) thus allows a great deal of flexibility in how a state shapes its new program, permitting wide latitude to develop stronger standards and more stringent enforcement than the federal program. A state, however, cannot avoid altogether the application of Section 18 by declaring that it will not seek federal approval for an occupational safety and health enforcement mechanism, but rather will apply general criminal law to workplace behavior covered by OSHA standards.

In turn, Section 18 and the legislative history of the act demonstrate clearly that Congress did not intend an exception to the broad preemptive effect of Section 18(a) to be made for state criminal sanctions. Congress carefully scrutinized the performance of the states with respect to workplace safety and health, including the record of applying criminal sanctions, and found it unacceptable.

Sen. Dominick's remarks about the difficulties of getting "an enforceable criminal penalty in these types of cases" is typical of the congressional reaction to the success of the states. As a result of this unacceptable state record, Congress chose to preclude assertions of jurisdiction by all state agencies and courts over safety and health issues not addressed by federal standards and to establish a comprehensive national program, relying on the development of uniform standards and a carefully chosen balance of civil and criminal sanctions.

At the same time, Congress provided states with the opportunity to re-establish their jurisdiction by setting comparable or more stringent standards and by adopting comparable or more stringent enforcement mechanisms.

The Act's Savings Clause — Section 4(b)(4)

Some people have argued, however, that Section 4(b)(4) of the act, 29 653(b)(4), provides an exception to the preemption rule of Section 18. This is simply not the case.

Section 4(b)(4) preserves suits under "workmen's compensation laws" and "any law with respect to injuries, diseases, or death of employees arising out of, or in the course of, employment."[36] General criminal laws are obviously not workmen's compensation laws. General criminal laws are also not laws "with respect to injuries, diseases, or death of employees arising out of, or in the course of, employment."

Under the common-sense maxim of statutory construction *ejusdem generis* ("of the same kind"), general words that follow a specific expression apply only to things akin to those of the specific expression.[37] Unlike workers' compensation laws, general criminal laws are not concerned with compensation for injured employees.

Moreover, Section 4(b)(4)'s legislative history makes it abundantly clear that it was intended to preserve only civil compensation remedies for injured employees. The Senate committee that drafted most of the act stated that Section 4(b)(4) would ensure that the act "does not affect any Federal or state workmen's compensation laws, or the right, duties, or liabilities of employers and employees under them."[38] The House committee described in almost the same terms the effect of an identical clause in a House bill.[39]

As the U.S. Court of Appeals for the First Circuit has pointed out, Congress feared that the federal and state courts would infer from the new legislation a private right of action for damages that would compete with or bypass state worker's compensation schemes.[40] Federal and state courts had long inferred such actions from remedial statutes, including workplace safety statutes.[41]

An exchange of letters among the American Society of Insurance Management, the solicitor of the U.S. Department of Labor, and the House Select Subcommittee on Labor shows that Section 4(b)(4) was intended to allay this concern. The insurance society stated that some of its members "are quite concerned that under proposed [occupational safety and health] legislation ... an injured employee could claim violation of the requirements and thus bypass the applicable state workmens' compensation benefits through an action in the federal courts."[42] The solicitor of labor assured the chairman of the House committee that:

> The provisions of S. 2788, the Administration's proposed Occupational Safety and Health Act of 1969, would in no way affect the present status of the law with regard to workmen's compensation legislation or private tort actions. Section 21(b) of S. 2788 [substantially identical to Section 4(b)(4)] is self-explanatory on this point ...[43]

The legislative history shows that Section 4(b)(4) served another function as well—to leave worker's compensation schemes untouched while they were studied by a commission established by the new law. Congress had come under some pressure to supplement or federalize worker's compensation, which had been widely attacked as inadequate and giving employers too little economic incentive to provide safe and healthful workplaces.[44] Congress decided to leave the situation as it was until the Workmen's Compensation Commission established by Section 27 prepared a report that would "provide a basis for an informed decision by Congress of this question [of federal involvement] in the future."[45] Succinctly stated, Sections 4(b)(4) and 27 show that Congress intended to preserve only the compensation remedies available to injured employees. In short, Section 4(b)(4) has nothing to do with state criminal laws.

Much Ado About Little

On Dec. 9, 1988, an assistant attorney general in the Department of Justice's Office of Legislative and Intergovernmental Affairs wrote a letter to a House subcommittee that had produced a report critical of the department's efforts to prosecute criminally violators of the act. (See note 3 above.) This letter touched briefly on the preemption issue:

> As for the narrower issues as to whether the criminal penalty provisions of the OSH Act were intended to preempt criminal law enforcement in the workplace and preclude the States from enforcing against employers the criminal laws of general application, such as murder, manslaughter, and assault, it is our view that no such general preemption was intended by Congress. As a general matter, we see nothing in the OSH Act or its legislative history which indicates that Congress intended for the relatively limited criminal penalties provided by the Act to deprive employees of the protection provided by State criminal laws of general applicability.

This passage has been widely publicized and was relied on by the court in *Chicago Magnet Wire*. What has escaped attention, however, is how little it says. The letter says only that the act's "criminal penalty provisions" were not intended to preempt state criminal prosecutions. The letter says nothing about the preemptive effect of the state plan provisions in Section 18.

CONCLUSION

In the final analysis, the arguments of prosecutors, unions, and other opponents of preemption rest upon what they believe to be sound public policy. They are convinced that because of OSHA's statutory and resource limitations, the ability of a state without a state plan to prosecute employers for crimes such as murder or manslaughter, when the elements of proof for such crimes are present, will have a significant added deterrent effect on unsafe conduct by employers in the workplace. Congress could have added a provision to Section 18(a) allowing assertions of jurisdiction under criminal laws of general applicability, irrespective of the existence of a federal standard, if it had found state criminal sanctions to have been effective in deterring and punishing criminal workplace conduct. Furthermore, armed with such a finding, Congress could have put broader criminal sanctions in the act itself. But Congress took neither step.

In fact, only recently have state and local prosecutions arising out of workplace safety and health conditions resulted in convictions. Perhaps if Congress were to review the current record of state criminal enforcement activity, it would re-evaluate the positive and negative impact of these recent developments and reconsider the balance that was struck almost 20 years ago. But, as the *Kelran* court pointed out, that is a decision for Congress, not the courts. Although state prosecutors may have "set forth valid public policy considerations in arguing that state and local prosecution should not be preempted by OSHA ... Congress and not [the courts] must determine the merits of these public policy considerations."[46]

ENDNOTES

[1] 13 OSHC 2001 (Ill. S.Ct. Feb. 2, 1989), rev'g 157 Ill.App.3d 797, 510 N.E.2d 1173, 13 OSHC 1337 (1987).

[2] See, in addition to *Chicago Magnet Wire*, *Wisconsin ex rel. Cornelier v. Black*, 425 N.W.2d 21, 13 OSHC 1761 (Wis.Ct.App. 1988)(no preemption); *Sabine Consolidated Inc. v. Texas*, 756 S.W.2d 865, 13 OSHC 1881 (Tex.Ct.App. 1988)(preemption); *Colorado v. Kelran Construction Inc.*, 13 OSHC 1898 (Colo.D.Ct. 1988)(preemption). In addition, a trial court in New York held, in an unpublished ruling, that such prosecutions are preempted; the ruling has been appealed. *See* 18 OSHR 1670 (1989), and 17 OSHR 955 (1988), discussing *Pymm Thermometer*, No. 93086 (N.Y. App. Div.), appeal filed, No. 1971 E-88 (App.Div., 2d Dept. Nov. 24, 1987).

[3] Petition for Leave to Appeal, pp. 14-15 (July 31, 1987). The specter of unpunished murderers also has been raised by nearly identical titles of a House committee report, "Getting Away With Murder in the Workplace: OSHA's Nonuse of Criminal Penalties for Safety Violations," HRep 100-1051, (discussing preemption), and a *Harvard Law Review* article, "Getting Away With Murder: Federal OSHA Preemption of State Criminal Prosecutions for Industrial Accidents," 101 *Harvard L. Rev.* 535 (1987).

[4] *See Oil, Chemical and Atomic Workers v. American Cyanamid Co.*, 741 F.2d 444, 449 (D.C. Cir. 1984). We use the term "purposeful" here in the sense used in the Model Penal Code 2.02(2)(a) (1962) — that it was the defendant's "conscious object" to bring about the death of an employee.

[5] *See United Steelworkers v. Auchter*, 763 F.2d 728, 733-4 (3d Cir. 1985).

[6] *New Jersey State Chamber v. Hughey*, 774 F.2d 587, 592 (3d Cir. 1985); *see also Environmental Encapsulating Corp. v. The City of New York*, 855 F.2d 48 (2d Cir. 1988).

[7] OSHA standards such as 29 CFR 1910.94, 1910.132, 1910.133, 1910.134, 1910.1000, and 1910.1200 require personal protective equipment, ventilation, and employee training.

[8] In *Chicago Magnet Wire*, the prosecutor told the court that "the proof at trial may very well show a violation of OSHA standards as being probative of the *mens rea* [state of mind] possessed by Defendants ...." Brief for Prosecution at 25.

[9] *See, e.g.*, Preamble to the Formaldehyde Standard, 52 Fed. Reg. 46168 (1987), esp. 46173-46237 (findings on health effects and assessment of risk), 46237-46245 (findings on feasibility and similar issues), 46250-46252 (discussing why OSHA chose the eight-hour, time-weighted average exposure limit), 46252-46254 (same as to short-term exposure limit), and 46254-46290 (discussing other issues). In rulemaking dealing with toxic substances, OSHA determines the exposure level that creates a "significant risk" of material impairment of health, whether it is "feasible" to bring employee exposure below that level, and whether the standard is cost-effective and otherwise "reasonably necessary." *E.g., Building and Construction Trades Dept. v. Brock*, 838 F.2d 1258 (D.C. Cir. 1988)(asbestos standard).

[10] Unofficial transcript of hearing on HR 2664, Corporate Criminal Liability Act of 1987, 100th Cong., 1st Sess. 6 (Nov. 19, 1987). *See also* press release from Ms. Holtzman's office, dated March 4, 1987, announcing the formation of the Environmental Crimes and Worksite Safety Unit.

[11] *Id.* at 36-37.

[12] *Motor Coach Employees v. Lockridge*, 403 U.S. 274, 292 (1971).

[13] The court also reasoned that OSHA does not punish an employer for his or her culpable state of mind because "for the most part OSHA imposes strict liability for violation of its standards ...." The court contrasted this with Illinois criminal law, which, the court believed, imposes an additional sanction on the defendants because of their attitude toward safety.

The court incorrectly characterized liability under the Occupational Safety and Health Act. The act never imposes strict liability. *E.g., Pennsylvania Power & Light Co. v. OSHRC*, 737 F.2d 350, 354 (3d Cir. 1984); *Brennan v. OSHRC (Alsea Lumber Co.)*, 511 F.2d 1139, 1144-5 (9th Cir. 1975). To prove a violation, OSHA always must show that the employer either knew of a violation or was not reasonably diligent in preventing it. *Alsea Lumber*, 511 F.2d at 1143-5; *Prestressed Systems, Inc.*, 9 OSHC 1864 (1981). *See also* 29 CFR 2200.35(b)(5) (1988) (codifying knowledge requirement). In addition, penalty assessments take into account the good faith of the employer and can be increased 10-fold if the violation is willful or repeated. Sections 17(a) and (j), 29 USC 666(a) and (j). Thus, the sanctions imposed on employers under the act are based, at least in part, on their attitude toward safety.

[14] V. Trasko, Bureau of Occupational Safety and Health, Dept. of Health, Education, and Welfare, *Status of Occupational Health Programs in State and Local Government* (January 1969), printed in *Occupational Safety and Health Act, 1970: Hearings on S. 2193 and S. 2788 before the Subcommittee*, Subcommittee on Labor of the Senate Committee on Labor and Public Welfare, 91st Cong., 1st and 2d Sess., Part 1, at 103-109 (1970) ("Senate Hearings"), and in *Occupational Safety and Health Act of 1969: Hearings on H.R. 843, H.R. 3809, H.R. 4294 and H.R. 13373 before the Select Subcommittee on Education and Labor*, 91st Cong., 1st Sess., Part 1, at 103-110 (1970) ("House Hearings"); Bureau of Labor Standards, U.S. Dept. of Labor, *Directory and Index of Safety and Health Laws and Codes* (undated, submitted Sept. 25, 1969), printed in Senate Hearings at 1327-1443 and in *House Hearings at 243-358; A. Hosey and L. Ede, Bureau of Occupational Safety and Health, Dept. of Health, Education, and Welfare, A Review of State Occupational Health Legislation* (undated, submitted Sept. 24, 1969), printed in House Hearings at 74-103, and in Senate Hearings at 100-137.

[15] *See, e.g.*, Senate Hearings at 123-124 (quoting state laws providing for fines, imprisonment, or both, and making violations misdemeanors). In addition, at least one state, New York, had a provision in its labor code making state general criminal law applicable to violations of its "general duty" provision, akin to that contained in 5(a)(1) of the act. N.Y. Lab. Law 200, 214 (McKinney 1986). More than 36 States had similar "general duty" provisions. SRep 1282, 91st Cong., 2d Sess. 9-10 (1970)(hereinafter "Senate Report"), *reprinted in* Senate Subcommittee on Labor, *Legislative History of the Occupational Safety and Health Act of 1970*, 92d Cong., 1st Sess. at 141, 149-150 (1971)("Leg. Hist.").

[16] Senate Hearings at 1585, 1588, 1595, 1602, reproducing coroner's verdicts noting the possibility that several workplace deaths might be attributable to "criminal negligence."

[17] Legis. Hist. at 425 (remarks of Sen. Dominick, reproduced below at text accompanying note 31).

[18] *See, e.g.*, Legis. Hist. at 1049 (remarks of Rep. Karth regarding California and New York); House Hearings at 258-324, 395-6, 441-486, 491-497 (testimony of C. Hagberg of Wisconsin and M. Catherwood of New York); Senate Hearings at 258-324 (testimony of C. Hagberg of Wisconsin) and 413 (testimony of F. Barnako).

[19] Senate Report at 4, Legis. Hist. at 141, 144. *See also* HRep No. 1291, 91st Cong., 2d Sess. Legis. Hist. 831, 845; (1970), *reprinted in* Legis. Hist. at 1003 (remarks of Rep. Daniels that state laws are "woefully lacking in protection for employees"), and at 1049 (remarks of Rep. Karth).

[20] Legis. Hist. at 513 (remarks of Sen. Muskie); *see also* House Hearings at 494-5.

[21] 30 USC 801 et seq. (1976) (amended 1977).

[22] Section 506, 30 USC 955.

[23] *Compare* HR 13373, 91st Cong., 1st Sess. 14(b) (1969), reprinted in Legis. Hist. at 679, 706; *and* S 2788, 91st Cong., 1st Sess. 14(b) (1969) (bill favored by administration, *reprinted in* Legis. Hist. at 31, 58; and S 2193, 91st Cong., 2d Sess. 17(h) (1970) (as passed by Senate), *reprinted in* Legis. Hist. at 529, 572; with Section 18(h) of the act. That the Senate receded on this point was expressly noted in the conference report. HRep 1765, 91st Cong., 2d Sess. 42 (1970), *reprinted in* Legis. Hist. at 1154, 1195. The wisdom of such a scheme also was discussed during committee hearings. *See, e.g.*, House Hearings at 396-397, 487-8, 490-492.

[24] Senate Report at 1, Legis. Hist. at 141.

[25] Senate Report at 18, Legis. Hist. at 158. *See also* Legis. Hist. at 525 (remarks of Sen. Saxbe that the Senate bill "does not take away forever the power of the States to regulate their health and safety .... The bill does contemplate ... that the States can regain their control ....") and *id.* (remarks of Sen. Dominick that a provision in the bill "permit[s] the States to regain some administrative control, but I do not think we should be under any illusion. The Federal Government is going to be setting standards for safety and the health criteria for all industries and businesses ....").

[26] HR 4294, 91st Cong., 1st Sess. 9 (1969), *reprinted in* Legis. Hist. at 659, 668.

[27] HR 16785, 91st Cong., 2d Sess. (1970), *reprinted in* Legis. Hist. 721, 744-7; HR 19200, 91st Cong., 2d Sess. (1970), *reprinted in* Legis. Hist. at 763, 803-805.

[28] S 2788, 10, Legis. Hist. at 54-55; S 4404, 91st Cong., 2d Sess. 17 (1970), reprinted in Legis. Hist. at 73, 113-114.

[29] For a complete review of the legislative history of the act's penalty provisions, see Levin, *OSHA and the Sixth Amendment: When Is a "Civil" Penalty Criminal in Effect*, 5 Hastings Const. L.Q. 1013, 1014-1020 (1978).

[30] See note 15 above.

[31] Legis. Hist. at 425.

[32] Ariz. Rev. Stat. Ann. 23-418.

[33] Cal. Labor Code 6423 (West 1971 & Supp. 1989).

[34] *See, e.g.*, Haw. Rev. Stat. 396-10; Ind. Code 22-8-1.1-49; Minn. Stat. 182.667.

[35] *E.g.*, North Carolina may prosecute for willful homicide: N.C. Gen. Stat. 95.139. *See also Granite Construction Co. v. Superior Court*, 149 Cal. App. 3d 465, 197 Cal. Rptr. 3 (1983), in which the court upheld a manslaughter indictment. *But see Michigan v. Hegedus*, 169 Mich.App. 62, 425 N.W.2d 729 (1988), leave to appeal granted, 431 Mich. 870, 429 N.W.2d 593 (Sept. 28, 1988)(No. 83601), in which manslaughter charges were

quashed. The solution to the perceived problem in this case would be to reference such general criminal laws in the state plan as being applicable sanctions in appropriate cases.

[36] Section 4(b)(4) states in its entirety:

> Nothing in this chapter [the Occupational Safety and Health Act] shall be construed to supersede or in any manner affect any workmen's compensation law or to enlarge or diminish or affect in any other manner the common law or statutory rights, duties, liabilities of employers and employees under any law with respect to injuries, diseases, or death of employees arising out of, or in the course of, employment.

[37] *E.g., Harrison v. PPG Industries, Inc.*, 446 U.S. 578, 588 (1980).

[38] SRep 1282, 91st Cong., 2d Sess. 22 (1970), *reprinted in* Senate Subcommittee on Labor, *Legislative History of the Occupational Safety and Health Act of 1970*, 92d Cong., 1st Sess. at 162 (1971)("Leg. Hist.").

[39] HRep. 1291, 91st Cong., 2d Sess. 34 (1970), reprinted in Leg. Hist. at 864 ("nothing is intended to affect state or Federal workmen's compensation laws, or the rights, duties, or liabilities of employers and employees under them.").

[40] *Pratico v. Portland Terminal Co.*, 783 F.2d 255, 256 (1st Cir. 1985). *See also United Steelworkers v. Marshall*, 647 F.2d 1189, 1235-6 (D.C. Cir. 1981); *Frohlick Crane Service v. OSHRC*, 521 F.2d 628, 631 (10th Cir. 1975).

[41] *See, e.g., J.I. Case v. Borak*, 377 U.S. 426 (1964)(damages could be awarded for use of false proxy material in violation of federal securities law); *Texas & Pacific Railway v. Rigsby*, 241 U.S. 33, 39 (1916)(employee damage remedy implied under Federal Safety Appliances Acts because "safety of employees and travelers is their principal object"). *See generally* Prosser and Keeton, *The Law of Torts* 36 (1984); Harper, James and Gray, *The Law of Torts* 17.6 (1986).

[42] *Occupational Safety and Health Act of 1969: Hearings on H.R. 843, H.R. 3809, H.R. 4294 and H.R. 13373 before the Select Subcommittee on Education and Labor*, 91st Cong., 1st Sess., Part 2, at 1592 (1970) ("House Hearings") (letter of J. Bailey, legislative counsel).

[43] *Id*. at 1592-93 (letter by L. Silberman, now a judge of the U.S. Court of Appeals for the District of Columbia Circuit).

[44] *See, e.g., Occupational Safety and Health Act, 1970: Hearings on S. 2193 and S. 2788 before the Subcommittee*, Subcommittee on Labor of the Senate Committee on Labor and Public Welfare, 91st Cong., 1st and 2d Sess., Part 1, at 218-9 (1970) (testimony of Jerome B. Gordon) ("Senate Hearings"). Another witness during the Senate committee hearings even urged that federal legislation allow juries to assess damages in favor of injured workers, arguing that awards under state workers' compensation schemes provided too little incentive for employers to provide safe workplaces and that supplementary tort awards would be more effective than federal regulation. *See id*. at 190-1 (testimony of George Daugherty). *See also* Legis. Hist. at 522 and 524 (remarks of Sens. Cranston and Javits).

[45] Senate Report at 25, Legis. Hist. at 165. The report of the commission can be found at 4 Larson, *Workmen's Compensation Law* Appendix G (1988).

[46] 13 OSHC at 1901.

II. High-Risk Notification

About 68,000 chemicals are registered by the Environmental Protection Agency under the Toxic Substances Control Act, and approximately 7,000 more are added to the list every year, the agency says.

The degree of toxicity of 80 percent of these substances is unknown, the National Research Council said in 1984, a fact that makes it difficult to determine how they affect people who work with them.

For some health and safety experts, the proliferation of chemicals in the workplace has created a crisis in which many workers slowly are being robbed of their health and, in some cases, their lives.

Union officials and some health and safety advocates say the federal government is not doing enough to protect the labor force from potentially dangerous chemicals.

Although the Occupational Safety and Health Administration's (OSHA) hazard communication standard requires employers to inform employees about the current dangers of working with toxic chemicals, employees may not be aware of the serious health risks they face from past exposures, according to critics.

These concerns led to the introduction of bills in Congress in 1988 to require the federal government to identify and notify workers who may be at high risk of disease from past chemical exposure. In addition, at least one state and several federal agencies have or are in the process of implementing high-risk notification programs.

'THOUSANDS BEING MADE SICK'

As many as 70,000 deaths a year may be caused by exposure to toxic substances in the workplace, and 350,000 new cases of chemically induced workplace illness occur annually, according to Philip J. Landrigan, director of environmental and occupational medicine at Mount Sinai School of Medicine in New York City.

"This is not an obvious crisis ... but behind the headlines disease is quietly occurring and thousands of workers are being made sick," Landrigan said at a May 18-19, 1989, health and safety conference sponsored by BNA.

Landrigan obtained his figures by estimating the number of occupational disease cases in New York and multiplying it by 10, since New York has nearly 10 percent of the U.S. work force.

In addition to the human toll, chemically caused illnesses may be costing U.S. business $6 billion a year in lost time and health care costs, Landrigan claimed.

Government Urged to Reveal Risks

Many in the business community, however, claim the problem can be addressed best by strengthening and enforcing existing regulations in agencies established to make workplaces safe, such as the Occupational Safety and Health Administration. Business claims workers would be told many situations are dangerous when in fact they are not, creating undue stress and worry for thousands of employees and their families.

Many health and safety advocates say it may take a chemically caused disaster, such as the discovery that a compound has caused cancer in thousands of workers, before the government adopts a program to tell workers about the dangers they face from chemicals in the workplace.

Other labor issues, such as minimum wage and parental leave legislation, will divert the 101st Congress's attention from high-risk notification legislation, leaving states and some federal agencies to take action designed to assure that workers know about workplace chemical exposure risks, observers say.

State Initiatives

If Congress does not enact high-risk notification legislation, Landrigan told BNA, states would:

- Strengthen the occupational safety and health sections of their health departments to provide better monitoring and regulation of the workplace;
- Launch worker education programs; and
- Help start clinics staffed by physicians trained to recognize and treat occupational disease.

Similar action occurred in the early 1980s with right-to-know legislation. When OSHA failed to act quickly to issue a hazard communication standard, 25 states adopted right-to-know laws.

If states also begin adopting high-risk notification laws, Congress probably will follow suit and pass high-risk notification legislation, some observers say.

"First, some states will adopt programs, then industry will do it, then somebody will say 'this should be a federal program'," former Acting OSHA Administrator Patrick R. Tyson, an attorney with Constangy, Brooks and Smith in Atlanta, Ga., said at the BNA conference.

SOME PROGRAMS ALREADY PLANNED

Some federal agencies already are drafting notification programs. The Department of Energy (DOE), for example, says it plans to notify former workers who may face a high risk of respiratory disease from exposure to beryllium in nuclear plants owned by the federal government.

The scope of the risk notification effort has not been determined, but it could cover thousands of people who worked in government-owned, privately operated nuclear facilities from the early 1940s through the 1980s, the department says.

DOE announced the proposal in the summer of 1988, saying it planned to send letters to former workers who were exposed to beryllium, advising them to obtain medical examinations. DOE also plans to hold technical sessions to provide these people with additional information.

Reasons for DOE Concern

Since World War II, beryllium has been used widely in nuclear plants and research laboratories to moderate fission reactions and reduce leakage of neutrons from the cores of nuclear reactors. Exposure to beryllium dust can lead to acute and chronic respiratory disorders. Severe cases can be disabling and sometimes fatal, studies indicate.

Plans for the notification effort were spurred by concern about two incidents, DOE says: the identification of a case of beryllium disease at a DOE-owned facility in Rocky Flats, Colo., and the filing of a lawsuit by a metallurgist who claimed that he developed respiratory disease after being exposed to beryllium dust while working in the top-secret Manhattan Project in the 1940s.

DOE spokesman Will Callicott told BNA in May 1989 the Bush administration is committed to the project and the department is "fine-tuning" the proposal, but no date had been set for starting it.

The National Institute for Occupational Safety and Health (NIOSH) is designing a program for notifying workers found in past NIOSH studies to be at risk of disease, and hopes to begin the program by October 1989, an institute official told BNA. In Minnesota, state health authorities are seeking additional funds from the state legislature to extend a program under which 1,200 former workers were notified in 1988 about a potential risk of disease from possible past exposure to asbestos.

LEGISLATIVE OUTLOOK DIM

A major high-risk notification bill (HR 162), introduced by Rep. Joseph M. Gaydos (D-Pa), passed the House 225-186 in 1988, but its Senate counterpart (S 79), sponsored by Sen. Howard Metzenbaum (D-Ohio), was withdrawn when Democrats failed to end an industry-backed Republican filibuster against it.

Metzenbaum on March 15, 1989, introduced S 582, the "High-Risk Occupational Disease Notification and Prevention Act." (See Appendix F.) But the measure, nearly identical to S 79, probably does not have enough support to pass this session, despite changes Metzenbaum made to accommodate business concerns, observers told BNA.

No Position Yet by Bush

Like its predecessor, S 582 would require the federal government to notify workers at risk of illness from occupational exposures to hazardous materials or chemicals. Workers would be told of the nature of their risk, the diseases or conditions associated with it, and their option to seek medical monitoring to detect any disease symptoms.

"Thousands of workers are dying each year from occupational cancers and other diseases that could have been prevented," Metzenbaum said in his statement to the Senate when he introduced the bill. "I believe that workers have the right to know whether they are at risk of life-threatening illness. This bill grants them that basic civil right."

Opponents of the measure say it would duplicate existing federal programs and would be too costly for employers.

The Reagan administration strongly opposed S 79 and HR 162. The Bush administration was not expected to take a position on S 582 until late summer 1989, but it probably will be much like that of its predecessor, Mark Bien, a congressional liaison officer at the Labor Department, told BNA.

'Unnecessary, Costly Bureaucracy'

Business claimed in 1988 that S 79 and HR 162 would produce a stream of unfounded worker's compensation and tort claims after employees were told that they are potentially at risk of disease from current or previous exposure to a workplace hazard.

S 582 and its anticipated House counterpart will face strong industry opposition, and few business or labor union representatives expect S 582 to come to a vote in the Senate.

"We oppose the bill wholeheartedly," Phyllis Eisen, director of risk management for the National Association of Manufacturers (NAM), told BNA. "We're surprised the senator put it back in."

The Coalition on Occupational Disease Notification (CODN), a group of nearly 200 manufacturing associations led by NAM, sent a letter to members of Congress April 17 urging them not to support S 582.

"For a variety of reasons, including the creation of an unnecessary and costly federal bureaucracy, the federal mandating of health benefits, and the triggering of many unwarranted lawsuits which would be filed against employers, the vast majority of the business community opposed" S 79, the letter said. "The business community still opposes the needless and costly high-risk legislation."

Existing Standards Said Sufficient

Business says concerns about worker exposure can be met by existing OSHA standards that require employers to notify workers of hazards from toxic substances used in the workplace and to provide medical and exposure records on request to employees, worker representatives, and OSHA inspectors. Concerns also can be met by the standard issued by OSHA Jan. 13, 1989, which tightened exposure limits for more than 400 substances, industry argues.

"These regulations, along with a myriad of other existing laws and regulations, result in a safer and healthier workplace and eliminate any possible need for the high-risk legislation," CODN said in its letter.

"This kind of thing we don't believe should be legislated," Thomas F. Evans, director of safety and environmental health at Monsanto Co., told BNA. "It's adding another level of bureaucracy to an already overcrowded situation."

Unions See Gaps in Current Regulations

The bill's supporters say OSHA regulations do not cover former employers and that OSHA has issued 26 health standards in its history. Only a small percentage of recognized carcinogens are regulated by the agency, they say.

More importantly, advocates say, a worker-notification law would extend the OSHA communication standard by providing a means to help workers in potentially dangerous industries.

"Notification is not enough," Eric Frumin, safety and health director at the Amalgamated Textile and Clothing Workers Union, told BNA. "If all you wanted to do was notify people, the Republicans would be right; the hazard communication standard would be enough. But worker notification also means taking care of people."

Eisen says OSHA regulations could be strengthened in lieu of legislation, but she predicted nothing will be done until a U.S. General Accounting Office (GAO) study on the Occupational Safety and Health Act is completed.

"A lot of concerns about notification are being looked at through the regulatory periscope," she said. "If they fail, then you go to statute."

The GAO study is scheduled for release in February 1990, according to William Gainor, director of education and employment issues for the accounting office. The office will analyze all components of OSHA and will propose options for Congress to improve the OSH Act, Gainor said.

'PIECE' WORK PROGRAM

Organized labor, which strongly backs the risk-notification legislation, has adopted the position that it probably will not be adopted in its entirety in a single federal act, but instead will be created in steps by states adopting NIOSH research recommendations and notification plans such as those contemplated by DOE.

Worker notification may not emerge "in one bite," David L. Mallino, legislative director for the Industrial Union Department of the AFL-CIO, told BNA. "We may very well do it in pieces."

Unions hope to combine worker notification legislation with a larger health and safety measure or make it part of an omnibus OSHA reform bill, but they say it is too early to determine what shape those bills may take.

"It's hard to say how it's all going to fit together," Michael Silverstein, a physician with the safety department of the United Auto Workers (UAW), told BNA.

With that in mind, and given industry and Republican opposition to worker notification measures, labor is adopting a wait-and-see attitude toward Congress on the issue.

"We're kind of sitting and watching the whole issue percolate," Mallino said. The AFL-CIO plans "no massive campaign" to change the thinking of senators opposed to the measure in the 100th Congress, Mallino said.

'Different Priority' by Labor

Labor has assigned the legislation a lower priority so it can concentrate on minimum wage and parental leave legislation, a decision that may doom any worker notification bill this session, Leslie Cheek, senior vice president for federal affairs at Crum and Foster Insurance Cos., told BNA.

"Labor has given [the bill] a different priority in its universe," Cheek said. "Until labor begins to push it, it probably will languish."

Business is fighting the bill because it fears an avalanche of lawsuits from workers already exposed to potentially dangerous chemicals, Cheek said. "There's so much concern about that aspect of the bill that no one's focusing on the future."

PROVISIONS OF METZENBAUM BILL

Like its predecessors, S 582 would create a seven-member risk assessment board in the Department of Health and Human Services (HHS) responsible for reviewing medical and scientific data on the incidence of diseases linked with exposure to occupational health hazards and for identifying and designating those populations to be notified.

The panel also would develop an appropriate form of notification and determine the type, if any, of medical monitoring or health counseling to be provided on the disease associated with the risk.

Each employee in the designated population at risk who was exposed to the occupational health hazard within the last 30 years would be notified, under to the bill.

For each designated population, the board would publish a notice of proposed determination in the *Federal Register*, provide 60 days for the public to submit written comments, and hold a public hearing within 45 days of the notice. The board would be required to issue a final determination on the population to be notified within 60 days of the hearing, according to the bill.

Elements of Notification

Notification would include identification of the occupational health hazard, including the names, composition, and properties of known chemical agents; the disease associated with the exposure; the risk for the workers compared to the general population; the latency period from time of exposure to time of clinical manifestation of the disease; information on non-occupational contributing factors to the disease; and advice on medical or health counseling available to the worker.

The bill also would establish a toll-free long distance telephone hot line for notified employees to call for medical and scientific information associated with the risks and diseases involved.

Many in industry are uncomfortable with the role science and the board would play under the bill. For example, Evans says Monsanto fears the government will rely on a few studies—or ill-conceived research—to frighten workers unnecessarily about occupational dangers.

Board Membership

Bill proponents, however, say the composition of the assessment board will ensure the most careful consideration of available information. The board would comprise four career or commissioned members of the Public Health Service and three members from outside government.

Board members would include two physicians specializing in occupational medicine, an epidemiologist, a toxicologist, an industrial hygienist, an occupational health nurse, and an occupational biostatistician.

The measure would authorize $25 million for each of the fiscal years 1990 through 1992, an amount opponents claim is not enough to operate an effective notification program. The money would enable the federal government to notify 300,000 workers a year, according to NIOSH.

To prevent further hazardous occupational exposures, the measure states that employees who are members of a designated at-risk population with evidence of a disease should be able to transfer to a less hazardous job without losing benefits, seniority, earnings, or other employment rights under "medical removal protection programs."

The measure provides a maximum civil penalty of $10,000 for each violation of the act.

MEETING CONCERNS ABOUT LAWSUITS, COSTS

The issue of worker lawsuits stemming from a worker notification program is of prime concern to business. In a report analyzing HR 162, however, GAO said the measure would not "change existing laws that permit workers to sue for damages or file claims for compensation."

Even if a bill is not enacted, GAO said, "an increase in lawsuits and claims might occur because workers are generally becoming more aware of their rights to compensation for health problems caused by workplace exposure to hazardous substances."

Gaydos and organized labor point to small-scale past-worker notification efforts in Augusta, Ga., Port Allegheny, Pa., and among U.S. automobile and agricultural equipment pattern makers as evidence that massive tort and compensation settlements resulting from notification are unlikely.

Nevertheless, Metzenbaum's 1989 bill would prohibit employee tort or worker compensation claims based on mental or emotional harm, fear of disease, or stress resulting directly or indirectly from any "report, notice, medical evaluation decision, or monitoring decision."

Exemptions for Small Employers

To quell complaints by small businesses that S 582 would cost them too much money, Metzenbaum added provisions to exempt employers with fewer than 100 employees during 1989 and 1990, and fewer than 50 in subsequent years, from providing medical removal protection. The measure also would not cover seasonal agricultural workers employed for less than six months.

Industry representatives said HR 162 and S 79 would have hit small businesses particularly hard. The National Tooling and Manufacturing Association, for example, estimated last year that HR 162 and S 79 could have cost each small business as much as $50,000 a year. The Small Business Legislative Council estimated the bills would cost employers $3,500 per year for each worker.

ISSUE 'HAS NOT GONE AWAY'

Rep. Paul Henry (R-Mich), ranking minority member on the House Education and Labor Subcommittee on Health and Safety, told BNA in an interview that congressional interest in high-risk notification is being upstaged by whistleblower protection legislation and other measures to protect and benefit workers.

"The issue has not gone away, although it is receiving less attention this year," he said.

Continuing questions about the extent of employer liability for workplace hazards, the relationship between chemical exposure and illness, and the effect of worker lifestyles on disease will make it difficult for Metzenbaum's bill to succeed in the 101st Congress, Henry said.

Likelihood of Floor Vote Questioned

An aide to Sen. Orrin Hatch (R-Utah), who led the fight against S 79, told BNA Hatch also opposes S 582. But the aide doubted the bill will get to the floor for a vote.

High-risk notification still "is a legitimate issue ... but this bill isn't the answer" for addressing it, the aide said.

A Metzenbaum aide said the senator "is moving slower" with the current bill than its predecessor to give business more time to scrutinize it and recommend changes. Hearings were expected to be held in the summer of 1989, and Metzenbaum is contemplating several new strategies to reduce business opposition, the aide said. He would not elaborate.

"I hope that this time around, members of the business community who [did] not support the bill last year will reconsider their position," Metzenbaum said when he introduced S 582. "We are ready and willing to discuss the concerns of the business community."

House Measure Expected

A Gaydos aide said the congressman will not introduce a House counterpart of S 582 until the Senate acts on that bill. The bill will contain no significant changes from HR 162, and the aide predicted the House will pass it.

Metzenbaum will need a herculean effort to convince opposition senators and business of the merits of S 582, however, Eisen told BNA. In addition to the coalition members opposed to the bill, several business groups that backed the measure in the 100th Congress, such as the Chemical Manufacturers Association (CMA), have withdrawn their support, she said.

Randal P. Shumacher, director of health and safety for CMA, told BNA the association will neither support nor oppose S 582 until several issues surrounding the bill are resolved, such as employer liability questions and the role of OSHA in the notification process.

"Basically, we're keeping our options open," Shumacher said. "We'll consider [the bill] once the contentious issues are resolved."

NIOSH RISK NOTIFICATION

Regardless of the outcome of high-risk notification legislation in Congress, NIOSH expects to launch a notification program on its own by October 1989, according to Paul Schulte, chief of screening and notification for NIOSH's Industrywide Studies Branch.

NIOSH's goal is to notify workers found to be at risk of disease in epidemiological studies it conducted from 1968 to 1988, Schulte told BNA.

"With these studies, we analyzed death and disease records for a company's, or an industry's, work force without the individual's knowledge," he said. "We now want to go back later and inform these workers of the risks."

Two groups studied in past NIOSH epidemiology research—cadmium-recovery workers and workers involved in processes that left them exposed to chloromethyl ethers—probably will be the first to be notified by the institute, Schulte told BNA.

He declined to identify the companies where the workers were employed, saying final notification plans have not been approved. "We prefer not to initiate notification through the media," he told BNA June 12, 1989.

The notification process planned by NIOSH seldom has been attempted, Schulte said. Consequently, he told BNA, the institute has spent a great deal of time developing criteria for deciding which workers to notify about potential risks and a method for contacting them.

34 Candidate Studies

Since 1982, Schulte and his staff have been evaluating the results of 132 epidemiological studies to determine which ones involve workers who need to be notified of disease risks. From this core group, 34 studies have been selected for initial consideration; of these, nine are being evaluated in-house and 25 are being examined by outside contractors.

These 34 studies will be evaluated according to criteria approved by NIOSH's Board of Scientific Counselors, the institute's independent technical advisory board. Before a group of workers from a particular study is considered to be candidates for risk notifications, the methodology of the study must be reviewed rigorously and the results of the study evaluated to determine if they are biologically plausible and consistent with the findings of other similar studies, Schulte told BNA.

Before NIOSH can proceed with a specific notification effort, a worker notification profile (WNP) must be developed for the at-risk group. A prototype WNP has been devised, Schulte said, which calls, in part, for the institute to state the rationale behind any notification decision.

This rationale might include data from studies conducted outside of NIOSH that examined the level of risk posed by the hazardous substance to which the workers in the proposed notification effort were exposed and information on whether the exposures described in the NIOSH study were still occurring in that workplace.

Because risk notification is "a contentious issue," the Board of Scientific Counselors felt a defensible rationale was needed to defuse criticism of this effort in every case, Schulte said, calling the WNP "a unique document in government evaluation of research."

Means for Notifying Workers

Once a WNP has been devised, the notification effort should take nine months to complete, said Schulte. NIOSH's current plans call for a WNP to be followed by a draft notification implementation plan, then a final plan prior to notification; after the notification is complete, the effort will be evaluated and a final report written.

A major component of any notification plan is the means by which workers will be contacted, according to NIOSH officials. The risk notification project's procedural criteria include an algorithm to determine how workers will be notified in accordance with the level of risk a particular compound poses.

If a NIOSH study found that a compound posed a low risk to a particular group of workers exposed to it, NIOSH would notify the company and the union representing the workers, and then post a general notification report, he said. For compounds posing a moderate risk, these same steps would be taken, and NIOSH might also consider issuing a public health alert.

For compounds posing the greatest risk, NIOSH would take all the preceding steps, and also send a letter to each worker in the group exposed. The letter would explain the potential disease risks faced by the worker, suggest appropriate medical intervention, and provide a contact person's name and telephone number for workers wishing additional information.

Reasons for Not Notifying Workers

Because the concept of worker risk notification involves important psychological, legal, and social considerations, NIOSH has been striving to ensure that this project proceeds "in a conscientious and comprehensive manner," Schulte said. The institute wants to include companies and unions in the process, believing that the most meaningful notification will occur within a "whole support framework," he said.

In deciding whether to notify a particular group of workers of a potential risk of job-related disease, NIOSH will consider such variables as whether the employer already has such a program in place, and how much time has elapsed since the study involving the workers was performed. If most of the workers are dead or if a company has taken steps to reduce occupational risks in the intervening years, NIOSH might not proceed with an intensive notification effort, he said.

NIOSH believes workers who may be at risk have a right to know this fact, even though it may upset them, Schulte told BNA. "Our job is to soften any disruption by clarifying the information workers receive and helping them to deal with the risks they face."

Research into psychological stress resulting from notification indicates "no short-term pathology," he said, because most workers appreciate receiving information that gives them better control over their lives. A NIOSH notification program at a Georgia chemical plant showed the benefits of notification far outweighed any mental stress it caused, he said, noting that an asbestos risk notification project recently described in the *American Journal of Industrial Hygiene* reported a similar psychological outcome.

By the time it starts its first worker notification effort, NIOSH hopes to have a special telephone hot line in place that workers can call for additional information, Schulte said.

He emphasized that the risk notification program pertains only to workers involved in past NIOSH studies. Future studies will have a notification component built in, he said. "This project is addressing a historic situation. It is a systematic means of handling a backlog of NIOSH studies. This type of situation will not exist in the future."

GOING TO COURT WITH NOTIFICATION

While NIOSH prepares to launch its first efforts, the Minnesota Health Department is working on a follow-up to a 1988 project in which it located, contacted, and examined a group of former workers thought to be at serious risk of asbestos disease. The workers were employed from the late 1950s to the mid-1970s at a plant in Cloquet, Minn., owned by Conwed Corp., a manufacturer of ceiling tile and wallboard.

As part of the project, which cost the state $326,000, the department successfully filed suit in state court to order Conwed to turn over detailed records needed to locate the more than 3,500 former employees believed to be at risk.

The study found "strong evidence" of asbestos-related disease in a "substantial" number of former employees who volunteered for testing after they were notified of their potential risk, the health department announced in a March 1, 1989, preliminary report to the state legislature.

More former Conwed workers still need to be notified and screened, the report said. The department said it plans to ask the legislature in the coming year for additional funds to conduct such efforts and to establish a larger program to monitor work-related disease within the state.

The results also improve chances for legislation to create a national worker notification program, according to Steve Trawick, safety and health director for the United Paperworkers International Union (UPIU), at whose request the Minnesota Health Department launched its study. Such programs are "a clear indication that there are workers who need to be notified, and that there's a need for" federal legislation, Trawick offered.

'An Adversarial Relationship'

UPIU's Local 158 represented workers at the Cloquet plant until August 1985, when Conwed sold the facility to U.S. Gypsum Co. The paper workers union became aware of a risk of asbestos disease among employees at the plant in late 1985, when an international representative worked with the local to answer members' questions about health insurance, benefits, and pension plans in the wake of the sale, according to Clare Sullivan, an occupational health specialist with the union.

The international representative found that "four or five" of the former employees were having trouble breathing, and one had been diagnosed as having mesothelioma, a malignancy caused by asbestos, Sullivan told BNA.

UPIU interviewed about 30 former workers who had worked in different areas of the plant, asking them about their exposure to asbestos. From their responses, it became evident that a large portion of the work force had been exposed to asbestos from 1958 to 1974, the period in which the material was used at the plant, Sullivan recalled. The union's executive board set aside $50,000 to initiate efforts to screen the former Cloquet employees for asbestos diseases.

The union had the names of some workers it could contact to advise medical screening, but it needed Conwed records to reach as many employees as possible who had worked at the Cloquet plant during the 16 years asbestos was used there, Sullivan told BNA.

On April 3, 1986, UPIU asked for the records under the authority of an OSHA standard, 29 CFR 1910.20, that requires employers to provide employees or a designated representative with copies of the employees' medical and exposure records. The company provided some information, she said, but not nearly as much as the union needed.

"We would have periodic conversations with their attorney," Sullivan said, "and they got more and more panicked as time went on. It became just an adversarial relationship. They did nothing to assist and a lot to make things more difficult."

Robert Brownson, an attorney with the law firm of Stitch, Angell, Kreidler & Muth, Minneapolis, which represents Conwed, told BNA that the company had no comment on

Sullivan's remarks because it believes that the Minnesota Health Department will pursue further litigation in the case.

State Authorities Intervene

According to an affidavit filed by Sullivan Dec. 10, 1987, with the Carlton County (Minn.) District Court, Conwed submitted information to the union on June 5, 1986, and Sept. 3, 1986, after UPIU asked OSHA and the Minnesota Department of Labor and Industry to intercede in the case.

However, Sullivan said in her affidavit that the company "withheld" detailed records that the union needed for determining whether individual employees had worked in jobs that exposed them to asbestos. "It was clear to me that Conwed's intent was not to cooperate but to delay production of documents," Sullivan told the court.

In the meantime, the union arranged to have researchers from the University of Michigan's Department of Environmental and Industrial Health conduct medical screening for 283 former Conwed workers, including 214 who had worked in departments where a "high risk" of asbestos exposure was believed to have existed, according to an Aug. 31, 1987, report by the researchers.

The screening found asbestos-related lung abnormalities in about 40 percent of the former employees, the report said.

The Minnesota Department of Health, which had put UPIU into contact with the university, concluded that the former Conwed employees might be facing serious health hazards, so it asked the company for the same information sought by the union, Alan Bender, chief of the department's chronic disease and environmental epidemiology section, told BNA.

The attempt was "unsuccessful," Bender said.

Minnesota Assistant Attorney General Scott R. Strand said his office entered the picture when the health department began encountering resistance from Conwed. At first, he said, litigation was not involved. The attorney general's office acted as negotiator for the department, asking Conwed attorneys if the company could produce personnel records for the 17 years in question. For a time, he said, it appeared that Conwed would comply with the state's requests.

"Conwed, prior to the time we filed the lawsuit, made a number of promises that they'd turn over the information we were seeking. They never really did it, though, and that's why we went to court," Strand said. Conwed attorney Brownson declined to comment on the statement.

Basis for Court Action

According to court records, Conwed supplied the state with some information on its former employees. The material provided, however, did not give the state enough information on which to base a screening, Strand said. In July 1987, according to records, the company sent the health department a handwritten list of approximately 3,260 Conwed employees identified by name or partial name only. No other information was provided.

Although the state could have petitioned the Social Security Administration for the names, Strand said his office believed it would be more efficient to use the company to obtain the records. The company was more likely to have the records and the information the state wanted, he told BNA.

The state filed its suit against Conwed on Jan. 15, 1988, in Carlton County District Court, using three different theories, Strand said, as to why the court should order the company to produce employee records:

- Under statutory and common law, the attorney general's office had the authority to protect public safety and health in the state.

- Under a then relatively new Minnesota statute, the state commissioner of health had the power to enjoin as a public health nuisance any activity or failure to act that adversely affected the public health. Conwed's failure to provide all the information sought by the state was defined as a public health nuisance.

- Under the OSH Act, the state's Department of Labor and Industry had the authority to obtain employer records pertaining to workers' medical histories and on-the-job exposures in order to determine if Conwed had violated any standards requiring that such records be maintained.

Conwed's position on each argument was that the state did not have the authority to demand that its records be produced, Strand said. It also argued that it had concerns about employee privacy and the confidentiality of its records, and that it had complied with requests for information on its former employees. Conwed attorney Brownson declined to comment on Strand's statement.

If it committed any "wrongdoing," Conwed argued in briefs submitted to District Court Judge Dale A. Wolf, it was through its failure to "comply with [the state's] demand in the manner and the time frame dictated by [the state]." It said it had tried to cooperate and provide the state with employee records, "even in light of [its] unreasonable deadlines."

'Delay ... Doesn't Wash With This Court'

On May 10, 1988, Wolf ordered Conwed to provide the Department of Health with the records it sought. "I am ... assuming [Conwed] is sincere in all of its representations. ... But much of the admitted delay in providing the needed information just doesn't wash with this court... There has been no good faith effort shown by [Conwed] to comply voluntarily."

On July 15, 1988, the health department began mailing letters to 1,200 former Conwed workers, notifying them that they may have been exposed to asbestos and offering them the opportunity to undergo free medical tests designed to detect asbestos-related health effects. The screening program began on Sept. 7, 1988.

The screening paints a generally "ominous" picture for male Conwed employees who worked with the company for long periods when the plant was using asbestos in its products, Bender told BNA.

According to the March 1, 1989, preliminary report, the longer an employee had worked at Conwed and the longer it had been since he or she allegedly had been exposed to asbestos, the greater the chance that he or she had developed asbestos-related abnormalities. The report indicated that among workers who started at Conwed more than 28 years before the study was conducted, 27.5 percent of the men tested and 7.7 percent of the women tested had lung abnormalities.

Of employees who started work with Conwed 25 to 28 years before the study, 23.7 percent of the men tested showed lung abnormalities. Clinical symptoms suggesting asbestos exposure were found in about half of all workers who began their jobs with Conwed more than 28 years before the study and who also worked at the plant for 15 years or more.

Workers who began their jobs with the company less than 25 years before the study had relatively few symptoms of asbestos exposure, the report said, and few indications of lung abnormalities.

For workers who were employed by Conwed anytime during the 16 years asbestos was believed to have been used at the plant, 18.9 percent of the men and 3.3 percent of the women tested showed signs of lung abnormalities.

Mesothelioma Found

Twelve percent of the former employees screened were referred for medical attention, although not all of the referrals were for asbestos-related problems, Bender said. The screening did, however, identify six confirmed or suspected cases of mesothelioma, a rare and fatal cancer of the lung lining that is related to asbestos exposure, he said. That number is at least 70 times greater than that which would be expected in the general population, he said.

Additional mesothelioma cases probably will be found in other former Conwed employees in the future, the report said, because mesothelioma typically has a very long latency period. Former workers are only now entering their higher risk years, the report said.

Bender said former workers in each of the plant's nine to 12 departments had lung abnormalities, an indication that the disease risk was not confined to certain departments at the plant. When the screening was begun, the health department predicted that any asbestos effects found among the workers probably would be found in employees who had worked in six departments where asbestos was thought to have been used.

'How Much Do You Spend to Inform?'

Trawick told BNA that the outcome of the Conwed case represents a "positive scenario" for risk notification. But Donna Costlow, associate director of risk management for NAM, questioned whether the study proved that a national notification program would be useful in identifying and treating work-related diseases.

"There's a difference between asbestos and other hazards," Costlow told BNA. "It's easier to identify diseases related to asbestos."

Bender told BNA that he is not sure the procedure used in the Minnesota study should be used as a model for other states to follow. Although the state health department did all it could to locate and notify workers, "there are a number of people out there who probably still don't know they were exposed" to asbestos, he said.

Two points could be raised to justify the time and expense involved in the Conwed screening, Bender said: it confirmed suspicions that the former employees faced a significant risk of asbestos-related disease and it identified the conditions from which many of the workers suffered.

However, even if physicians are able to identify cases of asbestos-related disease, they can do little to treat the victims, except advise them to stop smoking to avoid aggravating their condition, Bender said. "How much do you spend to inform and notify people, when there's so little you can do?"

In offering free screening to former workers, the health department may have relied too much on the written word, Bender offered. Written communication often is "not a modality of the blue-collar world," he told BNA.

Even with the screening, Bender said he is concerned that former employees either are unaware of the risks they face, or are unconcerned about it. Bender said his audience was very stoic at a town meeting in late February 1989 to announce the preliminary findings of the screening. Hundreds of residents were there to learn the findings, he said, but their reactions to the report were very subdued.

Six Lawsuits Filed, More Expected

Michael Polk, an attorney with the firm of Hertogs, Fluegel, Sieben, Polk, Jones & LaVerdiere of Hastings, Minn., told BNA that six lawsuits stemming from alleged asbestos exposure at Conwed had been filed as of late May 1989, and that 200 more were expected soon.

The suits charged or are expected to charge 20 companies alleged to have manufactured and supplied asbestos used at the Conwed plant with product liability, negligence, and breach of warranty, Polk said. As the former employer, Conwed is shielded from lawsuits under the state worker's compensation law and it will not be named as a defendant, he said.

He said while little treatment is available for most of the diseases associated with asbestos, the workers are going to court to obtain compensation for the injuries alleged in the suits.

'Possibility of Going Back to Court'

Bender told BNA that the health department wants to continue studying the former employees from the Cloquet plant. But how much more it can do, he said, depends on how much money the state legislature can provide.

The department wants to complete the analysis and interpretation of the data collected, and then continue its efforts to identify other former Conwed employees who were exposed to asbestos, Bender said. Those workers should be notified and counseled of the potential health risks and how to reduce those risks, he said.

The department also would like to determine the feasibility of an ongoing surveillance program to communicate new health information to former Conwed workers and to conduct follow-up studies. The total cost of these efforts would be about $270,000, Bender said.

"Those who worked at Conwed a long time ago for a very short period of time seem to be missing [from the records]," Strand said. "There's nothing unusual about that, but we are still exploring other ways of making sure the list is as complete as possible. There's a possibility of going back to court and asking for more."

According to the preliminary report to the legislature, the department also wants to evaluate the feasibility of a statewide occupational disease surveillance system. The Conwed findings indicate the necessity of developing a system that can guide business, labor, and health professionals in the general "prevention and amelioration of workplace illness and injury," the report said.

■ ■ ■ ■

III. The Construction Industry

A major construction accident thrusts the issue of construction safety into the public spotlight about every five years:

- March 2, 1973: The Skyline Plaza condominium tower in Fairfax County, Va., near Washington, D.C., partially collapses, killing 14 workers.
- April 27, 1978: The scaffolding inside the giant concrete cooling tower of a power plant in Willow Island, W.Va., collapses, hurling 51 construction workers 160 feet to their deaths.
- April 15, 1982: An elevated highway ramp being built in East Chicago, Ind., collapses, killing 13 workers.
- April 23, 1987: Concrete slabs of the L'Ambiance Plaza apartment project in Bridgeport, Conn., collapse, killing 28 workers and injuring 16 others.

Such tragedies capture headlines and heighten public awareness of the dangers construction workers face. They often generate demands from labor unions for tougher regulation of construction safety. Unfortunately, critics say, the initial concern and interest usually fade, leaving lax enforcement of safety standards.

However, this did not occur after the latest construction catastrophe—the accident at L'Ambiance Plaza.

Instead, the Occupational Safety and Health Administration (OSHA) proposed a new standard to tighten its regulation of the construction technique used in the L'Ambiance project. OSHA also established an office to help the agency better investigate future construction disasters.

In addition, the L'Ambiance accident spurred the 101st Congress to consider legislation introduced by Connecticut senators and representatives that would create a construction safety division in OSHA, require formal safety programs at construction sites, and increase criminal penalties for violations of safety regulations.

In a development unrelated to the L'Ambiance disaster, but which also imposes new duties on construction employers, OSHA on March 17, 1989, began enforcing its hazard communication standard in the construction industry, intensifying its scrutiny of chemical hazards on job sites.

Nevertheless, some critics in organized labor and Congress say these efforts are insufficient, and that an agency separate from OSHA should be established to regulate construction safety and health.

Contending in turn that most workplace accidents occur because employees fail to observe safety precautions, management officials say that any new laws to strengthen construction safety requirements should include provisions to hold employees directly accountable for unsafe work practices and to permit employee drug testing on the job.

NEW CONSTRUCTION OFFICE

Construction led all industries in the number and rate of work-related employee fatalities in 1987, the most recent year for which data are available, according to the Bureau of Labor Statistics (BLS). In that year, 830 construction employees were killed on the job, 40 more than in mining. This equaled 25 workers killed for every 100,000 full-time employees in the industry, compared with 23.2 in mining, BLS found.

"Almost every day," then-OSHA Administrator John Pendergrass told a House Education and Labor Subcommittee on Health and Safety hearing on construction safety Feb. 2, 1988, "I receive across my desk reports of construction accidents involving deaths and multiple injuries which have just been brought to the attention of OSHA's field offices."

Representatives of labor and management say the construction industry has unique features such as constantly changing work sites and a largely transient work force that make it difficult for OSHA to regulate like the agency regulates the manufacturing and service industries. OSHA often fails to consider construction's unique features when it tries to regulate the industry, they say.

"A fundamental lack of understanding" exists "on the part of the agency that the construction industry cannot be treated the same way as every other industry," Rick Palmer, vice president of Palmer Painting Co. of Amarillo, Texas, told the House subcommittee hearing. Testifying on behalf of the American Subcontractors Association (ASA), Palmer charged that "since its inception, OSHA has tried to put a square peg in a round hole and it just won't work."

The L'Ambiance Plaza accident refueled criticism that OSHA does not give enough attention to construction and that the agency lacks the engineering expertise needed to assess construction safety.

As a result of this criticism and congressional pressure, OSHA created an Office of Construction, Maritime, and Health Engineering Support on Sept. 26, 1988, to investigate accidents, provide day-to-day technical expertise to the agency's field offices and national office, and conduct in-depth studies of recurring problems.

Workings of the New Office

The office comprises three teams of engineers who determine the probable causes of accidents, conduct engineering tests and other technical studies, and answer technical questions from OSHA regional offices on construction, health, and maritime industry safety issues. The construction component has seven professionals, compared with six each for health and maritime. The construction and maritime segments are located in Washington, D.C., while the health unit, which was established earlier, remains in Salt Lake City, Utah.

Charles G. Culver, director of the office, headed the L'Ambiance investigation in 1987 and early 1988 as chief of the Structures Division of the National Institute of Standards and Technology (formerly the National Bureau of Standards). Culver provided OSHA with technical findings on the probable cause of the accident.

The office enables OSHA to avoid hiring consultants, such as NIST, when it investigates accidents in the construction, maritime, and health industries, agency officials explained.

"The office would do what NIST did," Culver told BNA. He said the office is not "sitting around waiting for the next accident to happen," but rather is responding to regional offices' questions on such topics as work platforms in shipbuilding, fall protection, and steel erection.

The office will cull the agency's inspection-data computer bank, the Integrated Management Information System, for "trends and characteristics of accidents to identify ways to improve safety in ... construction," Culver said.

The office also is helping agency safety-standards staff members write a standard to regulate the lift-slab construction method used in the L'Ambiance Plaza project, Culver said. OSHA decided to issue such a standard as a result of the L'Ambiance Plaza catastrophe.

Under the lift-slab method, concrete floor slabs for new buildings are cast at ground level and then lifted into place with hydraulic jacks. The NIST concluded that the L'Ambiance Plaza collapse resulted from a failure in the jacking equipment used on the project.

A proposed lift-slab standard was issued for public comment on Sept. 15, 1988. It would apply to the only two U.S. companies known to use the lift-slab method, OSHA said in the proposal. Among other provisions, the proposed standard would prohibit the lifting of slabs weighing more than the rated capacity of the jacks, require that jacks and other lifting components be capable of supporting two and a half times the weight of the load, require that lift-slab operations be designed by a "qualified" person, and prohibit employees not "essential to the jacking operation" from being in the building during a lifting operation.

'After the Fact'

Most industry representatives are deferring opinions about the new office, saying they do not want to prejudge its performance.

Some construction industry officials told BNA that the office could help most by increasing OSHA's understanding of the special nature of the construction industry. Other industry representatives said they hope that the creation of the office will result in OSHA regulators working cooperatively with industry instead of adopting an adversarial enforcement stance, as it often has in the past, they said.

President Bush cited the new office as evidence that his administration wants to ensure that "every person who takes a construction job is as safe as we can make them," in an

April 18, 1989, speech to the AFL-CIO Building and Construction Trades Department (BCTD) annual conference in Washington, D.C.

Bush called on union members to rely on the office for technical advice and support. The new unit wants "to work more closely with you for better accident investigation and prevention," he told the labor gathering.

Secretary of Labor Elizabeth Dole told the same conference that the office will do much to prevent catastrophes like that at L'Ambiance Plaza, and she praised Culver for his technical expertise and his efforts in coordinating the L'Ambiance investigation.

Several union officials told BNA they were less enthusiastic than Bush and Dole about the new office. "It's after the fact," said Jim Lapping, director of safety and health for the Building and Construction Trades Department. The office is not an adequate alternative to a separate construction safety agency within the Department of Labor, but independent of OSHA, which the building trade unions want, Lapping said.

"The office has no authority, ... no supervisory or coordinating capabilities," Lapping observed.

Better Inspection Targeting Urged

Industry and union groups say OSHA can make better use of its limited enforcement resources by concentrating its inspections on companies with histories of flagrant abuse of safety standards.

Construction union officials say OSHA should require employers to inform the agency when each construction project begins and ends, and to notify it of any injuries during the project. This would enable OSHA to focus on the more potentially dangerous projects and to monitor companies with poor safety records, union leaders say.

OSHA currently inspects construction sites randomly, using a computerized tracking system. The agency contracts with a private company that produces the *Dodge Reports*. These reports track construction activity nationwide and include descriptions of each project, when it is expected to start, and its owner, architect, and contractor. This information is processed through a computer at the University of Tennessee that determines when the most dangerous work is likely to be done and then randomly selects projects for OSHA to inspect.

OSHA says this system enables it to avoid charges that it singles out certain employers or that it focuses on unionized or non-unionized firms to inspect.

Supporters of the unions' proposal say OSHA might have prevented the L'Ambiance Plaza collapse if it had known that in 1987 a worker was injured when a lifted floor dropped 15 inches at another project managed by one of the L'Ambiance contractors.

"This accident should have served as a warning of impending tragedy," Rep. Christopher Shays (R-Conn) said at the 1988 House subcommittee hearing. "Had OSHA been notified of what happened ... they would have put a stop to the engineering and construction practices that later caused the death of 28 workers in Bridgeport."

Industry does not support the building trade unions' proposal that every project be reported to OSHA, according to Palmer. However, he told the House subcommittee hearing, "there are serious deficiencies in OSHA's inspection system which are crying out for change because of the uneven and sporadic approach that OSHA takes on its inspections."

OSHA inspectors usually visit larger projects because they have the highest profile in the public eye and are easily accessible, Vic DiGeronimo, president of the National Utility Contractors Association, told the House panel. The agency does this despite the fact that "these bigger companies ... usually have the best safety programs," he said. " As a result, the unscrupulous contractors, more often than not, go undetected."

Rep. Paul B. Henry (R-Mich), the ranking Republican on the subcommittee, said at the 1988 hearing "there is no consistent pattern that [OSHA is] going after the sleaze bags out there .... Ill-trained and unknowledgeable inspectors are going, in some respects, after safe job sites rather than really focusing on the known repeat violators."

Most labor and management officials agree with Henry. OSHA inspectors often lack technical expertise about construction operations, they charge. As a result, different inspectors often interpret OSHA construction regulations differently, creating inconsistent enforcement and confusion for employers, industry representatives charge.

Pendergrass admitted during the 1988 House subcommittee hearing that "most" compliance officers do not have construction experience. However, he insisted that the agency is "continually striving to improve the caliber of [its] investigations" by providing better training to inspectors and issuing guidelines from headquarters on interpreting and enforcing standards.

Employers and workers may comment on inspections by participating in confidential surveys conducted by compliance officers after each visit, OSHA officials told BNA. The surveys invite comment on the overall effectiveness of the inspection and the inspector and additional observations. According to Joe Nolan, chief of OSHA's Division of Maritime and Construction Compliance, only 3 percent to 5 percent of the surveys are completed and returned to the agency, and 80 percent of those who do respond comment favorably on the inspections, he said.

NEW OSHA CONSTRUCTION UNIT PROPOSED

While labor and industry debate the merits of OSHA's new construction-support office, members of Congress are introducing companion Senate and House bills that would far exceed the administrative changes made by the agency.

The Construction Safety, Health, and Education Improvement Act of 1989 (S 930, HR 2254), introduced April 24, 1989, by Connecticut's delegation on the second anniversary of the L'Ambiance Plaza accident, includes some but not all of the changes recommended by organized labor and industry.

The bills would establish an OSHA Office of Construction Safety, Health, and Education that would:

- Develop mandatory criteria for comprehensive safety and health programs on individual construction sites;
- Assume control of construction sites where accidents have occurred, including oversight of rescue operations;
- Help OSHA's Advisory Committee on Construction Safety and Health develop training courses and curricula for employees who would be designated as construction safety specialists on individual work sites, with the authority to identify hazardous conditions;
- Consult with and advise employers, employees, and labor organizations on preventing occupational injuries and illnesses;
- Increase awareness of construction site safety through education, training, and outreach programs; and
- Provide technical experts to answer contractors' technical questions.

The new office proposed in the bills is designed to replace the support unit headed by Culver; it would be stronger and would perform more functions, according to Jeff Anders, a legislative aide to Sen. Christopher J. Dodd (D-Conn). Dodd thinks the current office "does not go far enough" toward improving OSHA's construction safety programs, Anders told BNA.

Recommendation for 'Safety Specialist'

The bills would not create the construction safety agency in the Department of Labor, apart from OSHA, which construction union officials had sought even before the L'Ambiance Plaza catastrophe. However, the bills include several construction union proposals, such as requiring contractors to hire "construction safety specialists" or "other employees" to be responsible for the overall safety of projects.

The bills also include labor's proposal that employers be required to have written safety and health programs that would include emergency evacuation plans, work site safety instruction, and "regular" safety meetings. In addition, all supervisors and employees would have to obtain general safety and health training.

The bills would require OSHA to develop procedures for determining whether the required training had been provided.

On-site construction safety and health plans also would be required. The plans would describe construction processes used during projects; list inspections and tests; identify hazards; establish benchmarks by which on-site construction safety specialists would monitor contractors' compliance with the plans; and allow such specialists to stop work or remove workers from areas where an "imminent danger" was present.

Bills introduced by Dodd, Shays, and others during the 100th Congress, but not passed, would have required employers to hire "professional engineer-architects" to supervise each project. Construction, engineering, and contractor groups contended that few engineers and architects have much background in safety concerns.

THE CONSTRUCTION INDUSTRY

The current bills would create an OSHA Construction Safety and Health Training Academy similar to the National Mine Health and Safety Academy, which instructs Mine Safety and Health Administration inspectors and trains miners in safe working practices. The new academy would certify construction safety specialists for individual construction projects; train OSHA construction inspectors; and enter into cooperative agreements with educational institutions, state governments, labor unions, and employers to train workers to be work site construction safety specialists.

OSHA would be required to provide manuals, videos, audio tapes, and other materials to help employers comply with construction standards.

Stronger Reporting Requirements, Penalties

S 930 and HR 2254 would strengthen current regulations under which employers must report accidents directly to OSHA by requiring them to notify OSHA, within 24 hours, of any accident resulting in one or more deaths or serious injuries, and of any structural failure leading to the collapse or near-collapse of a building under construction.

Currently, employers have 48 hours to notify OSHA and they are required to report only those accidents that result in a fatality or the hospitalization of five or more employees.

The two bills also would increase the maximum jail sentence for a criminal violation of OSHA regulations from six months to 10 years. Additionally, the measures would allow federal prosecutors to seek criminal penalties for violations that result from "criminal negligence." Under current guidelines, such action can be taken only when violations are "willful."

Legal experts say such changes are needed to encourage federal prosecutors to pursue criminal charges under the Occupational Safety and Health Act (OSH Act) more vigorously. (See Chapter I.)

The Connecticut delegation's bills also would require OSHA to establish an inspection-scheduling system that would assign priority to sites with the greatest potential for serious accidents.

Individual construction projects, operations, or contractors would be exempted from any provisions OSHA found not to be feasible for them, or not useful in promoting "a significant increase in employee safety and health." OSHA would have to publish notice of such exemptions in the *Federal Register*.

UNIONS 'PLEASED,' BUT SEPARATE AGENCY URGED

Construction union officials told BNA they generally support the bills, but they will continue to lobby for more comprehensive legislation that imposes tougher controls on construction industry safety.

Les Murphy, assistant director of health and safety for the BCTD, described the bills as a "giant step forward," but he said the department "is hopeful" that they will be amended to establish a separate agency in the Labor Department.

"There is no way to recast" OSHA to make the agency more effective in the construction industry, Murphy maintained. "Nothing works over there."

Joe Adam, director of safety and health for the United Association of Journeymen and Apprentices of the Plumbing and Pipe Fitting Industry, told BNA that he is "pleased" with the bills, but he added that his union was "not giving up the fight" for a separate agency.

George E. Smith, safety director for the International Brotherhood of Electrical Workers, echoed Murphy's and Adam's comments about a separate agency, but said a construction office within OSHA nevertheless "would be a lot better than what we have now." Such an office "should be sufficient" if top OSHA management supports it, he said.

The building trade unions' proposal for a separate construction agency is not supported by all AFL-CIO member unions, labor officials told BNA. The AFL-CIO's position on the issue is that "all workers should be under [the jurisdiction of] OSHA," according to Margaret Seminario, AFL-CIO associate director for occupational safety, health, and social security.

The AFL-CIO plans to lobby for a legislative overhaul of OSHA, focusing on requiring all workplaces to have safety programs, plus improvements in the agency's training, record keeping, and inspection efforts, Seminario said. The proposal also may include provisions that apply only to the construction industry, she said.

Murphy said the OSH Act should be amended to allow inspectors to enter construction sites without a warrant. However, Bill Henry, director of communications for the Associated General Contractors (AGC), told BNA the AGC is "pleased" that S 930 and HR 2254 do not allow "warrantless searches."

EMPLOYERS SAY WORKER ACCOUNTABILITY NEEDED

Construction industry employers fault the Dodd and Shays legislation for not including provisions that would hold workers accountable for unsafe behavior and permit employers to test workers for drugs.

The OSH Act "removed all accountability for accidents from the worker and made the employer the only party responsible for accidents and accident control," Gilbert Czaplewski, vice president and general superintendent of Klug and Smith, Milwaukee, Wis., said at the 1988 congressional hearings on behalf of AGC.

According to AGC, unsafe work practices by employees, not unsafe working conditions, cause most accidents on work sites. At the 1988 hearings, Czaplewski noted studies by E.I. du Pont de Nemours & Co. which estimate that 90 percent of all workplace injuries are caused by "unsafe acts" by workers; a BLS study showing that nearly 70 percent of all chemical burns were classified as injuries caused by worker errors; and statistics from the Construction Safety Association of Ontario, Canada, which indicated that 68 percent of all work injuries involve unsafe employee behavior.

Drug use is a contributing problem, according to industry groups. Companies that conduct drug tests on employees after accidents have found that 40 percent of those employees tested positive, according to AGC. A 1987 survey by the Construction Industry

Institute found that some 10 percent of the employees of one construction company had some form of substance abuse problem, according to AGC.

Industry Opposes Safety Specialist Idea

Officials with AGC, the Associated Builders and Contractors Inc. (ABC), ASA, and other industry groups also criticized the bills' requirement that construction safety specialists be at individual job sites. Small contractors with limited budgets and work forces would find it especially difficult to comply with such a provision, they told BNA.

The bills would require construction safety specialists to complete a 40-hour training course and 18 months of on-the-job training. Charles E. Hawkins, ABC's vice president of government affairs, called the proposal "impractical" for small businesses.

The idea "won't work for mom-and-pop operations," agreed Michael P. O'Brien, ASA's assistant director of government relations. Employers would have difficulty finding workers to serve as safety specialists because the job would be "laced with liability," O'Brien offered.

OTHER LEGISLATIVE PLANS

Studies of OSHA's construction safety programs by several House members in early 1989 were expected to spawn additional legislation, according to congressional staff members.

Rep. Joseph M. Gaydos (D-Pa), chairman of the House subcommittee, and Michigan's Rep. Paul Henry, the ranking Republican member of the panel, have asked the U.S. General Accounting Office (GAO) to examine the overall effectiveness of the agency.

Gaydos said he would hold hearings after the GAO findings are released. Sources told BNA the report would be finished in September or October 1989.

While the report was not expected to focus on OSHA's role in construction safety, it is to make broad recommendations affecting all sectors that come under the agency's jurisdiction, BNA was told. The unions' proposal to create a separate agency was among the issues the panel was scheduled to discuss, according to a subcommittee spokeswoman.

The GAO audit parallels work being conducted by a five-member task force informally convened by Henry in 1988 to investigate OSHA's effectiveness and to recommend OSH Act amendments. The task force is trying to determine whether the act needs a general overhaul, including additional safety requirements for all industries, not just construction.

The task force has drafted several recommendations addressing issues pertinent to construction safety, such as increasing penalties for health and safety violations, alternatives to federal inspections, and improving OSHA's training methods, but an aide to Henry said it probably will not offer "construction-specific" language.

According to Henry, the task force will offer a "balanced" alternative to legislation proposed from labor unions, providing a "straight-on proposal with a breadth of support ... that would carry the day over splintered groups."

The task force originally was expected to suggest draft legislation to Henry by January 1989, but members of the group were required to deal with other legislative issues, such as proposals to increase the minimum wage, according to Henry's aide.

HAZARD COMMUNICATION

In addition to its demands for stronger safety regulation in construction, organized labor has pressed OSHA to devote more attention to health hazards on construction sites.

Samuel S. Epstein, a professor at the University of Illinois Medical Center, told the BCTD's annual conference in Washington that stronger health regulation is needed because many construction workers are unaware of the potentially lethal consequences of inhaling fumes from solvents, chemicals, and other toxic substances used on the job, including coal tar pitch, asbestos, benzene, lead, cyanide, and mercury.

Construction workers live an average of eight years less than employees in other professions, a difference that cannot be attributed solely to risks from such off-the-job factors as smoking, alcohol consumption, and poor diet, Epstein said.

Unions applauded OSHA's extension of its hazard communication standard to the construction industry as a major step toward providing employees with more information about health hazards on the job. But employer representatives say the standard, developed in 1983 for the manufacturing industry, will be generally unworkable on construction sites.

OSHA extended the standard to construction and other industries in August 1987 as a result of a federal court order. Construction employer groups appealed the decision, but the U.S. Court of Appeals for the Third Circuit upheld the ruling.

Although that decision is under review by the U.S. Supreme Court in response to petitions by AGC and ABC in December 1988, OSHA started enforcing the standard on March 17, 1989.

Requirements for Employers

Under the standard, construction employers are required to inform workers about the hazards of specific chemicals by:

- Training employees before they work with hazardous chemicals and when new hazardous chemicals are introduced into the workplace, and making sure they know how to protect themselves from chemical exposure;
- Labeling chemical containers;
- Making material safety data sheets (MSDS) with detailed information about individual chemicals readily available; and
- Having a written hazard communication program that also is accessible to all workers.

The construction industry's opposition to the standard coalesced in late 1987 when 33 trade associations formed the Construction Industry Hazard Communication Coalition.

THE CONSTRUCTION INDUSTRY

The coalition said the standard would require employers to brief each crew coming onto the site about the hazards posed by chemicals there, and to find out which substances other employers on the site were using.

"These 'HazCom breaks' will interrupt the normal work habits of even the most dedicated crew and force delays which will add to the cost of the project," AGC's Dennis Neal told OSHA on behalf of the coalition during a December 1988 public hearing on the standard.

Training Requirement Problems Seen

Training requirements often overlap at multi-employer project sites, and the high turnover rate among construction workers presents "incredible problems" in complying with such requirements in the standard, according to Neal.

Companies must teach their workers about the hazardous materials that their employer and other contractors bring to the work site. When subcontractors discover that other subcontractors have brought hazardous materials to the site, they must train their own workers about those materials, Berrien Zettler, deputy director of the OSHA Office of Compliance, explained at a March 8-9, 1989, meeting of the OSHA Advisory Committee on Construction Safety and Health.

Neal predicts that "employers will be required to spend inordinate amounts of time training new employees for what may amount to only a few weeks of work." As a result, he added, employers may begin to under-staff their jobs rather than spend the time and energy necessary to hire and train a sufficient number of new workers.

Unions reject such arguments. The BCTD maintains that the standard's requirements "mesh well with the processes that go on at a construction site."

"Just as subcontractors must coordinate with regard to each other's operations, they can coordinate with regard to hazardous chemicals to which employees might be exposed," BCTD said at a December 1988 public hearing on the standard.

Data Sheets: 'Regulatory Twilight Zone'

Employers also object to the standard's requirements for MSDSs. The standard requires subcontractors using a hazardous chemical at a work site to give an MSDS for the chemical to the primary contractor, and to notify other subcontractors that the data sheet is available so those subcontractors, in turn, can tell their employees about the availability of the sheet.

Whether any employee ever asks to see any MSDS is irrelevant, according to OSHA. The standard's goal is to provide workers the opportunity to know immediately what chemical hazards they face.

Because of the large number and wide variety of chemicals covered by the standard and the complexities of multi-employer work sites, requiring that an MSDS collection be maintained on every construction site "simply cannot work," according to Martin Perlman, vice president and treasurer of the National Association of Home Builders, a member of the industry coalition opposing the standard.

"It's one thing to label products with appropriate precautionary measures," Perlman told the December 1988 OSHA hearing; "it is quite another to require all construction companies in the country, no matter how small, to maintain technical libraries on all job sites consisting of hundreds of Material Safety Data Sheets that do little more than tell us what the American Conference of Government Industrial Hygienists thinks about the hazards of a particular chemical."

Industry also says OSHA does not have a uniform MSDS system. The standard "leaves the employer in a regulatory twilight zone," Thomas J. Godwin Jr., second vice president of ASA, told OSHA. "Either give us the freedom to determine what's best for our own company or just tell us what to do and how to do it."

The BCTD says unless data sheets are present and employers share hazard information where risks of exposure exist, "the standard is simply inadequate in construction."

The BCTD argues that MSDSs can be kept at project sites just as easily as plans, purchase orders, delivery schedules, and material specifications, which construction firms maintain there.

What Inspectors Will Seek

To determine whether hazard communication programs on construction sites are adequate, inspectors will ask employers certain questions, OSHA said in a program directive:

- Does a list of hazardous chemicals exist in each work area on a construction project or at a central location?
- Did you tell your workers where MSDSs are located and how they can get them?
- What kind of training did you provide to employees?
- Did you train your employees on the hazards of using the materials you brought to the site?
- Did you train your employees on the hazards of using the materials other contractors brought to the site?
- Did you tell other employers about the chemicals you brought to the site and which precautionary measures to use?
- Who is responsible for obtaining and maintaining MSDSs?

Industry officials told BNA that construction employers remain confused about the requirements of the standard. Few are seeking OSHA's help because they do not think that the agency knows how to apply hazard communication concepts to construction sites properly, they said.

Several documents designed to help employers comply with the standard are available at no charge from OSHA, the agency announced. These include *Chemical Hazard Com-*

munication, a booklet describing the rule's requirements in simple terms, and *Hazard Communication Guidelines for Compliance*, a booklet providing further clarification on various provisions.

However, representatives from both industry and labor say these documents are inadequate. Several construction-related organizations have developed their own guides for complying with the standard.

Under an OSHA grant, the National Constructors Association (NCA) has been sponsoring a series of three-day lectures and workshop training sessions on the standard's application in the construction industry. The program, held in various parts of the country, is presented by safety and health professionals from NCA, National Erectors Association members, and the BCTD and its affiliated members. The last of the 1989 programs was scheduled to be held in September in San Diego.

Other industry groups that have helped their members create a basic hazard communication program include the AGC, whose 72-page booklet contains samples of a written hazard communication plan and MSDSs, as well as where employees can purchase generic MSDSs. The guide outlines employee training requirements, including advice for "toolbox talks," and community right-to-know requirements. The guide also suggests how to inventory chemicals and lists standard chemicals common in construction operations.

IV. New Chemical Exposure Standards

By considering more than 400 industrial chemicals and substances in one regulatory proceeding, we were able to make a 20-year leap forward in the level of worker protection in a relatively short period of time.
—John A. Pendergrass, former assistant secretary of labor for occupational safety and health.

Completion of the air contaminants standard makes it credible for future OSHA administrators to do even more.
—Michael Baroody, Labor Department assistant secretary of policy.

OSHA is merely rubber-stamping the chemical industry's own recommendations.
—Jack Sheinkman, president, Amalgamated Clothing and Textile Workers Union.

The broad scope of the rulemaking process ... raises questions about its scientific soundness, and about industry's ability to comply with so many changes at one time.
—Susan Spangler, director of occupational safety and health, National Association of Manufacturers.

In a sweeping action at the close of the Reagan administration, the Occupational Safety and Health Administration established more stringent permissible exposure limits (PELs) for almost 400 chemicals and other air contaminants, such as dusts, fumes, and vapors.

The rulemaking was completed in near-record time, pushed through in 18 months by then-OSHA Administrator John A. Pendergrass and a team of agency standards writers assigned to the effort.

The agency called the rule its "most significant workplace exposure action ever."

Although the rule is expected to improve protection for 4.5 million workers, it is likely to strain the capabilities of some employers and fall short of labor union expectations, according to sources contacted by BNA.

Industry is divided on the new rule—larger companies, especially those in the chemical manufacturing industry, say they already are complying voluntarily with the lower limits or comparable corporate guidelines. Smaller companies and industries that deal with a handful of controversial substances, however, maintain that the new rule will force them to install costly engineering controls and take other steps to meet the lower limits. Most companies contacted by BNA in the spring of 1989 were still assessing the possible impact of the rule, which will go into full effect in 1992.

Unions contend that the new limits generally are a rubber stamp of industry recommendations that will have little impact on industry. Organized labor was awaiting the appointment of a new OSHA administrator before setting strategies for getting the agency to adopt new and more stringent regulations.

Publication of the new exposure-limit table also raised the prospect that OSHA would address—on its own or at the behest of outside groups—chemicals that need more extensive attention or better cost and feasibility information, or those that were bypassed when the agency tightened the standards.

GENERIC STANDARDS

Until recently, OSHA—with limited funds and personnel—regulated hazards on a priority basis. Standards were issued for those substances that posed the greatest risk to workers at any given time.

Over almost 20 years, however, the agency has issued only 24 health standards covering specific substances, such as asbestos, lead, benzene, formaldehyde, vinyl chloride, and cotton dust.

The agency has been unable to regulate the thousands of substances commonly found in the workplace or to review scientific information to determine whether different limits were required for the more than 400 substances regulated before 1988 under exposure limit tables contained in Subpart Z of the air contaminants standard (29 CFR 1910.1000).

It became clear to OSHA officials during the Reagan administration that the agency's traditional substance-by-substance rulemaking procedure was ineffective. The lengthy and complicated process could not keep pace with the number of new chemicals and other hazards occurring in the workplace. Moreover, very specific standards are less likely to be applicable to all employers or work sites, some agency staffers and industry officials argued.

Revision of 20-year-old PELs contained in Subpart Z ("the Z-tables") was seen by some at OSHA and its parent agency, the Department of Labor (DOL), as a way to reduce worker exposures to chemicals in a single action.

The new final rule is only a "starting point" to establish more up-to-date PELs than those contained in the previous Z-tables, the agency acknowledged. OSHA said it plans to develop and implement a system by which PELs can be revised periodically to keep them abreast of current information on the substances.

Generic Efforts To Continue

The new standard will not end agency efforts to develop substance-specific standards when necessary or appropriate, OSHA said in its final rule. In addition, the agency will continue to develop other generic standards—including exposure monitoring, medical surveillance, respiratory protection, and methods of compliance—to complement protection from the reduced PELs, according to OSHA.

The reduced PELs also will be coordinated with OSHA's hazard communication standard to provide further worker protections, the agency said. The hazard communication standard requires that employers provide hazard labeling on chemical containers; maintain material safety data sheets that detail chemical hazards, health effects, and precautions; and train and educate their workers.

By revising the exposure limit tables and setting generic standards for exposure monitoring and medical surveillance, the agency hopes to fulfill requirements of the Occupational Safety and Health Act (OSH Act). The act mandates that the agency, "where appropriate," regulate employer monitoring of employee exposures and prescribe the frequency and type of medical examinations or other tests.

Through generic standards, some safety and health professionals argue, the agency will have a far-reaching effect on worker safety and health and on companies and the way they operate.

THE PEL PROJECT

The massive OSHA rulemaking changes PELs in some way for 376 chemicals and other air contaminants. The rule sets more stringent PELs for 212 substances listed in Table Z-1 of OSHA's air contaminants standard and establishes PELs for 164 substances not previously regulated by the agency.

The agency's proposed rule, published in June 1988, contained PELs for 428 substances.

The updated table (Table Z-1A), in combination with two unchanged exposure-limit tables, Tables Z-2 and Z-3, now regulates a total of more than 600 air contaminants. Tables Z-2 and Z-3, which contain PELs for beryllium, cadmium, mercury, toluene, silica, coal dust, graphite, and other substances, were retained unchanged. According to the agency, limits in those tables could not easily be included in the format of the new Table Z-1A.

Short-Term Limits Imposed

The final rule also imposed 15-minute short-term PELs for 116 substances, including such commonly used ones as acetone, carbon dioxide, carbon disulfide, chlorine, gasoline, ozone, propyl alcohol, toluene, and toluene-2,4-diisocyanate.

Short-term limits are used to control high bursts of chemical exposures over short periods of time. The American Conference of Governmental Industrial Hygienists (ACGIH) established short-term limits in 1974, which were published in 1976, to protect workers against chemicals' acute effects, such as irritation, chronic or irreversible tissue change, or narcosis severe enough to increase the likelihood of accidents, impair self-rescue, or reduce work efficiency.

The previous Z-tables did not regulate for short-term exposures because that exposure-limit category did not exist when the threshold limit values (TLVs) were adopted by the agency in 1971. The original tables, however, contain some "acceptable ceiling concentrations" that act as short-term limits, according to the agency. The new short-term limits were adopted as a result of toxicological evidence of recognized acute effects, OSHA said.

Skin Notations Added

"Skin notations"—indications that workers should wear gloves, coveralls, goggles, or other appropriate protective gear to prevent skin absorption—were added for 49 substances on the exposure-limit table. Substances for which skin notations were added include acrylic acid, hexaflouroacetone, methyl acrylonitrile, methyl alcohol, and *p*- and *o*-toluidine.

The decision to add skin notations in those cases was based on the substances' ability to be absorbed through the skin in adequate amounts to cause systemic toxicity, the agency said.

And for four substances—DDT, dipropylene glycol methyl ether, hexachloroethane, and respirable paraquat dust—the agency retained skin notations that had been dropped by ACGIH. The agency contended that deleting those designations would increase worker exposures and decrease worker protections.

Deferred Decisions

OSHA declined to act on four substances for which it proposed to revise limits, pending further rulemaking activity, such as health and feasibility studies.

In addition, the agency's final rule raised the PEL proposed for 12 substances, although final limits will be lower than current limits.

Moreover, the rule does not cover nine substances that already are undergoing rulemaking under Section 6(b) of the OSH Act. Those substances are 1,3-butadiene, cadmium dust and fume, four glycol ethers, ethylene dibromide, methylene chloride, and 4,4-methylenedianiline.

Finally, the agency retained previous PELs for three substances—benzene, formaldehyde, and cotton dust—where particular operations or industries are not covered under comprehensive standards.

No changes were considered for the 24 substances already regulated under comprehensive standards.

'Most Commonly Used Substances'

According to the agency, the new final rule addresses many of the most commonly used substances in industry, including chloroform, carbon monoxide, hydrogen cyanide, perchloroethylene, and wood dust.

NEW CHEMICAL EXPOSURE STANDARDS

The substances range from chemicals like acetone, to gases like ammonia, to welding fumes and mercury vapors, and to dusts created by woods and grains.

The substances are grouped into 18 categories, for the most part based on the adverse health effects they pose. The categories include carcinogens, irritants, and respiratory, cardiovascular, or systemic effects.

The chemicals are found in a wide range of industries, from semiconductor manufacturers to meat casing producers. One witness at a hearing on the proposed rule suggested that it could be considered "un-American." Lower limits proposed for carbon disulfide, a chemical used in food processing, could drive meat-casing manufacturers overseas and put hot dog producers out of business, he said.

Limits Selected

"Industrial experience, new developments in technology, and scientific data clearly indicated that in many instances [the previous exposure] limits had become inadequate and failed to consider many toxic materials commonly used in workplaces," Pendergrass stated in announcing the final rule.

"The inadequacy of the existing OSHA Z-tables was demonstrated by the fact that many large industrial organizations felt obligated, for worker safety, to supplement those OSHA PELs with their own internal guidelines," he added.

Although the ACGIH threshold limit values are criticized by many labor union officials, the agency used them for most of the substances in the final rule.

Almost all of the TLVs recommended in 1968 by ACGIH were adopted by OSHA in its Z-tables when the agency was established in 1971.

OSHA also considered a broad selection of data bases and PELs in developing the new rule, including limits set by foreign governments.

Generally Accepted Limits

Agency officials decided that the ACGIH limits covered the largest number of substances, were supported by written documentation, and were generally recognized by health professionals.

But critics from organized labor contended that the ACGIH limits are overly influenced by industry, that evidence and reasons used to set those limits are not made public, and that the limits do not protect workers sufficiently.

For about 35 substances, OSHA adopted the PELs recommended by the National Institute for Occupational Safety and Health. In those cases, OSHA argued that the NIOSH limits provided more appropriate protection to exposed workers.

In several cases—including acetone, acetonitrile, and trichloroethylene—the agency rejected the PELs recommended by NIOSH and adopted less-stringent ACGIH levels.

160 Substances Not Addressed

In response to criticism at the public hearings, OSHA defended its decision in the final rule not to change PELs for substances where OSHA's current PEL and ACGIH's TLV are the same.

OSHA argued that considering those additional 160 chemicals would further overwhelm the rulemaking, making the effort too complex and slowing it down. Moreover, the agency can initiate Section 6(b) rulemaking for those substances in the future, it noted.

Substances for which limits were not proposed for change included:

- Dimethylformamide, a solvent used in leather tanning and aircraft repair industries, and linked to testicular cancer;
- Methyl isocyanate, the toxic gas involved in the December 1984 accident in Bhopal, India, that killed more than 3,000 people;
- Phosgene, a substance used as a poison gas, in the synthesis of organic chemicals, and in making dyes. Phosgene is associated with illnesses ranging from nose and eye irritation to pulmonary edema, sometimes resulting in death; and
- Respirable silica dusts, found in mining, stone, ceramic, glass, and foundry operations. The dusts are associated with various forms of lung disease, including silicosis and, in the case of quartz, evidence increasingly indicates that the dust may cause cancer, according to OSHA's data.

Completion In 'Near Future'?

The agency also contended that its decision not to include substances currently undergoing full Section 6(b) rulemaking in the air contaminants standards was appropriate. It stated that rules for those nine substances would be completed in the "near future." Only one of those rules was in the final stages of rulemaking in mid-1989, however. Standards normally take at least two years to complete from that stage, according to an informal review of past rulemaking timetables.

Since 1985, the United Auto Workers Union has been trying to get OSHA to adopt a standard to reduce exposures to methylene chloride, a common industrial solvent. No proposed standard has been issued, and union officials doubt that such a rule will be promulgated in the near future. The OSHA April 1989 regulatory agenda said a proposed standard will be issued in September 1989, but it mentioned no date for issuing a final rule.

OSHA also acknowledged that certain new PELs would not eliminate "significant risk," primarily because data were not available to justify lower limits at the time. But PELs may be changed as more information becomes available, the agency said, noting that was the case with asbestos, for which PELs were set in 1971 and revised in 1972, 1976, and 1986.

Pendergrass contended that the agency's approach in this rulemaking "saved American taxpayers tens of millions of dollars that otherwise would have been spent over a long period of time in individual rulemakings on updating the PELs for each of the substances."

Industries Affected

Although the new air contaminants standard covers only employers in general industry, the agency indicated in the final rule's preamble that the revised PELs will be applied eventually to the construction industry.

The agency also is considering applying the revised limits to the maritime and agricultural industries.

Pendergrass said in announcing the final rule in January 1989 that a proposed revision of the PEL tables for the construction industry would be presented to the Advisory Committee on Construction Safety and Health (ACCSH) "later" in 1989. The advisory committee, which the agency is required to consult before setting standards that will have a major impact on construction, has been informally consulted, OSHA said.

The agency acknowledged that application of the revised standard to the three industries may require some "modifications" because of differences in exposures, work situations, and feasibility.

In the preamble, OSHA said although it is appropriate at this time to limit the newly revised PEL table to general industry, the agency will pursue its expansion to other industrial sectors as a "second-stage rulemaking."

Health Benefits

OSHA estimated that more than 21 million workers work where they could be exposed to hazardous chemicals. Of these workers, 4.5 million are exposed to substances above the lower limits established by the new standard, the agency observed.

The revised PELs will protect workers against a wide variety of health effects that could impair their health or their ability to function, OSHA stated in papers explaining the new standard.

The agency said the lower PELs would prevent many adverse reactions, ranging from "catastrophic effects," such as cancer, lung disease, and cardiovascular, liver, and kidney damage, to "more subtle" effects, such as central nervous system damage, narcosis, respiratory effects, and sensory irritation.

The agency said the lower PELs could reduce the number of annual work-related illnesses—now more than 55,000; lost-work-day illness—now more than 23,000; and lost workdays due to illness—currently almost 520,000.

OSHA also estimated that compliance with the newly mandated PELs will reduce by 683 the number of deaths each year from exposure to substances that cause cancer, respiratory disease, cardiovascular disease, or liver or kidney disease.

Under the new standard, 4.5 million workers will receive additional occupational health protection at an annual cost of $150 per worker, according to OSHA's calculations.

1989 Compliance Date

The standard went into effect March 1, 1989, and employers must achieve the new limits by Sept. 1, 1989. Until Dec. 31, 1992, employers may use any combination of engineering controls, work practices, and respiratory protection to achieve and maintain the new limits. After that time, they must use engineering controls and work practices to comply.

However, if OSHA has not published a final rule on compliance methods by Dec. 31, 1991, employers will have until Dec. 31, 1993, to meet the new PELs through engineering controls and work practices.

The agency determined that existing analytical or sampling methods cannot detect seven substances at the lower limits. These are aluminum alkyls, ethylidene norbornene, hexaflouracetone, mercury (alkyl compounds), oxygen difluoride, phenylphosphine, and sulfur pentaflouride.

Because developing sampling and analytical procedures is a "dynamic, rapidly progressing technology," it is appropriate to establish lower limits for these substances without adequate procedures, the agency reasoned. However, it explained that it will not enforce PELs for these chemicals until such methods are developed.

OSHA did not indicate how long this is expected to take. But it noted in the final rule that when the original PELs were adopted in 1971, analytical methods existed for only a few of the hundreds of chemicals. "In the intervening years," it said, "OSHA, NIOSH, and the private sector have developed and tested hundreds of methods and have made these available to the industrial hygiene community."

As new analytical methods are developed, OSHA will announce them in the *Federal Register* and propose dates for enforcing the relevant PELs for the seven substances.

OSHA said it anticipated no problems enforcing the lower limits established for the other substances by the new air contaminants standard.

In April 1989, the agency directed states operating their own safety and health plans to adopt an air contaminants standard "at least as effective as" the new OSHA standard, as required by the OSH Act.

Agency Assistance Available

OSHA is advising employers on how to comply with the lower PELs established by the air contaminants standard, according to OSHA spokesman Akio Konoshima and the agency's consultation program guidelines.

OSHA helps employers identify potential hazards and provides advice on engineering controls, work practices, and personal protective equipment, such as respirators, they can use to meet the standard.

The agency also tells employers, during inspections, about the new standard and the Sept. 1, 1989, compliance date, one staff member in OSHA's health compliance assistance office told BNA.

OSHA also offers employers a booklet that summarizes the rule and contains the new PEL tables, according to an agency spokesman. The booklet contains a nine-page summary of the standard's provisions and compliance schedules, and reprints both "transitional," or previous, limits as well as the new limits for all 600-plus substances in the PEL projects.

Also, OSHA published an index and table of contents in the March 28, 1989, *Federal Register* that is expected to make it easier for employers to use the final rule and its preamble, the spokesman said.

Transitional Limits Enforced

Because employers were not required to meet the new limits until Sept. 1, 1989, OSHA inspectors would not cite employers under the standard until then, the health compliance staffer told BNA.

Until then, agency compliance officers would continue to enforce the old, transitional limits and the required engineering controls for those limits, she said.

A directive to regional offices instructing inspectors how to enforce compliance with the new standard was to have been sent out before the Sept. 1 compliance date, she added.

Section 21(c) of the OSH Act directs the agency to provide training and consultation with employers and employees on preventing occupational injuries and illnesses.

The agency provides free on-site consultation to private-sector employees, according to OSHA's "Consultation Policies and Procedures Manual." At the request of employers, agency consultants identify safety or health hazards, provide guidance on how to correct them, help employers develop or improve their safety and health programs, and provide training and education to employers, workers, and supervisors.

In most states, the service is provided through annual cooperative agreements with a state government agency under Section 7(c)(1) of the act. In a few states, according to OSHA, the service is offered through the state's occupational safety and health program under Section 18(b) of the act.

Although no trade associations or other professional groups have issued formal guidelines for complying with the new rule, most are distributing information about the new PELs among their members and clients and are helping these companies informally.

Legal Challenges

Although the rule was not challenged immediately, legal activities began soon after it appeared in the *Federal Register*. Challenges to OSHA rulemakings must be filed in court within 60 days after the rule is published.

By Jan. 23, 1989, five petitions for review of the standard had been filed by four industry groups (the Inter-Industry Committee on Carbon Disulfide, Courtaulds Fibers Inc., the American Iron & Steel Institute, and the Corn Refiners Association) and the AFL-CIO in federal appeals courts in four separate circuits.

In February, the cases were consolidated in the U.S. Court of Appeals for the Eleventh Circuit, under provisions of 28 USC 2112(a)(3).

In addition, two industry groups, the International Fabricare Institute and the National Lime Association, asked OSHA to delay enforcement of some provisions of the standard pending court review of the final rule, and several other groups, including the Styrene Information and Research Center, Hoechst Celanese Corp., and Tennessee Eastman Co., asked the agency to reconsider other requirements.

As of mid-June 1989, the agency had issued no administrative stays of the standard—or parts of the standard—based on petitions for review it received.

Legal proceedings were being conducted in mid-1989, although a schedule for a first round of briefs had not been set as of mid-June 1989.

In addition to the five groups that had filed initial challenges in January 1989, other groups challenging the standard included the International Fabricare Institute, the Enzyme Technical Association, the Fertilizer Institute, the Fiberglass Fabrication Association, the Interstate Natural Gas Association, the National Grain and Feed Association, and the Society of the Plastics Industry.

Challenges also were filed by Inco United States Inc., Hoechst Celanese Corp., Caterpillar Inc., the American Mining Congress, ASARCO, Brush Wellman Inc., the Institute of Makers of Explosives, and the International Institute of Ammonia Refrigeration.

Other challenges were filed by Kennecott Corp., the National Lime Association, NGK Metals Corp., the Salt Institute, the Scientific Apparatus Makers Association, and United Technologies Corp.

Attorneys for various challengers expressed some concern about specific PELs, but they admitted to BNA that in such an early stage of the litigation their objections to the rule were broad-based.

Motions to intervene in the court challenges were filed by the American Petroleum Institute, the Chemical Manufacturers Association, the Furniture Workers Division of the Electronic and Electrical Workers, IUE, Local 800, the Halogenated Solvents Industry Alliance, Inco United States Inc., the Inter-Industry Wood Dust Coordinating Committee, the Polyurethane Manufacturers Association, the Thermal Insulation Manufacturers Association, and United States Gypsum Co.

New Limits

The new OSHA standard tightens PELs for some chemicals and sets limits for the first time for other others.

Among the 212 substances for which PELs were tightened are acetaldehyde, acetic acid, acetone, carbon monoxide, chlorine, chloroform, nitrotoluene, octane, perchloroethylene, portland cement, styrene, sulfur dioxide, and toluene.

NEW CHEMICAL EXPOSURE STANDARDS

In the final rule, the agency also set PELs for the first time for 164 previously unregulated chemicals, including captafol, captan, carbofuran, fenthion, formamide, gasoline, grain dust, indene, indium and compounds, iodoform, iron salts, isoctyl alcohol, methylacrylonitrile, MBOCA, methyl parathion, nonane, paraffin wax fume, platinum (metal), resorcinol, *m*-toluidine, *p*-toluidine, tungsten (insoluble and soluble compounds), *n*-valeraldehyde, vinyl acetate, vinyl bromide, welding fumes, and wood dust.

Early Reaction

"I am very proud of what the agency has done, and the position it is in for the future," Pendergrass said at a Jan. 13, 1989, news conference announcing the rulemaking action.

Initial reactions to the final rule were generally positive, but some industries and unions said they were concerned about certain PELs and other aspects of the new rule. Criticism soon built, and group after group challenged the standard in federal appeals courts around the nation.

Tim O'Leary, a spokesman for the Chemical Manufacturers Association (CMA), told BNA that the chemical industry supports the revision of the PELs and the generic approach taken by the agency. The exposure-limit tables were "badly in need of an overhaul," he said.

The CMA official added, however, that the association "fervently hopes OSHA doesn't find itself in the same position 20 years from now." He said CMA supports an automatic review mechanism to keep limits up-to-date.

The AFL-CIO said the final rule did not include any of the labor federation's suggestions, such as using NIOSH's recommended PELs or including methylene chloride and cadmium—currently undergoing the full rulemaking process—in the PEL update, AFL-CIO industrial hygienist Diane Factor told BNA.

The labor federation was pleased, however, by OSHA's further reduction of the proposed PEL for perchloroethylene, a solvent used in the dry-cleaning industry, she said.

The Synthetic Organic Chemical Manufacturers Association (SOCMA), a trade group representing companies involved in the manufacture, distribution, and marketing of synthetic organic chemicals, said it was pleased with the revised PELs. Calling the new rule a "historic step for worker safety," SOCMA Chairman James Mack said the action is "a major step forward for every industry, a new landmark in cooperation between the private sector and the federal government."

Mack, however, criticized the agency's adoption of PELs recommended by NIOSH for some substances. Those limits do not "necessarily represent the most realistic basis for judging safe PELs," Mack said.

'Rubber-Stamping' Industry Levels

The Amalgamated Clothing and Textile Workers Union (ACTWU) criticized the new OSHA standard as "merely rubber-stamping the chemical industry's own recommendations," a reference to the ACGIH limits.

ACTWU President Jack Sheinkman called the new limits for perchloroethylene and other chemicals "simply unacceptable." The new limits, he contended, will leave workers at risk and will allow cancer to occur in 2,000 of the 100,000 workers exposed to perchloroethylene in the dry-cleaning industry.

The union "will continue to fight to protect dry-cleaning and other workers from occupational disease due to toxic chemicals on the job," Sheinkman said.

NAM applauded OSHA's effort, but said it was concerned about the magnitude of the new regulations.

"The broad scope of the rulemaking process ... raises questions about its scientific soundness and about industry's ability to comply with so many changes at one time," NAM said in a statement on the rule. NAM said it favored a "more staggered approach" to setting chemical standards.

Controversial Substances

The new rule's controls on styrene, perchloroethylene, grain dust, and wood dust received the most comment. It included the following:

STYRENE

Styrene, a chemical commonly used in the plastics industry, was classified as a carcinogen based on ACGIH information.

However, the Styrene Information Research Council and other groups objected to that classification. SIRC argued that nine animal studies on styrene showed either no evidence of cancer or were inconclusive due to study limitations, such as faulty study design, high background tumor incidence, or high morbidity among test and control animal groups.

OSHA also acknowledged that studies on workers exposed to styrene were not reliable for demonstrating the chemical's carcinogenicity because the workers also were exposed to other chemicals, including 1,3-butadiene.

Consequently, the agency classified styrene as a narcotic, based on demonstrated central nervous system dysfunction—headache, fatigue, dizziness, confusion, malaise, and a feeling of intoxication.

The agency adopted its proposed PEL, reducing maximum styrene exposures to 50 ppm for any eight-hour work shift and short-term exposures to 100 ppm for any 15-minute period. The previous limit prohibited exposures above 100 ppm per work shift, or 200 ppm for five minutes in any three-hour period.

OSHA noted in its preamble that large chemical companies, including Rohm and Haas Co. and Dow Chemical Co., already had established limits at their plants ranging from 25 ppm to 50 ppm.

The new limits can be achieved through engineering and work-practice controls in all industries that use styrene, with the exception of two operations in the boat-building industry, OSHA contended in the final rule. General dilution ventilation, local exhaust ventilation, and process enclosures are effective control measures for the manufacture of large reinforced-plastic items such as tubs, showers, and diving boards, the agency stated.

But in two operations—manual "lay-up and spray-up" in the boat-building industry—OSHA said insufficient evidence existed that engineering and work practice controls could be used to achieve the new levels. For those operations, the agency said, respirators may be used to reduce exposure levels from previous limits to the new limits until OSHA can demonstrate that engineering and work-practice controls are feasible.

PERCHLOROETHYLENE

In its proposed rule, OSHA had recommended that exposures to perchloroethylene, a chemical widely used in dry-cleaning operations and as an industrial degreaser, be reduced to 50 ppm, with a short-term limit of 200 ppm for any 15-minute period. Those limits were based primarily on evidence of the chemical's narcotic effects. Previously, the agency had required that exposures to the chemical not exceed 100 ppm in an eight-hour work shift or 200 ppm for more than five minutes in any three-hour period.

But after emotional and extensive testimony at its public hearings on worker over-exposures and health effects, as well as NIOSH's disapproval of the proposed limit, the agency established a much lower limit of 25 ppm in the final rule.

NIOSH also disagreed with the agency's original classification of the chemical as a narcotic. The research institute said the chemical should be regulated as a carcinogen and that exposures be held to the "lowest feasible limit."

The Halogenated Solvents Industry Alliance (HSIA), however, argued during the rulemaking process that one study of workers exposed to perchloroethylene—and to no other dry-cleaning solvents—showed no increased cancer risk. The group also contended that some of the cancer found in rodents used in other studies was due to physical characteristics that mice and rats have but humans do not.

HSIA also stressed that the Environmental Protection Agency and the International Agency for Research on Carcinogens had concluded that evidence on perchloroethylene carcinogenicity is "inadequate."

Citing testimony presented by the Amalgamated Clothing and Textile Workers—which represents dry cleaning workers—and the American Public Health Association, OSHA concluded that perchloroethylene is a "potential human carcinogen that presents a significant risk of material health impairment to workers exposed to it."

The agency said the perchloroethylene study cited by HSIA reviewed too few workers to demonstrate the "noncarcinogenicity" of the chemical.

Moreover, although EPA and IARC considered inadequate evidence that perchloroethylene is a cancer-causing agent, that description fits evidence about all but a few chemicals, according to OSHA.

OSHA relied heavily on a risk assessment which concluded that among every 1,000 workers exposed to perchloroethylene, 45 more die from cancer than would die from the disease if none of the 1,000 were exposed to the chemical. "Clearly this high risk of mortality represents a significant risk" justifying the further reduction of perchloethylene exposures, OSHA said.

The standard's four-year phase-in period should enable small and larger dry cleaning operations to achieve the lower 25 ppm limit, the agency added.

OSHA said it was "sympathetic to the circumstances of small businesses," and if, after three years, it appears meeting the 25 ppm limit will cause small dry cleaning operations significant economic harm, it would consider extending the compliance period for them.

GRAIN DUST

Previously unregulated by OSHA, grain dust was classified as dust generated by oats, wheat, and barley, and including such extraneous components as husk particles, starch granules, fungi spores, insect debris, pollens, rat hair, and mineral particles. Many opponents of the proposal maintained that no evidence exists that workers face a significant risk of being exposed to more grain dust than the 4 milligrams per cubic meter of air allowed under OSHA's proposed limit.

In addition, grain-elevator operators, especially small and family-run facilities, and grain cooperatives said it would cost too much to reduce exposures to the proposed level.

The National Grain and Feed Association said a NIOSH-sponsored study that OSHA used to set the proposed grain dust limit was based on "workers' arbitrary interpretations of 'average' exposures" at 4 mg/m^3, rather than actual, accurate measurements of exposures.

OSHA said the study found a significant excess of respiratory symptoms even among workers found through air sampling to be exposed to less than or equal to 10 mg/m^3.

Responding to charges that the study was biased, OSHA said the research institute reviewed many study populations, including eight grain elevators, state grain inspection agencies, and longshoring companies, and it is unlikely that answers to questionnaires distributed at that many locations would be biased as a result of employee dissatisfaction with working conditions at any one of them.

The American Feed Industry Association said a PEL for grain dust is "unwarranted and unnecessary" because the substance is a nuisance dust, and should be regulated instead under the 15 mg/m^3 limit for particulates. Although grain dust may have "some effect on some individuals' health," no evidence exists that those effects are "anything more than reversible and non-serious," the industry group stated.

The Food and Allied Trades Department, AFL-CIO, however, argued that such symptoms materially impair health. Deborah Berkowitz, director of safety and health, said "living with chronic bronchitis [resulting from grain dust exposure] is not a hazard that should go unchecked. In fact, study after study point to the possibility of very real long-term damage from chronic cumulative effects of exposure to grain dust."

OSHA said it was unable to establish a specific threshold level for respiratory effects or the feasibility of the proposed 4 mg/m^3 limit, especially for smaller grain elevators.

Nevertheless, the agency decided that a PEL should be established to reduce the significant risk of adverse respiratory effects, including "grain fever," wheezing, chest tightness, cough, eye and nasal irritation, and symptoms of chronic respiratory disease. Grain dust also may cause asthmatic reactions, especially in individuals prone to allergies, the agency stated.

As a compromise, OSHA selected a 10 mg/m^3 limit, saying it is technologically feasible and would substantially reduce the chance that workers' respiratory systems would be harmed by grain dust.

WOOD DUST

OSHA previously regulated workplace exposure to wood dust under its 15 mg/m^3 nuisance-dust standard, but the Occupational Safety and Health Review Commission ruled in 1985 that the standard should not be applied to wood dust or grain dust because they are not mineral dusts.

To address hazards posed to workers, OSHA included wood dust in its PEL revision project, and proposed to set exposures at levels recommended by ACGIH: 1 mg/m^3 for hardwood dust and 5 mg/m^3 for softwood dust. Hard woods were defined as broad-leaf, flowering trees and soft woods as those that do not lose their leaves in the winter.

Exposure to dust during the processing or handling of wood is associated with a variety of adverse health effects, including dermatitis, allergic respiratory effects, mucosal and non-allergic respiratory effects, and cancer, according to the agency.

Many officials of various wood industries, including furniture and cabinet manufacturers, casket producers, and paper and pulp operations, told OSHA that achieving the proposed wood dust limits would not be technologically or economically feasible, or it would be extremely difficult.

Many of those that commented, including at least two unions, argued that two different PELs would be burdensome for employers and could create compliance problems. The United Brotherhood of Carpenters and Joiners of America said "given the frequent intermixture of wood types in the workplace, this would render OSHA's compliance efforts virtually worthless."

The agency's assumption, based on studies of British workers exposed to hardwoods, that a more stringent limit was required for that type of dust also was challenged. University scientists and industry representatives argued that the distinction between hard and soft woods is botanical, and does not warrant two separate PELs.

Consequently, the agency adopted a single eight-hour PEL of 5 mg/m^3 and a 15-minute 10 mg/m^3 limit for both hard and soft woods. Additionally, OSHA set a 2.5 mg/m^3 limit for Western red cedar wood dust, based on evidence in the rulemaking record that such dust can cause allergic sensitization and occupational asthma.

Mixed Compliance Success Expected

Companies affected by the OSHA's final rule lowering PELs for hundreds of air contaminants think they will achieve the required levels for some substances but not others, according to sources contacted by BNA.

Most chemical companies will not have to change their operations significantly to meet the new lower PELs, according to CMA's O'Leary. "To a large extent, industry is already is compliance with those limits," he stated, especially large chemical manufacturing companies.

Compliance will not be "severe by any stretch of the imagination" for most smaller chemical companies, except where OSHA adopted especially stringent limits from NIOSH recommendations, Eric Christensen, a spokesman for SOCMA, told BNA.

"We do have some concerns about the NIOSH limits," conceded Susan Akers Kernus, SOCMA's government affairs manager. Those limits will be difficult and expensive for some members to achieve, she said, "but we're just going to have to live with them."

Meeting the standard may be difficult for certain "downstream" companies, or companies that do not manufacture chemicals but use, process, or distribute them, such as dry cleaners, O'Leary said.

'Arbitrary Decision'

The new exposure limits for styrene and perchloroethylene will be costly and perhaps impossible for many companies to meet, industry representatives told BNA.

Jack Winnick, general manager of Gold Shield Fiberglass Inc., a Fontana, Calif., company that used styrene to manufacture fiberglass products such as recreational vehicles and plumbing fixtures, said the new styrene standard would affect his industry "rather adversely."

The new standard allows most companies that use styrene in fiberglass boat-building operations to meet the new exposure limit by having workers use respirators. However, companies like Gold Shield that use polyester resins containing styrene in a different kind of open-molding process will have to meet the standard through engineering controls.

This was an "arbitrary decision" by OSHA, Winnick said. "Boat manufacturers don't have any bigger problems than anyone else" in trying to meet lower PELs through engineering controls, he contended.

Other fiberglass industries besides boat-building, such as the manufactures of underground storage tanks, tubs, spas, and planters, have processes that may warrant the use of respirators, agreed Daniel P. Boyd, a spokesman for the Styrene Information and Research Center. "Conditions are virtually the same throughout the open-mold sector of the reinforced plastics industry," Boyd said.

'It Will Put Some ... Out of Business'

Stephen Risotto, communications director for the Halogenated Solvents Industry Alliance, told BNA that the final 25 ppm standard for perchloroethylene will cause "some hardship" among dry cleaners, the primary users of the solvent, as well as difficulties for operations that use the chemical in degreasing operations, such as the auto, aerospace, electronics, and semiconductor industries.

To meet the new standard, industry spokesmen said, dry cleaners will have to replace older equipment, which requires clothing to be cleaned and dried in separate machines, with more expensive "dry-to-dry" machines that clean and dry.

This change would eliminate worker exposure to perchloroethylene fumes during the transfer of clothing from one machine to another, but it also would reduce profits for facilities that would have to invest in more up-to-date equipment, according to Risotto. "I suspect it will put some dry cleaners out of business," he said.

Even using dry-to-dry machines, dry cleaners will have difficulty meeting the new 25 ppm limit consistently, according to John Meijer, a spokesman for the International Fabricare Institute (IFI), which represents 11,000 dry-cleaning establishments. Machines will still leak chemical vapors after normal long-term use, and chemical residues will continue to collect inside the equipment, he told BNA. Consequently, one-third of all dry cleaners "won't meet the standard at all," Meijer said.

IFI, however, is more concerned about the fact that perchloroethylene has been designated a potential carcinogen and the effect of this action on businesses that use the chemical, according to Meijer. Over time, "that will probably close more doors than the 25 ppm PEL," he said.

Unions To Press For More Standards

In addition to filing lawsuits to force OSHA to reduce further those exposure limits that they believe were not made stringent enough under the PEL project, union officials will lobby OSHA to set more stringent standards for other toxic substances not covered by the PEL project, several union officials told BNA. Labor officials said they were delaying their drive for new and tougher limits until President Bush named a new OSHA administrator.

"We will be knocking at OSHA's front door as soon as there is someone there to answer," advised Franklin E. Mirer, director of the UAW's health and safety department.

Some unions that petitioned OSHA to set more stringent standards for particular substances during the Reagan administration continued to press for those regulations after President Bush took office, according to Diane Factor, an industrial hygienist with the AFL-CIO. Factor noted a UAW request for a methylene chloride standard and a petition by the Public Citizen Health Research Group and the International Chemical Workers Union for a cadmium standard.

The UAW first petitioned OSHA for a methylene chloride standard in July 1985; after receiving no response it asked again in February 1986, Mirer told BNA. In November 1986, OSHA told the union that it expected to propose a regulation in early 1987 and to publish a final standard in the summer of 1988. No proposal had been published by late June 1989, however.

Mirer said it was too soon in the Bush administration to say whether UAW would ask for standards on other chemicals about which it has particular concerns, such as glycol ethers.

"We don't know whether we're facing a situation where we have to petition OSHA in the courts for every standard or whether the new administration will take a more reasonable approach to rulemaking," he said.

Labor Priorities

The AFL-CIO wants the agency to conduct more "grouped rulemakings"—adopt standards that address groups of chemicals or hazards—Factor told BNA. She said that solvents and isocyanates would fit that type of rulemaking.

The AFL-CIO opposed the PEL project because the rulemaking did not meet statutory requirements, such as relying on the best scientific evidence available, Factor said.

The adequacy of the new PEL for perchloroethylene is expected to be an issue in the Amalgamated Clothing and Textile Workers Union's petition for review of the air contaminants standard, according to Eric Frumin, safety and health director for the union. In addition, ACTWU intends to oppose the dry cleaning industry's stand against the lower limit established in OSHA's final rule, he said.

ACTWU is informing its members that perchloroethylene has been classified as a carcinogen and that some of their employers will be required to reduce current exposure levels when compliance with the new limits is required, Frumin said. In addition, the union is encouraging its members to pursue further reduction of chemical exposures with their employers through collective bargaining agreements, according to the union official.

EXPOSURE MONITORING, MEDICAL SURVEILLANCE

As OSHA prepares to respond to the lawsuits filed over the PEL standard, it is considering setting additional requirements for the chemicals included in the standard, agency officials said.

Under these additional requirements, employers would have to monitor workers' exposures to the PEL substances and conduct medical surveillance programs for these workers. Such requirements appear in most other OSHA health standards.

In two advance notices of proposed rulemaking (ANPRs) published in September 1988, OSHA asked for public comment on monitoring and surveillance to help it determine if it should set generic measures in these areas and, if so, what the measures should contain and how much complying with them might cost.

Exposure monitoring requires employers to take and analyze workplace air samples. By analyzing these samples to see whether a regulated substance exceeds its limit, employers can determine whether they need to take further steps to reduce workers' exposure to the substance.

Similarly, medical surveillance provisions require employers to conduct periodic medical examinations of employees in order to determine if the workers show any abnormal physical conditions that may be caused by exposure to a certain substance.

Generic regulations in these areas would set broad requirements that could be used in conjunction with the exposure limits for the hundreds of chemicals included in the PEL project, and with other standards on hazard communication and methods of compliance, the agency said.

Most companies, unions, trade associations, and other parties that commented on the ANPRs have advised OSHA to proceed with the rulemakings, OSHA Health Standards Director Charles Adkins told BNA.

Potential Model Provisions

In the ANPRs, OSHA did not propose any particular provisions for generic rules on exposure monitoring and medical surveillance. However, the agency said some measures in its past health standards for specific substances have "potential" for serving as models for such provisions.

These include provisions that:

- Require employers to perform initial exposure monitoring;
- Specify how frequently follow-up monitoring should be conducted;
- Indicate whether air samples should be taken generally in the workplace or within workers' personal breathing zones;
- Require employers to notify workers or worker representatives of the results of monitoring;
- Set procedures for allowing employees to observe monitoring;
- Require medical surveillance for employees exposed to substances above a certain level, usually one-half the exposure limit; and
- Call for medical examinations to be performed or supervised by physicians at no charge to the employee.

Doubts About Schedule

OSHA's most recent regulatory schedule, issued April 24, 1989, said the agency expected to propose the additional regulations in September 1989. Agency plans call for it to issue final rules in 1990, according to staff members in OSHA's health standards office.

However, some agency staffers and outside observers told BNA they doubt whether OSHA will be able to issue any proposals before the end of 1989, and that final rules are years away.

Realistically, the agency could complete a "good first draft" by the end of 1989, Darrell K. Mattheis, a health and safety specialist with Organization Resources Counselors Inc., a Washington, D.C., firm that advises companies on technical issues, told BNA.

Foreshadowing potential industry opposition to these rulemakings were comments received by the agency in response to its ANPRs.

Although the comment period ended Dec. 27, 1988, the agency said it would give "full consideration" to comments received after that date. As of April 1989, the agency had received about 80 comments on each proposed rule.

Small Businesses at a Disadvantage?

Most companies that commented on the OSHA notices agreed with the concept of or need for exposure monitoring and medical surveillance, but they disagreed sharply over how best to implement such programs and what the standards should require.

Officials from industry and OSHA told BNA that larger companies may be better prepared than smaller ones for the potential standards. "The great majority of companies [already] have monitoring programs to make sure they're meeting OSHA exposure limits," Adkins said. Sheldon Weiner, director of OSHA's Office of Standards Analysis and Promulgation, said most medium-sized and large companies have some type of monitoring program, but many smaller firms may not be conducting any monitoring.

However, all companies have access to OSHA's free consultation programs to help them set up monitoring programs, Adkins and Weiner said.

"The larger companies are going to find that any new generic standards for exposure monitoring and medical surveillance will not differ significantly from what they are already doing," ORC's Mattheis said.

Smaller firms will be forced to spend a lot to hire new experts and to implement new programs because they do not have industrial hygienists, safety engineers, safety and health consultants, and occupational physicians, Mattheis noted.

Some observers argue that companies already are required to conduct such programs when workers are exposed to specific hazardous substances. Generic rules would give employers more guidance, and they would be easier to comply with than an extremely large number of provisions under standards for separate substances, these observers contend.

'More Potential Harm Than Benefit'

However, others say new regulations will require changing existing monitoring and surveillance programs, forcing companies to spend more money and change their operations. It is impossible to say to what degree that may occur until OSHA issues more detailed proposals, according to O'Leary.

Because "there is no one way" to approach exposure monitoring and medical surveillance, industry should not "have to abide by one set of rules," O'Leary said. Such programs, he added, require flexibility and site-specific strategies. "This is not a cookie-press procedure," O'Leary offered.

Some industry representatives said OSHA should not develop generic standards for exposure monitoring and medical surveillance.

Pratt & Whitney, a subsidiary of United Technologies Corp., said in written comments to OSHA that the agency should let companies develop their own industrial hygiene programs, which the firm said would be cost-efficient and would meet each employer's needs. "OSHA should stick to developing performance standards and leave the practice of industrial hygiene to the professionals," G.W. Lancour, the company's manager of safety and health services, said.

"OSHA may be doing more potential harm than benefit by dictating to professional hygienists as to frequency of monitoring, appropriateness of sampling methods, and selection of a sampling strategy," Gail M. Driscoll, a health and safety information and regulation specialist for Merck & Co., said.

Driscoll said industrial hygienists are familiar with plants and their operations and they develop effective air sampling programs based on that knowledge.

Gulf Power, an electrical utility in Pensacola, Fla., told the agency that a medical surveillance rule could cause labor relations problems if doctors try to disqualify workers from certain jobs because they smoke or drink alcohol and those habits put them at higher risk of developing an illness.

Medical surveillance probably will not improve employee health unless OSHA can prove that chronic exposure to specific substances can cause serious illness or death, according to Fred R. Cully, Gulf Power's manager for organizational planning and development. "What is needed is an organized, prioritized research project to identify what chemicals cause irreversible adverse effects to humans and how the onset and progression of these illnesses can be tracked medically," Cully said.

Unions: 'Key to Protecting Workers'

Labor unions told OSHA that exposure monitoring and medical surveillance can supplement traditional substance-by-substance OSHA rulemaking.

"Monitoring and medical surveillance for exposed employees is key to protecting workers from workplace hazards" and can "greatly expand the effectiveness of OSHA's regulations," Margaret Seminario, associate director of occupational safety, health, and social security for the AFL-CIO, told the agency.

Exposure monitoring standards should contain criteria for monitoring exposed workers, including provisions for an overall exposure assessment, evaluation of chemicals and agents and conditions of use, evaluation of operations and workers at risk, and control measures used to limit exposures, the AFL-CIO said in comments to the agency.

"Exposure monitoring is the only acceptable means of determining overexposure and the adequacy of control methods; there are no acceptable alternatives," the American Federation of State, County, and Municipal Employees told the agency. Moreover, without mandatory requirements, employers will not conduct exposure monitoring, AFSCME contended.

Labor officials also said generic standards should not replace standards for individual substances. Matthew Gillen, executive director of the Worker Institute for Safety and Health, a non-profit organization, told BNA that the average company is small, has few if any safety or health personnel, and needs specific standards with which to comply. Also, without specific standards, "OSHA can't pin employers down" in citing alleged health or safety hazards, Gillen said.

■ ■ ■ ■

V. Recordkeeping

"We've largely made our points on recordkeeping ... I don't think we'll need the same intensive enforcement efforts," Frank A. White, then-deputy director of the Occupational Safety and Health Administration (OSHA), said Feb. 10, 1988, speaking at a conference of shipyard industry safety officials, convened in Washington, D.C., by the Shipbuilders Council of America.

White was alluding to the agency's recent emphasis on enforcing regulations that require companies to record work-related injuries and illnesses among their employees. Amid growing charges by labor unions that employers were under-reporting such cases, the agency had begun to examine injury and illness logs closely during workplace inspections and to seek record-high fines totaling millions of dollars against employers for alleged failure to record cases.

"It's no mystery why inaccuracies in records should be of concern to OSHA," White said. He noted that the agency usually relies on national statistics based on those records to determine which types of workplaces it needs to inspect regularly and which ones it need not inspect. "OSHA steers away from whole industries [on the basis of their reported injury] rates," White said.

However, OSHA's get-tough policy had come under widespread criticism from industry officials, who charged that the agency was over-enforcing "paperwork" regulations. Some industry representatives asked if the effort signaled a change from OSHA's policy under the Reagan administration of trying to work "cooperatively" with industry in regulating workplace hazards.

"Our industry sees, in the last couple of years, the pendulum swinging back to an absolute enforcement approach," said one member of White's audience at the Washington meeting.

By mid-1989, the controversy surrounding OSHA's recordkeeping enforcement policy had subsided. As White predicted, the number of citations issued declined from the peaks of 1986 and 1987, but it remained higher than it was in the early 1980s and OSHA inspectors were continuing to examine injury records closely, Frank Frodyma, OSHA acting director of policy analysis, told BNA.

INDUSTRY, LABOR COMPLAINTS

Nevertheless, industry and labor representatives told BNA that they continue to have serious concerns about the OSHA recordkeeping system. Those concerns have led OSHA and the U.S. Bureau of Labor Statistics, which assists the safety agency in designing its recordkeeping program, to plan major changes in recordkeeping forms and guidelines in the next two years.

Those changes may eliminate some of the complaints that employers have about OSHA's recordkeeping system, but they also may generate new questions and controversies, according to industry representatives. Union officials say they will try to get Congress to amend the Occupational Safety and Health Act (OSH Act) to eliminate what they say are basic flaws in the recordkeeping system.

Industry and labor have complained for several years about the OSHA recordkeeping system, each for different reasons. Industry officials charge that OSHA's regulations—and supplemental guidelines from the Bureau of Labor Statistics (BLS) designed to clarify them—often are ambiguous or over-inclusive in describing which injuries and illnesses must be recorded.

As a result, they say, employers must record injuries that they regard as trivial, like poison ivy irritations, or irrelevant to job conditions, like slips and falls in company cafeterias. Employers may be cited for recordkeeping violations if they fail to record these minor incidents, industry representatives complain.

"We will record any [injuries] we have any question about," Thomas F. Evans, director of safety and environmental health for Monsanto Co., told BNA. However, such all-inclusive injury records are "meaningless" for identifying serious hazards in the workplace, Evans said.

Union officials say regardless of how many work-related injuries are recorded on individual company logs, most such cases are never reported to OSHA. "Unless OSHA shows up at the plant gate, it doesn't know" how many injuries occur in a given workplace, according to Eric Frumin, director of health and safety for the Amalgamated Clothing and Textile Workers Union. Frumin also is chairman of the Labor Research Advisory Committee, a group of labor union officials that provides technical advice on government recordkeeping programs.

THE KEYSTONE REPORT

OSHA and BLS are seeking answers these concerns and others, officials from the two agencies told BNA. These efforts include two BLS projects that would result in major changes in the recordkeeping rules:

- New recordkeeping forms that would make information about worker injuries more readily available to government regulators, statisticians, health authorities, and researchers by including important data on one form, rather than on two forms as is currently the practice. These documents are expected to replace current forms on Jan. 1, 1991.

- New recordkeeping guidelines that may differ significantly from current ones. The bureau plans to seek comments from industry, labor, and other parties on what they think the new guidelines should contain.

OSHA expects to repeat a 1987 investigation in which inspectors examined injury records from 400 workplaces in two states to determine how accurate the documents were.

The framework for developing new forms and guidelines, officials told BNA, comes from a report issued Jan. 31, 1989, by a 43-member study group comprising representatives from government, labor, and industry. The group was convened in 1987 by the Keystone Center, a non-profit Colorado organization that brings differing parties together to draft consensus policy recommendations for the government on environmental and safety and health issues.

Pilot Studies on Forms

The Keystone group focused on OSHA Form 200, an annual log of injuries and illnesses that each workplace must maintain and make available to OSHA inspectors on request, and Form 101, a supplementary record used to provide further information about the circumstances of each injury.

The Keystone report said the forms should be changed to provide "more complete injury and illness data, including information on employment tenure, nature and source of injury and illness, and factors which contributed to the incident."

Bureau officials told BNA that they have launched four pilot studies in 10 states to test changes like those recommended by the report. BLS will use the study results to decide what type of form or forms will replace those currently used, they said.

In one of the studies, 1,050 employers in three states are using a modified Form 101 that asks the employer to record the number of lost workdays that result from a given injury or illness. To obtain this key information about the severity of injuries, BLS currently must examine both Form 200 and Form 101, according to William L. Weber, chief of program analysis in the BLS Office of Safety, Health, and Working Conditions.

In three states, 1,050 employers are using a single form that asks for the information now sought on Forms 200 and 101.

Some 700 employers in two states are using a similar consolidated form that also includes codes for various types of injuries or illnesses. BLS wants to see if the codes enable it to obtain information more quickly for additional studies of health and safety hazards, according to Weber.

These three studies began in April 1989 and are expected to continue through March 1990.

A fourth study in which 700 employers used the current forms was conducted in two states from January through March 1989,. The results of that study will give BLS information for comparing results from the three other studies, Weber said.

Avoiding 'Future Shock'

BLS will use the study results to determine which type of revised form best improves the current system, bureau officials told BNA. "We want to look at the resources required [from employers], in terms of differences in costs and time," Weber said.

Industry representatives said they are willing to cooperate with BLS on testing new ideas, but they added that the bureau must act now to avoid confusing employers with new forms in two years.

"There's no question that if there are changes in the forms, you're going to have confusion," said Kyle Olson, a health and safety specialist with the Chemical Manufacturers Association.

"BLS will have to be very careful in how it redesigns the forms," Olson said. "They have to make them more effective tools, but not give too great a sense of future shock. They will need to do a lot of communication and a lot of education."

BLS is explaining its plans in presentations to trade associations and discussions with the Business Research Advisory Committee, the industry counterpart of the labor advisory panel chaired by Frumin, Weber said.

"Larger employers are already collecting this information," he added. "Their attitude is, 'Why did it take the government so long to see the need for what they've already been doing for years?'"

PREPARATIONS ON NEW GUIDELINES

The bureau also wants to prepare employers for new recordkeeping guidelines scheduled to go into effect in January 1991, officials said. BLS probably will seek public review and comment on proposed changes in the spring of 1990, according to Robert Whitmore, acting chief of recordkeeping in the BLS Office of Occupational Safety and Health Statistics.

"We need time to receive comments, go through them, make final revisions, and get things signed off by OMB. It's a very ambitious schedule," Whitmore said.

The bureau had planned to issue new guidelines in mid-1988 and put them into effect in January 1989, but it decided to delay the guidelines until the Keystone report was released, and until it could obtain OSHA's input on the revision, officials said. BLS's current guidelines, issued in 1986, will remain in effect until final changes are made.

The Keystone report contained three recommendations for changing the guidelines:

- BLS should stop defining as "work-related" any injuries on company premises that involve "personal food preparation or eating," off-hour activities, pre-existing physical conditions, or "acts of violence unrelated to the work situation (e.g., domestic quarrels)."
- The guidelines should include a comprehensive list that distinguishes between injuries requiring only "first aid," which need not be recorded, and those requiring "medical treatment," which must be recorded.
- BLS should change its concept of "restricted work activity" for determining the seriousness of an injury. The guidelines say injuries must be recorded if the injured employee returns to work but cannot per-

form all the duties of his or her job. Instead, the guidelines should specify that injuries affecting employees' ability to do their job must be recorded only if the employee is not fully capable of performing the duties undertaken when the injury occurred and the duties scheduled for that week of work.

'Restricted Work Activity' Criticized

Industry officials told BNA they endorse the Keystone recommendations, especially the suggestion to revise the definition of "restricted work activity."

To avoid citations, employers are "recording every nick and scrape rather than focusing on ... the most serious injuries," Olson said. "If I got a paper cut, I guess I would have to record that because I wouldn't be able to shuffle papers."

Phyllis Eisen, director of risk management for the National Association of Manufacturers, said the changes suggested in the Keystone report are needed "so we can best do what we need to do" to monitor significant workplace injuries.

BLS probably will use the Keystone recommendations as a starting point in drafting proposed revisions to the current guidelines, bureau officials told BNA. "We aren't tied to their recommendations, but unless someone comes up with a better idea," the changes suggested by the Keystone group "will be in our draft revision," Whitmore said.

Some industry officials said they are eager to comment on the guidelines. Richard F. Boggs, vice president of Organization Resources Counselors Inc., a consulting group that advises companies on technical safety and health issues, noted that the planned revision project had dropped two years behind schedule.

"They have legitimate reasons why they're running behind," he added. "We're just saying, 'Don't run behind too long.'"

However, BLS will have to be careful to "minimize" any confusion or disruption caused by new guidelines, Boggs and other industry representatives added. Eisen said new guidelines and forms pose problems that are especially difficult for small businesses, which may "end up out of compliance because they're not sure what to do."

'OUTREACH' PLANNED

Bureau officials acknowledged that it will "take a large outreach effort" to educate employers about new forms and requirements. "We're telling ourselves to do a better job of getting to small employers," said William Eisenberg, director of the BLS Office of Occupational Safety and Health Statistics.

BLS plans to involve industry recordkeepers early in the process of revising the current forms and guidelines, Eisenberg said. The bureau plans ask a group of people who fill out injury records for their companies to practice using the proposed forms and guidelines after they are written, he said.

This will help the bureau identify "flaws which could lead to misleading or inaccurate information being reported," Eisenberg explained.

After the new forms and guidelines are issued, the bureau will explain employers' new obligations to them, Eisenberg and Whitmore said. The bureau plans to seek private organizations' help in conducting training seminars, as it did when the current guidelines were issued, they said.

In addition to using written educational materials, such as brochures, the bureau may produce a videotape to help employers understand the new requirements, Whitmore said.

Olson said BLS should provide employers a confidential toll-free hot line to help them interpret the new forms and guidelines.

CHECKING RECORDS' ACCURACY

OSHA enforcement statistics appear to support Frodyma's assessment of the agency's recordkeeping enforcement activities. Total dollar penalties sought against employers for alleged failure to record injuries and illnesses declined gradually from the beginning of fiscal 1987, from $8 million for fiscal 1987, to $3 million for fiscal 1988, and to $1.7 million for the first half of fiscal 1989, according to data provided to BNA. During 1984 and 1985 combined, only $17,000 in recordkeeping fines were proposed, according to other OSHA data cited by the U.S. General Accounting Office in a Dec. 30, 1988, report.

GAO said OSHA should examine injury records closely each time it inspects a workplace, and should routinely compare companies' injury logs with information from worker's compensation reports, daily injury reports, and other internal records.

The accounting office noted that OSHA conducted such audits at 400 workplaces in Massachusetts and Missouri in a 1987 pilot study, and found that about 20 percent of the injuries that should have been recorded on OSHA Form 200 had not been, according to BLS analysts who evaluated OSHA's data.

Frodyma said OSHA could not conduct such investigations routinely without reducing the time it needs to inspect working conditions. "We [told GAO] that we could do it, but at the expense of safety and health," he explained.

Inspectors try to determine the accuracy of the logs by questioning recordkeepers about their understanding of the OSHA requirements and BLS guidelines, and interviewing employees to obtain supplemental or unreported information on injuries, Frodyma said. He noted that these efforts take less time than investigations.

This information enables the agency to classify the alleged violations it finds, according to the OSHA official. "If the person keeping the records has a lot of training, and we find that injuries were not put down on the log in accordance with the requirements, that forms the basis for a willful [citation]," he said.

Frodyma and other officials said the agency wants to repeat the limited audit study periodically. In a Dec. 15, 1988, letter to GAO commenting on the preliminary findings of the accounting office's report, then-OSHA Administrator John A. Pendergrass said the agency planned to do so "every five years or so." Frodyma told BNA that the study might be repeated in 1992, after the new forms and requirements are expected to have gone into effect. "This puts us a little ahead of the time frame" mentioned by Pendergrass, he said.

MODELS FOR NEW LEGISLATION

Labor union officials told BNA they support a major redesign of the current recordkeeping system. The effort planned by BLS "clearly needs to move forward ... in a way that accommodates the needs of major users of the information, [such as] employers, individual workers, trade unions, OSHA, BLS, the National Institute for Occupational Safety and Health, public health authorities, worker's compensation agencies, Congress, and health care providers," Frumin offered.

He said that OSH Act amendments that would require employers to report workplace injuries directly to OSHA also are needed.

Frumin and other union officials noted that organized labor plans to push for a variety of revisions to the act during the 101st Congress. "Chances are quite good that any OSHA reform legislation would include changes on recordkeeping," the textile workers official said.

Frumin declined to say what specific changes labor might urge, but he said the "models are out there now":

- Mine Safety and Health Act requirements for mine operators to notify the Mine Safety and Health Administration of accidents;
- Provisions in "a lot of collective bargaining contracts," including those of the United Auto Workers, the United Steelworkers of America, and the Oil, Chemical, and Atomic Workers Union, that require employers to provide information on worker injuries and fatalities to the unions; and
- Public health laws that require doctors to report infectious diseases to state health agencies. "This sparks action," Frumin said. If "OSHA gets a report of an injury, that [would spur] action."

Frodyma said OSHA may already have the authority to demand reports on individual injuries. The OSH Act requires employers to provide OSHA with records "deemed necessary or appropriate" for enforcement, he observed.

If stronger measures are indicated, the agency might need only to amend its recordkeeping regulations to state, for example, that employers must send OSHA a copy of Form 101 each time a worker is injured, he said.

However, such a requirement would increase the controversy over how much more information OSHA needs and how it would handle the additional paperwork, according to Frodyma. "Do we have to know every time a secretary cuts their finger? What would we do with 5 million more pieces of paper?" he asked.

■ ■ ■ ■

VI. Regulating Blood-Borne Diseases

Between 1 million and 1.5 million people in the United States are infected with the human immunodeficiency virus (HIV), the blood-borne microorganism generally believed to be the cause of acquired immune deficiency syndrome (AIDS), according to an April 30, 1989, estimate by the U.S. Centers for Disease Control (CDC). As of that date, some 94,280 people were diagnosed as having AIDS, and 53,544 had died of the disease, the agency reported.

CDC predicts that by the end of 1991, these figures will more than double—270,000 AIDS cases diagnosed, 179,000 deaths.

In the area of occupational safety and health, these numbers are important beyond the immediate human tragedy they represent. Many health experts, government officials, and labor leaders say they show that the government should issue new regulations to protect millions of U.S. workers in health care, emergency service, and medical research occupations from an increasing risk of exposure to the AIDS virus on the job.

Some 5.3 million doctors, nurses, physician's assistants, laboratory employees, emergency rescue personnel, and others work in jobs that routinely bring them into close contact with human blood, according to the Department of Labor (DOL). As the number of AIDS cases grows in the general population, the risk also grows that these workers may be exposed to HIV-contaminated blood in ways that may introduce the virus into their own bodies, many experts say.

As a result of this concern and growing pressure to protect health care workers from an even more prevalent blood-borne disease—hepatitis B (HBV)—the Occupational Safety and Health Administration on May 30, 1989, proposed a standard that would require all U.S. employers with workers who are exposed to blood or potentially infectious materials on the job to use training programs, protective clothing and equipment, and other measures to safeguard such employees.

Government officials say the proposed standard would require employers to spend more money to comply with it than with any other regulation issued by OSHA. It also would mark the first time the agency tried to regulate a biological hazard, rather than a mechanical or chemical one.

OSHA issued administrative guidelines to its compliance officers in January 1988 instructing them to seek fines against hospitals, clinics, and industrial companies for alleged failure to protect workers from blood-borne disease risks. The standard would expand that effort. Labor leaders say the proposed rule is long overdue, but employers question whether the agency can deal effectively with infectious disease issues.

'NO KNOWN CURE'

"The gravity of becoming infected with the human immunodeficiency virus, a disease with no known cure and which is believed to lead inevitably to death, has raised a consciousness of personal risk in medicine that had been largely dormant," a February 1989 report by the Association of American Medical Colleges said.

HIV attacks the body's immune system and ultimately renders it unable to combat malignancies, infections, and other opportunistic diseases. According to the American Hospital Association (AHA), a person infected with HIV may experience a mild, flu-like illness within the first few weeks after the initial infection and then show no further symptoms for years. It may take up to five or six years to detect infection, according to the AHA.

CDC says the virus is most commonly transmitted through sexual contact with an infected partner and the sharing of contaminated needles by drug users. In health care settings, workers may be at risk of infection through direct contact with contaminated blood, semen, vaginal secretions, cerebrospinal fluid, synovial fluid secreted by the membranes of joints and tendons, pleural fluid from the membrane that covers the lungs and the lining of the chest cavity, peritoneal fluid from the membrane surrounding the heart, or amniotic fluid, according to CDC.

Workers are believed to be at risk if they are stuck with contaminated needles, if their mucous membranes come into direct contact with contaminated body fluid, or if they have an open wound that comes into contact with contaminated body fluid, CDC says.

Hepatitis B virus can be transmitted in the same ways. This virus, which causes inflammation of the liver, can lead to cirrhosis or liver cancer and may be fatal, according to CDC.

OSHA says workers who face a "high risk" of work-related infection from HIV and hepatitis B through routine exposure to potentially infected body fluids include doctors, dentists, and dental technicians; intensive-care, emergency-room, and operating-room nurses and technicians; X-ray technicians and pathologists; laboratory and blood bank technicians; and employees who draw blood from patients. Other workers who may be directly exposed to body fluids during certain work assignments include housekeeping personnel, laundry workers, and orderlies in health care facilities, morticians, research laboratory workers, paramedics, and medical examiners, the agency says.

'A Panic About AIDS'

As of Sept. 19, 1988, 169 health care workers with no identified risk for infection had AIDS, according to OSHA, and the agency says "it is reasonable to assume that at least some of [the cases] resulted from occupational exposures." CDC has documented only 25 cases involving health care workers in the United States who have work-related HIV infections. The cases are assumed to be work-related because other avenues of risk from personal behavior have been ruled out.

Only two laboratory workers, both employed in AIDS research by the National Institutes of Health, have reported testing positive for HIV infections apparently resulting from work-related exposures, according to Robert N. McKinney, head of NIH's safety division.

James M. Hughes, deputy director of CDC's Center for Infectious Diseases, estimates that the risk of transmission of HIV following a needlestick accident is less than 1 percent—or three to five infections per 1,000 people injured by contaminated needles, according to OSHA.

"The rate of occupational risk is expected to grow, but no one knows how much," then-Labor Secretary Ann McLaughlin told a Jan. 9, 1989, conference on AIDS sponsored by DOL and the Department of Health and Human Services (HHS). Because of this uncertainty and because no cure exists for the disease, "there is a panic about AIDS," McLaughlin said.

Risks from hepatitis B are easier to estimate, authorities say. OSHA estimates that direct worker exposures produce from 6,000 to 7,400 hepatitis B infections each year in the United States. According to Hughes, 12,000 new cases of hepatitis B infection in health care workers are reported each year. Some 500 to 600 of these workers are hospitalized, and between 200 and 300 die from cirrhosis or liver cancer caused by the disease. The risk of hepatitis B infection from needlestick injuries is between 6 percent and 30 percent, Hughes reported—and OSHA estimates that the annual HBV infection rate among health care workers who are frequently exposed to blood or other potentially infectious materials is between 4.89 and 6.63 per 1,000 exposed workers.

Unlike AIDS, hepatitis B can be prevented with a vaccine. The vaccine, which has been available since 1982, is about 90 percent effective in protecting against hepatitis B infection, OSHA says.

OSHA INSPECTION POLICY

Since 1983, when four unions cited the risk estimates for hepatitis B in petitioning for a standard to protect workers from exposure to that virus, OSHA has been under increasing pressure from organized labor and many public health advocates to regulate workplace conditions that pose a risk of exposure to blood-borne diseases.

In July 1987, OSHA announced that it would begin to develop such a standard. Although a final regulation had not been adopted by mid-1989, OSHA inspectors were inspecting health care facilities for blood-borne disease hazards, and were citing employers for alleged unsafe conditions. The agency authorized such inspections at hospitals, clinics, and other health care facilities in an administrative directive (CPL 2-2.44) issued Jan. 19, 1988, and updated on Aug. 15, 1988 (CPL 2-2.44A). (See Appendix J.)

The directive also instructs inspectors to examine clinics, health units, and nurses' stations at industrial work sites.

In the directive, inspectors are asked to determine whether employers are:

- Providing protective clothing to health care workers who "may reasonably be expected" to come into contact with blood or other body fluids on the job, regardless of whether the fluids come from patients known to be infected with HIV or hepatitis B;
- Warning these employees of possible infection hazards; and
- Keeping work sites clean and sanitary.

The directive is based primarily on "universal precautions" recommended by CDC to guard against transmission of blood-borne disease viruses in the workplace.

Protective Clothing

OSHA says employers must provide health care workers with disposable gloves, gowns, aprons, masks, and other items when the employee works with or handles blood and other body fluids.

Gloves must be made of latex, vinyl, or other "appropriate" materials, must fit the employee, and must be intact and not deteriorated, according to the directive.

Gloves also must be provided, and worn, whenever a health care worker takes a blood sample (phlebotomy). Employer may be cited for failing to make gloves available, or for discouraging employees from using gloves that are made available.

However, under the directive, an inspector cannot cite the employer if the employee taking the sample voluntarily declines to wear gloves, unless (1) the employee has a cut, scratch, or other break in the skin, (2) the inspector or the employee decides that the employee may come into contact with blood during the procedure, (3) the employee is drawing blood from the finger or heel of an infant or child, or (4) the worker is being trained in phlebotomy.

Protective gowns, aprons, or lab coats must worn be if the worker is likely to be splashed with body fluids. Gowns, including surgical gowns, must be made of or lined with impervious materials and they must protect all exposed skin.

Masks, protective eye wear, or face shields are required when a worker is likely to be splashed or sprayed in the eyes, nose, or mouth with body fluids—for example, OSHA said, during surgical or dental procedures. However, they are not required for routine care.

The agency also requires that pocket masks, resuscitation bags, or other such devices be provided in "strategic" locations and to key personnel, such as paramedics, who are likely to have to administer mouth-to-mouth resuscitation.

In dental procedures, employees must be provided with protective gloves, surgical masks, and protective eye wear, or chin-length plastic face shields if they are likely to be splashed with blood, saliva, or gingival fluids.

In laboratories or workplaces like funeral homes where postmortem procedures are performed, OSHA requires that gloves be used for processing body fluid specimens and that masks and protective eye wear be used when the worker's mucosal membranes may come into contact with body fluids. Also, OSHA requires workers performing or assisting in postmortem procedures to wear personal protective equipment to avoid exposure to body fluids.

Housekeeping Requirements

OSHA inspectors are instructed to evaluate housekeeping and waste-disposal requirements when they walk through workplaces and examine their infection control programs, according to the directive.

Under the directive, inspectors can cite employers for failure to provide a clean, orderly, and sanitary workplace. Rooms where body fluids are present must be cleaned as frequently as necessary based on the area of the institution where body fluids are present, the type of surface to be cleaned, and the amount and type of contamination present.

OSHA requires employers to use chemical germicides approved for use as hospital disinfectants to decontaminate blood and body fluid spills. A solution of 5.25 percent sodium hypochlorite (household bleach) diluted with water or another suitable disinfectant should be used for disinfection following the initial cleanup, according to the directive.

To avoid the risk of needlestick accidents, OSHA requires employers to ensure that needles are not recapped, purposely bent or broken by hand, removed from disposable syringes, or otherwise manipulated by hand.

In addition, disposable syringes and needles, scalpel blades, and other sharp objects must be placed in puncture-resistant disposal containers. The container must be within easy access to workers and located in all areas where needles are commonly used, including emergency rooms, intensive care units, and surgical suites. Containers must be constructed so they will not spill their contents or injure workers who handle them. They must be located on patient floors and in all other locations where blood is drawn and needles are used.

Tags and Warning Signs

The directive says that tags must be used to warn employees if a biological hazard such as the AIDS virus or hepatitis B virus is present or may be present in the workplace, and to identify equipment, containers, rooms, experimental animals, and other objects or areas that contain or are contaminated with such hazardous agents.

OSHA inspectors are instructed to check employers' warning policies when they conduct walk-around inspections and when they interview employees.

Under the directive, employers must use tags containing the word "BIOHAZARD" to label bags and other receptacles containing contaminated articles. Each tag must indicate that the bag could contain infectious wastes and must give additional instructions for handling the bag. For example, OSHA said, instructions might say that if the outside of the bag is contaminated with body fluids, a second outer bag must be used.

Employers also must explain to employees what the tags mean.

Citing Section 5(a)(1)

The directive allows OSHA inspectors to use four existing standards to cite alleged hazards involving blood-borne disease exposure:

- 29 CFR 1910.132, which requires employers to provide personal protective equipment to workers;
- 1910.22(a)(1) and (a)(2), which contain general requirements for proper housekeeping;
- 1910.145(a)(4)(i) and (ii), which contain safe waste disposal procedures; and
- 1910.145(f), which contains specifications for accident-prevention signs and tags.

Alleged hazards not covered by any of those standards may be cited under the general-duty clause of the Occupational Safety and Health Act (OSH Act), Section 5(a)(1), which requires employers to provide "a place of employment free from recognized hazards that are causing or are likely to cause death or serious physical harm to employees."

CDC's blood-borne disease guidelines may form the basis for a 5(a)(1) citation, because the health care industry "generally accepts and, therefore, recognizes" CDC's assessment of infectious disease hazards and recommendations for protecting workers from those hazards, according to the OSHA directive. The CDC guidelines were updated in February 1989.

Inspectors may base citations on Section 5(a)(1), the OSHA directive says, if the employer fails to:

- Develop a program for determining when hepatitis B vaccine should be provided to employees, or fails to offer the vaccine in amounts and at intervals prescribed by standard medical practice.
- Identify all laundry operations involving substantial risk of direct worker exposure to body fluids. Soiled linens must be handled and agitated as little as possible, put in leak-proof bags where they were used, and transported in such bags. Soiled linens may not be sorted or rinsed in patient-care areas.
- Use standard procedures for sterilizing instruments, devices, or other items contaminated with body fluids. Employers must follow CDC guidelines for cleaning, disinfecting, and sterilizing hospital equipment.
- Ensure that objects contaminated with potentially infectious materials are placed in impervious bags. If the outside of a bag is contaminated, it must be covered with a second bag.

- Ensure that workers who use gloves when exposed to body fluids wash their hands or other skin surfaces thoroughly and immediately after removing the gloves.

Testing After Possible Exposure

Section 5(a)(1) also may be used as the basis for a citation if the employer fails to monitor employees after they sustain possible exposure to the hepatitis B or HIV virus, according to the directive.

If a health care worker suffers a needlestick, cut, or mucous membrane exposure to body fluids or has a cutaneous exposure to blood when his or her skin is chapped, abraded, or broken, the patient from whom the blood came should be informed of the incident and tested, upon consent, for HIV and HBV infection, the directive said.

If the patient refuses to consent to an HIV test, or if the patient tests positive, the worker should be given an HIV antibody test as soon as possible and advised to seek medical attention for any acute febrile (fever-related) illness that occurs within 12 weeks after exposure.

If the worker tests negative, he or she should be retested six weeks after the exposure and thereafter at 12-week and six-month intervals.

Follow-up procedures for workers exposed to hepatitis B depend on whether the worker was vaccinated with the hepatitis B vaccine and whether the source patient had or has the HBV virus, according to OSHA.

If the health care worker refuses to take part in any of the above procedures, OSHA will not hold the employer responsible "since the procedures are designed for the benefit of the exposed employee," according to the directive.

Training for Workers

Employers also may be cited under Section 5(a)(1) if they fail to provide training and education to health care employees on precautionary measures, epidemiology, modes of transmission, and prevention of HIV and HBV, according to the directive. In addition, it says workers should be counseled regarding possible risks to the fetus from HIV and HBV and other associated infectious agents.

Workers also must be told about the location and proper use of personal protective equipment and trained on proper work practices and "universal precautions" for guarding against exposure. Moreover, workers should be trained about procedures to follow if they are exposed to needlesticks or body fluids, according to the directive.

PROPOSED FINES TOTAL $152,173

As of Jan. 30, 1989, OSHA had conducted 395 inspections under the directive, and had levied a total of $152,173 in proposed penalties, according to agency data provided to BNA.

The agency cited employers for a total of 1,001 alleged violations of the standards. Of those, five were classified as willful violations, four as repeat, 455 as serious, and 537 as non-serious.

Most of the citations involved alleged failures to provide personal protective equipment and warning tags. About 52 employers were cited for alleged violations under Section 5(a)(1), including alleged failure to protect workers from hazards posed by contaminated needles and infectious waste and alleged failure to provide protective equipment.

OSHA issued proposed penalties totaling $25,670 for the alleged willful violations; $4,400 for repeat; $120,223 for serious; and $1,880 for non-serious.

Of the 395 inspections, 141 were conducted as a result of employee complaints, 13 on the basis of referrals, and two as a result of an accident; the remainder were follow-up and unscheduled inspections.

Workplaces cited included blood banks, small health care clinics, hospitals, funeral homes, dental offices, dialysis clinics, and nursing homes, according to the OSHA data. Also cited were corporations in manufacturing and the service industry, including Trans World Airlines Inc., Sara Lee Bakery, and Wal-Mart.

OSHA'S BLOOD-BORNE DISEASE PROPOSAL

OSHA's proposed standard for blood-borne diseases is based on its current inspection program. Under the proposal, employers would be required to identify health care employees at risk of exposure and develop written programs to reduce such risks.

Although employers would be allowed to develop programs designed for their workplaces, every program would have to include requirements for protective equipment, housekeeping, waste disposal, and training similar to those currently enforced through the inspection program.

The OSHA proposal also would establish requirements for research laboratories and production facilities engaged in the culture, production, concentration, and manipulation of HIV and HBV. The proposal would require, for example, that laboratory doors remain closed when work involving HIV or HBV is in progress and that contaminated materials be placed in durable, leak-proof containers that are closed before being removed from the work area.

The proposal also would require that access to the work area be limited to authorized personnel and that all activities involving potentially infectious materials be conducted in biological safety cabinets or other physical-containment devices.

The proposed standard also would require employers to offer hepatitis B vaccination to employees occupationally exposed to blood or other potentially infectious materials on the average of one time or more per month, unless the worker had a previous HBV vaccination or was found through antibody testing to be immune.

Under the proposal, the employer would have to bear the full cost of complying with the standard, including the cost of the HBV vaccine. In its original Jan. 19, 1988, directive,

OSHA said that employers had to provide the vaccine free of charge to employees. But that requirement was relaxed in the Aug. 15, 1988, directive. Employers currently are required to make the vaccine available, but not to pay for it. That cost has to be borne by the employee if he or she wants to be vaccinated.

"In all previous OSHA health standards, the agency has required the employer to bear the cost for all provisions of the standard. This proposed standard would also require the employer to pay for all the provisions of the standard," OSHA said in the proposal. It asked for comment on the issue.

The employer further would have to make available to employees a confidential HBV medical evaluation and follow-up after any exposure to potentially infectious materials. This would have to include HBV antibody or antigen testing, counseling, instructions to the employee to report any illness, and "safe and effective" treatment.

Compliance with the standard would cost employers an estimated $800 million during the first year it was in effect, primarily to pay for protective clothing and equipment for employees. OSHA estimates employer compliance costs will average $1,373 a year, led by hospitals at $32,875, with funeral establishments spending the least, $141.

The proposed standard's first-year costs would be $200 million to $300 million higher than those under the hazard communication standard, now the most expensive regulation ever issued by OSHA, according to the agency. Most OSHA standards cost employers a total of about $100 million the first year they are in effect, agency officials say.

INDUSTRY QUESTIONS OSHA ROLE

Many health care employers think OSHA should not issue a blood-borne disease standard, and many feel the agency should not be involved in enforcing infectious disease control guidelines at all. OSHA inspectors, most of whom are trained to recognize chemical or mechanical hazards in industrial settings, lack the expertise to assess working conditions in hospitals and other health facilities, these critics say.

"OSHA is looking at [disease exposures] like they look at chemical [exposures], but it can't be handled the same way," Setsuko Nakahara, director of infection control/employee health for Valley Presbyterian Hospital, Van Nuys, Calif., told BNA. "The public is terribly fearful of AIDS, but the risk is minimal in the work situation."

OSHA officials say the agency trains compliance officers on inspecting health care facilities and on recognizing potential blood-borne infectious disease hazards.

OSHA Region II Administrator James D. Stanley admitted at the Jan. 9 AIDS conference that when OSHA started inspecting health care facilities, it was a "foreign inspection process." To remedy the problem in Region II (New York, New Jersey, and Puerto Rico), Stanley said his office enlisted physicians to answer inspectors' questions about the transmission of blood-borne diseases and proper handling of blood and body fluids.

Paperwork, Liability Concerns

Health care employers told BNA that OSHA's current inspection program and the prospect of a forthcoming standard have raised concerns about costs, paperwork, and liability, even though some facilities have conducted programs for many years to protect workers from infectious risks on the job.

Valley Presbyterian in Van Nuys has had a written infection control program since the late 1960s, and recently it also began providing special disposal containers for used needles in every patient's room and where needles are used, according to Nakahara.

Taking such steps to protect employees is an expensive process, according to Gina Pugliese, director of infection control and environmental safety for the AHA.

"The costs are prohibitive when they are not reimbursable," Pugliese said, adding that hospitals sometimes have trouble keeping enough protective gloves in stock and that some gloves are of poor quality.

Robin Rothrock, administrative director of the Hope Medical Group for Women in Shreveport, La., told BNA that her facility has always had an employee infection control program, and the center improved its employee education program in the past year, prompted by an OSHA inspection in August 1988 and the rising public awareness of AIDS.

Rothrock said the Hope center has developed a new employee safety manual containing the text of OSHA's enforcement guidelines. The center requires that all employees read the manual and sign a statement that they have done so, she said.

OSHA's inspection program creates a "real paperwork issue" for health care facilities, according to Rothrock. She said her facility now keeps written records to show which employees have refused vaccination against hepatitis B and which employees have been tested or not tested for hepatitis B infection. The facility also keeps records on all workplace accidents and potential exposures to viruses, she said.

Rothrock also said OSHA's involvement "brings the whole enforcement and liability concept into play …. Even if an employer has [personal protective] equipment and clothes, what can [it] do if the employee chooses not to use [them]? You can't fire them for refusal." Nevertheless, she noted, employers can be cited for non-compliance with OSHA guidelines if protective clothing is not worn.

Similar concerns exist about employers' liability for contract employees, Rothrock said. Because Hope Medical is a small facility, it contracts with a local firm to provide cleaning crews, she said.

The facility offers its staff health care workers free testing and vaccinations for hepatitis B, but it does not offer similar services for contract employees, Rothrock said. "That raises another question for employers. Who is responsible for protecting contractor workers?" she asked.

UNIONS SEEK MORE PROTECTION

Unions representing health care workers say a strong standard is needed to fill existing gaps in employee protection against HIV and HBV.

Although many health care facilities have written infection control programs, they often are not being conducted or are ineffective, according to Jordan Barab, director of occupational safety and health for the American Federation of State, County, and Municipal Employees (AFSCME). "Just because a policy is on paper" does not necessarily mean it is being used, Barab offered.

Although many health care workers who come into contact with blood and body fluids use protective clothing as recommended by CDC, employees "on the fringes"—such as laundry workers and housekeeping personnel—need more education about AIDS and hepatitis B and should be offered hepatitis B vaccine, Barab told BNA.

Several unions, including AFSCME, are using training programs, collective bargaining provisions, and other means to get employers to increase worker protection against exposure to blood-borne diseases, according to Barab and other officials.

Since 1985, the Service Employees International Union (SEIU) has answered more than 800 requests from its members for technical assistance on AIDS issues, conducted research, provided model contract language, reviewed infection-control policies at institutions where its members are employed, helped its locals get infection control practices implemented, and linked locals to community agencies that provide AIDS-related services, according to Bill Borwegen, director of SEIU's occupational safety and health department.

The union also provides educational seminars and training workshops on blood-borne infectious diseases nationwide. The day-long conferences enable local union leaders and members to learn about and discuss issues related to AIDS and its implications for workers, Borwegen said. SEIU has distributed more than 70,000 copies of its handbook, "The AIDS Book: Information For Workers," and more than 360,000 copies of the English- and Spanish-language versions of the pamphlet, "AIDS and the Healthcare Worker," he said.

In addition, SEIU has conducted two surveys, including an informal overview of infectious disease control practices in 40 hospitals in four metropolitan areas with high rates of AIDS infection. The other survey was a more extensive study of 100 departments in hospitals, nursing homes, blood banks, and mental health correctional facilities. Both surveys found that infectious disease control programs were "severely lacking," especially in the area of education of non-professional staff, Borwegen said.

Additional Union Funding

Borwegen said SEIU plans to conduct a larger, more extensive survey on health care facilities' compliance with the CDC and OSHA guidelines. The union decided to delay the study until OSHA publishes its proposed standard, so it could include questions whose answers would help OSHA develop a final regulation, he explained.

SEIU also plans to expand its AIDS educational efforts, helped by a recent $800,000 Robert Wood Johnson Foundation grant. The grant will enable the union to hire three regional AIDS coordinators—one each in San Francisco, Chicago, and Boston—as well as an AIDS project director and an AIDS writer and researcher for its Washington, D.C., office, Borwegen said.

Since 1983, AFSCME has developed publications, held conferences on AIDS, and has helped CDC develop AIDS guidelines. The union represents about 1.4 million public sector employees, many of whom work in hospitals, correctional facilities, and mental health institutions, Barab said.

In March 1989, AFSCME, SEIU, and the AFL-CIO received a total of about $240,000 from CDC to expand their AIDS educational programs. Barab said AFSCME will hire a specialist to work solely on AIDS-related issues and will expand its educational program to cover off-the-job risks of infection and provide on-site assistance to members with AIDS-related workplace problems, such as those who face discrimination because they have the disease, or who are afraid to work with an employee who has AIDS. SEIU intends to use some of the grant to reprint the third edition of its AIDS worker handbook and hire a writer to help author additional educational materials.

The three labor groups also plan to build community and national coalitions between unions and organizations engaged in AIDS research, education, and treatment, and develop a shop steward's manual on discrimination, confidentiality, and other workplace issues.

Workers To Monitor Employers' Programs

Union officials also told BNA that they are advising their members to monitor their employers' efforts to comply with OSHA and CDC guidelines. For example, Barab said one circular published by his union, "Forcing Compliance with AIDS and Hepatitis B Guidelines," instructs workers on how to evaluate their employers' infection control programs and how to request an OSHA inspection.

In the circular, AFSCME recommends that employees become familiar with CDC's guidelines for protections against blood-borne infectious disease exposures and with a 1987 advisory notice by DOL and HHS based on those guidelines.

Workers should ask their employers, in writing, to correct any unsafe conditions or work practices, AFSCME advises. Union locals should keep copies of complaints and records of management responses. Where employers do not comply and collective bargaining contracts include language to cover the situation, locals should consider filing a grievance, according to the union. If the grievance procedure is slow or unsuccessful, the local should consider requesting an OSHA inspection, the circular says.

In filing requests for inspections, workers should describe how the allegedly unsafe conditions violate the CDC guidelines and list the specific OSHA standards allegedly violated, AFSCME advises. The circular suggests wording that employees can use in their complaints, such as "Gloves not provided to employees who face frequent exposure to blood" and "Needles are left lying on table or thrown in the trash because no needle disposal boxes are available."

To reduce the possibility of retaliation, workers can have OSHA keep their names confidential, or complaints can be signed by union officers or staff representatives instead of individual workers, AFSCME advises.

When an OSHA inspector arrives at a workplace, a member designated by the union has a right to meet alone with the inspector and accompany him or her on the inspection. "It is very important that the union member be present to point out to the inspector where violations occur," the advisory says.

AFSCME recommends that photographs of the alleged unsafe conditions be given to the inspector, where possible, as well as written copies of the union requests to management to make the workplace safe. The inspector should examine records of needlesticks and cuts, and records of employees who have contracted hepatitis B or other infectious diseases on the job, according to the advisory.

Contract Provisions Recommended

To help members ensure employers' compliance with CDC guidelines, SEIU also advises its locals that represent health care workers to negotiate contract clauses under which employers agree to minimize the risk of HIV and HBV transmission in the workplace. According to the union, a contract clause can state that the employer will follow the CDC safety guidelines and abide by the 1987 DOL/HHS Joint Advisory Notice. Such contract language enables locals to grieve violations of the guidelines, the union notes.

AFSCME suggests that contracts state, "The employer shall comply with all CDC infectious disease guidelines and train employees about infectious diseases to which they are routinely exposed."

Other unions that have contract provisions on compliance with CDC and OSHA guidelines include the National Union of Hospital and Health Care Employees and Local 1199 of the Hospital and Health Care Employees Union, according to Barab.

Barab told BNA that although he does not know of any instance in which a union has filed a grievance against an employer under such contract language, in the absence of a standard such provisions encourage employers to comply with government guidelines.

Unions also are using contract language to obtain other protections for their members. For example, a three-year pact negotiated in 1988 between Local 3469 and Ohio Council 8 of AFSCME and the Mansfield-Richland (Ohio) County Board of Health requires the board to provide free hepatitis B vaccinations and AIDS testing to all of its employees who are regularly exposed to blood and body fluids on the job.

■ ■ ■ ■

VII. The Right to Act

Sweden has one. So does Quebec, Canada.

The United States should have one, too, according to union officials and some public-health specialists—a law that would require all employers to provide safety training to their workers and to correct hazardous conditions their workers identify.

Workers also should be given the right to refuse to do what they consider to be dangerous work assignments, without fear of being fired or otherwise disciplined by their employer, according to these labor advocates.

These ideas form the basis for a "right-to-act" movement that such labor officials as Anthony Mazzocchi, secretary-treasurer of the Oil, Chemical, and Atomic Workers International Union (OCAW), have been urging since the early 1980s.

Speaking at a May 19, 1989, conference sponsored by BNA on emerging occupational safety and health issues, Mazzocchi acknowledged that a system giving workers virtually equal authority with management over workplace safety and health programs would be "a radical change" from traditional practices in U.S. industry.

"But the mere fact we are talking about it today," he added, "means the issue has grown in quantum leaps."

Bills containing right-to-act provisions have been introduced in the U.S. Senate and in the Connecticut legislature. These measures have generated opposition from business officials, who argue that they would unnecessarily disrupt current worker-protection programs, infringe on management's traditional prerogatives, and lead to potential abuses by disgruntled workers.

According to observers, such opposition was instrumental in the defeat of the Connecticut bill in that state's most recent legislative session, which ended June 7, 1989. But right-to-act proponents say that defeat will not discourage them from pursuing further legislation in Connecticut and elsewhere.

These advocates say they will counter future attacks by emphasizing their view that right-to-act is a logical extension of existing requirements, such as the worker safety training laws in Sweden and Quebec, workplace and community right-to-know laws in the United States, and provisions in some union contracts.

BUILDING ON RIGHT-TO-KNOW

The bill (S 436) introduced in the U.S. Senate Feb. 23, 1989, by Sens. Howard Metzenbaum (D-Ohio) and Charles Grassley (R-Iowa) would prohibit employers from disciplining employees for refusing to perform a task when an employee has "a reasonable apprehension" that the activity would harm him or her, other employees, or the public.

The Connecticut bill that died in the 1989 legislative session—and a predecessor bill that met a similar fate in the 1988 session—contained provisions similar to those of S 436.

Advocates of right-to-act legislation say the concept expands the hazard communication or right-to-know regulations issued by the Occupational Safety and Health Administration (OSHA) and by some states in the early and mid-1980s. Those rules require employers to notify workers and residents of surrounding communities about the hazards of chemicals used in the workplace.

"Right-to-act is being treated seriously" because labor unions and community groups were successful in advocating those laws, Mazzocchi told BNA.

Right-to-know gave workers "the right to have good information about the workplace," Michael Wright, health and safety director for the United Steelworkers of America, told BNA. "But right-to-know by itself is meaningless unless you have a framework [for using] that knowledge," he added.

A task force of the National Association for Public Health Policy, a coalition of public-health professionals affiliated with the American Public Health Association (APHA), said in 1987 that the right to act, like the right to know, should be given to workers and to communities. The task force criticized regulatory cutbacks by the Reagan administration and said communities should have the right to inspect plants within their boundaries and to shut down operations that pose safety or health hazards to local residents. The comments were included in a statement presented at APHA's 115th annual meeting Oct. 20, 1987, in New Orleans, La.

"Only this sharing of power will enable workers and communities to protect themselves when government fails to act and protect the people," the statement said.

The change from the Reagan to the Bush administration has not reduced the need for right-to-act, according to Rick Melita, a spokesman for the Connecticut Council on Occupational Safety and Health, a worker-advocacy coalition that has supported the Connecticut legislation. "Even if the [Bush] administration viewed OSHA as its number one priority, there are not yet enough OSHA inspectors," he told BNA.

Current UAW/Ford Program

In some U.S. workplaces, another model for right-to-act exists in the form of joint labor-management committees established under collective bargaining agreements, according to labor officials. These committees—comprising equal numbers of employer and employee representatives—monitor safety and health conditions at individual work stations. Through these panels, employees can work with management representatives to correct unsafe conditions.

Such elements are present in a new joint program in which employees represented by the United Auto Workers International Union (UAW) at Ford Motor Co. plants are trained along with management representatives to recognize ergonomic problems on assembly lines, according to union and industry spokesmen. The program was launched in all of Ford's assembly and manufacturing plants Jan. 13, 1989, under a 1987 agreement between Ford and the UAW.

Under the program, workers and managers systematically try to identify and redesign jobs that pose a risk of physical stress, strain, and injury to the employee, labor and industry representatives told BNA.

In April 1989, a UAW-Ford National Joint Committee on Health and Safety brought together and trained representatives from each local representing employees at a Ford plant. These representatives, beginning in late May, returned to their respective plants to train other employees.

"In the last few days of the nationwide training effort we trained 600 hourly and management employees from every facility," Bill Stevenson, the union's health and safety director for Ford operations, told BNA. "We taught them how to analyze jobs, prioritize what needs to be done when, find out which jobs are bad and how they should go about fixing it, etc."

According to Franklin Mirer, director of health and safety for the union, joint safety programs under UAW contracts date from 1973, when extensive provisions were written into national agreements with automakers, first with Chrysler Corp. and then with the remaining auto, farm, and construction equipment companies.

Too few of these committees exist nationwide to serve the vast majority of U.S. workers, Mirer told BNA. "Most American workers are not represented by unions. They rely on government activity to establish their basic rights in the workplace."

Safety Committee Requirements Abroad

Margaret Seminario, associate director of occupational safety, health, and social security for the AFL-CIO, told BNA that workplace safety and health committees in Sweden and Quebec have a "more structured" role for identifying and correcting hazards than their U.S. counterparts.

She said laws in Sweden and Quebec provide "expanded opportunities" for worker-safety training.

Swedish law requires management to train one out of every 50 workers to recognize and evaluate workplace hazards, the AFL-CIO said in a discussion paper on right-to-act, presented at its November 1987 safety and health conference in Nashville, Tenn.

In Quebec, every workplace with 20 or more employees must have a joint safety and health committee. "In this system, the initial authority over safety and health in the plant rests with the joint committee," the discussion paper said.

The committees have the authority to choose a plant physician, approve safety training plans, select protective equipment, investigate accidents, and resolve employee complaints, the paper pointed out.

According to Melita, these laws may provide the basis for right-to-act in the United States. "A lot of [U.S.] companies [also] operate ... in Sweden and Canada. If these programs are practicable [there], how come they can't do it in the U.S.?"

WHISTLEBLOWER PROTECTION

Observers say a bill to mandate joint safety committees probably will be introduced in Congress later in 1989, perhaps as part of a larger measure to amend the Occupational Safety and Health Act (OSH Act). The AFL-CIO would like to see current law "expanded to focus more on developing health and safety programs in the workplace," Seminario told the May 19 BNA conference.

In the meantime, labor officials have voiced support for provisions in S 436, the Employee Health and Safety Whistleblower Protection Act.

Under the measure, any employee who believed that he or she had been fired or otherwise disciplined for refusing to perform hazardous work or for alerting federal authorities to an alleged hazard would be able to file a complaint with the Department of Labor (DOL).

The bill would establish an office in DOL to handle such complaints. The office would have the authority to investigate such allegations and to order the immediate reinstatement of the employee if it found "reasonable cause" to believe that the worker's rights had been violated.

If the employer challenged the order, the office would have the authority to convene a fact-finding hearing and to issue a final determination after the hearing. In conducting investigations and hearings, DOL would be able to subpoena witnesses and documents.

In announcing the measure Feb. 23, Metzenbaum noted that 16 existing federal laws prohibit employers from discriminating against employees for complaining about job-related safety hazards, including the OSH Act. Metzenbaum said S 436 would cover workers not currently protected by those statutes, would establish a uniform complaint procedure, and would give DOL new authority to issue subpoenas in attempting to resolve disputes over employee safety complaints.

Disrupt Existing Protections?

In written comments to the Senate Labor and Human Resources Subcommittee on Labor, the National Association of Manufacturers (NAM) said the measure might "go beyond its expressed scope and disrupt existing whistleblower protections."

For example, although the OSH Act allows employees to refuse to work in situations where they face an imminent danger of injury or death, it does not allow them to walk off the job in non-emergencies, NAM said.

"Under this scenario, [the bill] would create new rights and remedies for workers that are not available under the OSH Act," the association said. "At the very least, further clarification is necessary as to how [the] drafters envisioned the bill interacting with existing whistleblower provisions."

NAM also questioned the provision that would require an employer to reinstate a worker immediately, even if the employer challenged DOL's finding that the worker had been fired unjustly. The provision "raises constitutional due process questions," NAM contended.

S 436 is "totally unnecessary," Thomas F. Evans, director of safety and industrial hygiene for Monsanto Co., told BNA. "We already have [the] Taft-Hartley [Act] and other labor laws. Cases like this go up to the National Labor Relations Board all the time. The OSH Act also [protects employees from retaliation for safety-related activities]."

Metzenbaum, who chairs the labor subcommittee, hoped to hold a subcommittee mark-up session on the bill in summer 1989, the first step in seeking Senate passage of the measure, a congressional aide told BNA June 12, 1989.

The senator plans to work with industry critics to resolve their concerns about the bill, the aide said. "This bill is designed to protect workers from retaliation. It's not designed to be a right-to-refuse bill," the aide told BNA.

'MANAGEMENT HAS THIS RESPONSIBILITY'

Critics use some of the same arguments advanced against the Metzenbaum-Grassley bill to challenge the general concept of right-to-act.

Putting right-to-act programs into law would amount to "giving 100 million individuals the authority now vested in [the National Institute for Occupational Safety and Health], OSHA, and business," Mark D. Cowan, chief executive officer of the Jefferson Group, a Washington, D.C., consulting firm, said. Cowan, an OSHA deputy administrator during the Reagan administration, offered this view during a debate on the issue with Mazzocchi at the May 19 BNA conference.

Right-to-act has "tremendous potential for abuse" because it would enable "any worker in any situation ... to just say 'I'm not going to do that, boss'," Cowan said.

"If a company wishes to negotiate [labor-management] relationships, they should be free to do so within their system of collective bargaining, if they have one," Monsanto's Evans said at a May 23, 1989, roundtable on right-to-act at the American Industrial Hygiene Conference in St. Louis, Mo. But "employee relations and management relationships [cannot] be mandated by law," he contended.

"No outside entity should develop a process and criteria for selection of [labor-management safety] committees," Evans told the roundtable meeting. "Management has this responsibility and must do it."

CONNECTICUT BILL STALLED

Such objections from industry are a main reason the right-to-act bills failed in Connecticut in 1988 and 1989, labor and legislative spokesmen told BNA.

The bills' advocates said they had expected a right-to-act measure to find support in the state in the wake of the April 23, 1987, collapse of the L'Ambiance Plaza apartment building in Bridgeport, and a February 1987 study by a New Haven clinic that found liver ailments linked to alleged exposure to a solvent, dimethylformamide, at a small factory in Hartford.

However, a right-to-act bill introduced in the 1988 session was derailed by opponents, and a follow-up measure died when the legislature closed its 1989 session June 7 without taking action on it.

State Rep. Joseph Adamo (D), who sponsored both measures, told BNA June 2, 1989, that he intends to introduce the legislation again in 1990. Adamo and labor officials said they remain hopeful that a bill will be enacted then. Melita said he is concerned that the sense of urgency created by the L'Ambiance Plaza collapse may be lost if the bill is not passed in the next session.

Inspection, Hearing Process

Under the 1989 measure, HB 7346, and its 1988 predecessor, HB 5096, employees would have been given conditional rights to refuse to perform hazardous duties, and to inform other employees that they were "working in or exposed to a hazardous condition," without fear of being discharged, disciplined, or "otherwise penalize[d]."

Employees could have exercised the right only if the employer had refused or been unable to correct the hazard and the employee did not have enough time to use existing regulatory channels, such as OSHA, to address the problem.

The bills would have created a system for handling employee complaints similar to that proposed in the Metzenbaum-Grassley bill. Workers who believed they had been disciplined for refusing to perform hazardous work would have been allowed to file a complaint with the state Labor Department. The department would have had the authority to inspect the workplace to determine if the complaint was valid, to convene a hearing if the employee's concern was found to be justified, and to resolve the dispute.

The measures would not have shielded employees who refused to work "without a reasonable belief that a hazardous condition existed." In those instances, employers would have been allowed to take disciplinary action, "up to and including dismissal" of the worker.

The earlier bill would have set minimum requirements for mandatory workplace safety programs, required employers to establish labor-management safety committees in individual workplaces, and created a state-funded network of occupational health clinics. Those provisions were not included in the 1989 measure, Melita said. "We decided to take it one step at a time."

'Existing Laws Protect Employees'

Union officials in Connecticut supported the measures. "It is simply common sense to allow workers to walk away from a life-threatening situation without losing their jobs," Connecticut AFL-CIO President John Olson said at a labor-sponsored ceremony April 28, 1989, in Bridgeport honoring workers killed in job-related accidents.

"Workers resent having to gamble daily with their lives," Ron Nobili, business agent of Laborers Union Local 655 in Bridgeport, Conn., said at the same gathering.

But industry representatives questioned the need for a new law. "Because existing federal and state laws, regulations and court decisions all protect employees against an employer's retaliatory action ... HB 7346 [was] unnecessary," Kyle Ballou, an attorney with the Connecticut Business and Industry Association, the state's largest business lobby, told BNA.

The federal OSH Act already protects employees from retaliation or discrimination for exercising their rights under that statute, Ballou said. She also said the Connecticut right-to-act bills were "overly broad and vague," and failed to articulate the specific conditions and circumstances under which an employer would have been prohibited from taking adverse action against an employee.

But Melita argued that if the OSH Act was actually protecting employees, "we wouldn't have to be talking about [right-to-act]...This bill gives the employee the right to protect himself on the job."

Plans for 1990

According to the bill's supporters, the measure died because of industry opposition and because it would have required additional funds for the state Department of Labor at a time when the legislature is trying to reduce a budget deficit.

Adamo told BNA that he intends to reintroduce the bill early in the legislature's next session, which is scheduled to convene in February 1990. To reduce possible opposition, he said he will drop the $24,000 appropriation in the 1989 measure that would have paid the salary of an additional Labor Department staff person to help the department investigate and resolve employee complaints.

The measure stalled in the closing days of the 1989 session primarily because it "got caught up in the terrible onslaught of bills" on the final day of the session, Adamo told BNA.

Melita agreed with Adamo, but noted that the bill was ranked 118 of 120 pending in the legislature's Committee on Labor and Public Employees. "The [committee] leadership obviously didn't think worker safety was a high priority," Melita told BNA.

Olson said he hopes the measure will be adopted in 1990. The 1989 bill had been "moving along smoothly" until it hit "the fiscal wall," he told BNA June 2, 1989.

Building Grassroots Support

Melita said he is guardedly optimistic about the bill's chances in the upcoming session. "You can look at [the bill's lack of success so far] in two ways," he told BNA. "You can either say we've been butting our heads for two years, or you can say that we're just starting to get the concept out there."

Melita said his organization will try to "build some kind of groundswell" among workers for a future bill, similar to the strategy used by unions in the early 1980s to generate support for right-to-know standards.

Moreover, by focusing on concerns that hazardous substances used in the workplace also may be emitted into the outside air and pose possible health hazards to residents in surrounding localities, labor activists may try to enlist community groups as allies "down the road," according to Melita.

"It doesn't make sense to fight like hell not to be exposed [to hazardous substances] on the job and not protect [the worker's] family at home," he told BNA.

Meanwhile, labor unions in New Jersey hope to have a right-to-act bill introduced in the state legislature within the next several months, Rick Engler, an assistant to Archer Cole, president of the New Jersey Industrial Union Council, AFL-CIO, told BNA.

Engler said right-to-act supporters have not drafted the details of a bill, and it is unclear how successful advocates will be in lobbying for such legislation next year. "I wouldn't bet that it will pass, but I wouldn't bet that it won't," he said.

"In our favor," he said, is the fact that labor and community groups in New Jersey were successful in joining to support enactment of a state right-to-know law in 1983. "There's a history of organizations working together," Engler noted.

■ ■ ■ ■ ■

APPENDIX A

**"Getting Away With Murder in the Workplace:
OSHA's Nonuse of Criminal Penalties
For Safety Violations"**

**Report by the House Government Operations Committee
Oct. 4, 1988**

APPENDIX A

Union Calendar No. 619

100th Congress, 2d Session — — — — — — — — — — House Report 100-1051

GETTING AWAY WITH MURDER IN THE WORKPLACE: OSHA'S NONUSE OF CRIMINAL PENALTIES FOR SAFETY VIOLATIONS

SIXTY-SIXTH REPORT

BY THE

COMMITTEE ON GOVERNMENT OPERATIONS

OCTOBER 4, 1988.—Committed to the Committee of the Whole House on the State of the Union and ordered to be printed

---

U.S. GOVERNMENT PRINTING OFFICE

88-982   WASHINGTON : 1988

Published by The Bureau of National Affairs, Inc.

COMMITTEE ON GOVERNMENT OPERATIONS

JACK BROOKS, Texas, *Chairman*

JOHN CONYERS, Jr., Michigan	FRANK HORTON, New York
CARDISS COLLINS, Illinois	ROBERT S. WALKER, Pennsylvania
GLENN ENGLISH, Oklahoma	WILLIAM F. CLINGER, Jr., Pennsylvania
HENRY A. WAXMAN, California	AL McCANDLESS, California
TED WEISS, New York	LARRY E. CRAIG, Idaho
MIKE SYNAR, Oklahoma	HOWARD C. NIELSON, Utah
STEPHEN L. NEAL, North Carolina	JOSEPH J. DIOGUARDI, New York
DOUG BARNARD, Jr., Georgia	JIM LIGHTFOOT, Iowa
BARNEY FRANK, Massachusetts	BEAU BOULTER, Texas
TOM LANTOS, California	DONALD E. "BUZ" LUKENS, Ohio
ROBERT E. WISE, Jr., West Virginia	AMORY HOUGHTON, Jr., New York
MAJOR R. OWENS, New York	J. DENNIS HASTERT, Illinois
EDOLPHUS TOWNS, New York	JON L. KYL, Arizona
JOHN M. SPRATT, Jr., South Carolina	JAMES M. INHOFE, Oklahoma
JOE KOLTER, Pennsylvania	CHRISTOPHER SHAYS, Connecticut
BEN ERDREICH, Alabama	
GERALD D. KLECZKA, Wisconsin	
ALBERT G. BUSTAMANTE, Texas	
MATTHEW G. MARTINEZ, California	
THOMAS C. SAWYER, Ohio	
LOUISE M. SLAUGHTER, New York	
BILL GRANT, Florida	
NANCY PELOSI, California	

WILLIAM M. JONES, *General Counsel*
DONALD W. UPSON, *Minority Staff Director*

EMPLOYMENT AND HOUSING SUBCOMMITTEE

TOM LANTOS, California, *Chairman*

BILL GRANT, Florida	JOSEPH J. DIOGUARDI, New York
TED WEISS, New York	JON L. KYL, Arizona
NANCY PELOSI, California	CHRISTOPHER SHAYS, Connecticut
BARNEY FRANK, Massachusetts	

EX OFFICIO

JACK BROOKS, Texas   FRANK HORTON, New York

STUART E. WEISBERG, *Staff Director and Counsel*
LISA PHILLIPS, *Professional Staff Member*
JUNE SAXTON, *Clerk*
MATTHEW BEHRMANN, *Minority Professional Staff*

LETTER OF TRANSMITTAL

House of Representatives,
Washington, DC, October 4, 1988.

Hon. Jim Wright,
Speaker of the House of Representatives,
Washington, DC.

Dear Mr. Speaker: By direction of the Committee on Government Operations, I submit herewith the committee's sixty-sixth report to the 100th Congress. The committee's report is based on a study made by its Employment and Housing Subcommittee.

Jack Brooks, *Chairman.*

CONTENTS

	Page
I. Introduction	1
II. Background	2
III. Findings and conclusions	4
IV. Recommendations	5
V. Discussion	6
A. Pymm Thermometer	6
B. The need to beef up criminal provisions of the Federal OSH Act	7
C. OSHA and the preemption question	8
D. Cooperation with the States	10
VI. Summary	11

Published by The Bureau of National Affairs, Inc.

Union Calendar No. 619

| 100TH CONGRESS 2d Session | HOUSE OF REPRESENTATIVES | REPORT 100-1051 |

GETTING AWAY WITH MURDER IN THE WORKPLACE: OSHA'S NONUSE OF CRIMINAL PENALTIES FOR SAFETY VIOLATIONS

OCTOBER 4, 1988.—Committed to the Committee of the Whole House on the State of the Union and ordered to be printed

Mr. BROOKS, from the Committee on Government Operations, submitted the following

SIXTY-SIXTH REPORT

BASED ON A STUDY BY THE EMPLOYMENT AND HOUSING SUBCOMMITTEE

On September 27, 1988, the Committee on Government Operations approved and adopted a report entitled "Getting Away With Murder in the Workplace: OSHA's Nonuse of Criminal Penalties for Safety Violations." The chairman was directed to transmit a copy to the Speaker of the House.

I. INTRODUCTION

At a March 19, 1987, hearing on the Occupational Safety and Health Administration's [OSHA] policy of exempting companies from inspections based on injury records, the Employment and Housing Subcommittee heard testimony about the tragedy of Stefan Golab, a 61-year-old immigrant worker who in 1983 died from inhaling cyanide fumes while working at the Film Recovery Systems plant in suburban Chicago.

Two months prior to Mr. Golab's death, an OSHA inspector visited the Film Recovery plant where the company was engaged in the business of reclaiming silver from used x-ray film, reviewed the company's injury records in the plant's front office, determined that the company's injury rate was below the national average, and left. Had the OSHA inspector come in and observed the conditions on the plant floor, he would have seen 70 boiling vats full of used film from which lethal cyanide vapors were being released, the floor covered with cyanide-contaminated solutions, warning labels on cyanide containers painted over, and immigrant workers, many unable to speak English, unaware of the unsafe conditions. Had the OSHA inspector conducted a wall-to-wall inspection of the plant, the deadly cyanide exposure would easily have been detected.[1]

However, the story does not end there. Upon investigating Mr. Golab's death, OSHA inspected the plant and discovered numerous health and safety violations. OSHA issued a citation and fined the company $4,855, which was subsequently bargained down to less than $2,400.

By comparison, the Cook County State's attorney charged the company and its officials with criminal conduct. Company officials were convicted of murder and 14 counts of reckless conduct and were sentenced to 25 years in prison. The company was convicted of manslaughter and reckless conduct and fined $24,000.

During the May 1986 Senate confirmation hearings on John Pendergrass to be Assistant Secretary of Labor for OSHA, Senator Paul Simon referred to a worker's death at the Film Recovery Systems plant and stated that in a precedent-shattering case the four owners of that plant had been found guilty of murder. Mr. Pendergrass commented:

> You bring up a case that brings forth a passion, I think, in many people, it certainly does in me. . . . It was an inexcusable set of circumstances that would allow people to be exposed to things that would damage their health, such as happened there. As a personal opinion, I think the owners and managers got what they deserved.[2]

The criminal convictions of Film Recovery officers are now under appeal in the Illinois courts on the issue of whether State criminal prosecutions for workplace safety violations are preempted by Federal OSHA law.

The *Film Recovery* case, and others like it, prompted the Employment and Housing Subcommittee to hold a hearing on February 4, 1988, to examine OSHA's apparent nonuse of criminal sanctions for workplace safety violations and efforts by State and local prosecutors to fill the vacuum by utilizing historic police powers and enforcing State criminal laws against employers who knowingly and recklessly expose workers to toxic substances and dangerous working conditions, causing them serious injury or death.

II. Background

The Federal Government has been involved in worksite safety since the first Congress. In 1790, Congress passed legislation which allowed merchant seamen to refuse to serve on unsafe ships.

---

[1] In response in part to this committee's report, "Here's the Beef: Underreporting of Injuries, OSHA's Policy of Exempting Companies From Programmed Inspections Based on Injury Records, and Unsafe Conditions in the Meatpacking Industry," H. Rept. 100-542, Mar. 30, 1988, Forty-second report by the Committee on Government Operations, OSHA has changed its inspection policy. OSHA investigators are now required to inspect "high hazard" areas of the workplace even though an examination of the employer's injury and illness records shows a below average rate.

[2] Hearing on Nomination of John A. Pendergrass to be Assistant Secretary of Labor for Occupational Safety and Health before the Senate Committee on Labor and Human Resources, 99th Cong., 2d sess., S. Hrg. 99-864 at 27-28 (1986).

Through the decades, Federal laws evolved to address workplace safety issues, often targeting particular groups such as child labor, occupations such as railway workers, and industries such as mining.

In 1970, Congress passed the Occupational Safety and Health Act (OSH Act, 29 U.S.C. 651–668) to bring Federal interests in worksite safety standards and their enforcement within one new agency at the Department of Labor. The Occupational Safety and Health Administration [OSHA] was charged with assuring, "... so far as possible, every working man and woman in the Nation safe and healthful working conditions and to preserve our human resources." [OSH Act, Sec. 2(b).] To accomplish this goal, OSHA was given broad authority and responsibility to establish health and safety standards, and to enforce them through civil and criminal penalties.

Since the establishment of OSHA, more than 100,000 workers have lost their lives because of unsafe working conditions. It is estimated that annually 7,000–11,000 workers are killed on the job [3] and thousands more die from the long-term effects of occupational illnesses.

OSHA is charged with inspecting worksites and identifying unsafe practices and equipment to ensure that they meet industry-specific and general health and safety standards. OSHA has the authority to issue citations for violations and assess penalties. In addition, the Secretary of Labor is authorized to seek an injunction in U.S. district court when a condition of immediate danger is clear to OSHA inspectors.

OSHA may assess civil penalties of $1,000 for a serious violation (defined as one where there is a substantial probability that death or serious physical harm could result and the employer knew or should have known of the hazard) and up to $10,000 for a willful violation (defined as one where an employer knew that a hazardous condition or violation existed and made no reasonable effort to correct it).

OSHA has authority to seek criminal prosecution for workplace violations in three situations:

(1) a willful violation of a specific OSHA standard that results in death to an employee (maximum penalty is a $10,000 fine plus 6 months imprisonment); [OSH Act, Sec. 17(e).[4]

(2) giving advance notice of an OSHA inspection (maximum penalty is a $1,000 fine plus 6 months imprisonment); [OSH Act, Sec. 17(f)].

(3) knowingly making a false statement or supplying false documents to OSHA (maximum penalty is a $10,000 fine plus 6 months imprisonment). [OSH Act, Sec. 17(g).]

Cases are referred by OSHA to the Department of Justice for possible criminal action. The criminal prosecution of cases requires the recommendation of the Justice Department and the agreement

---

[3] Preventing Illness and Injury in the Workplace, Office of Technology Assessment, Washington, D.C. (1985); Report of the National Safety Workplace Institute, Chicago, Illinois (1987).

[4] For a second conviction the maximum fine and term of imprisonment are doubled.

of the local U.S. attorney's office who is responsible for prosecuting the case.

Since the establishment of OSHA some 18 years ago, only 42 cases have been referred by OSHA for criminal prosecution.[5] Only 14 of those cases were prosecuted, resulting in 10 convictions, with fines or suspended sentences. Tr. 75, 83–88.[6] In the 18-year history of OSHA, no one has ever spent one day in jail for a criminal violation of the OSH Act.

During the 1980's, State and local law enforcement officials have with increasing frequency used the historic police powers of the State to prosecute company officials for knowingly and recklessly exposing employees to toxic substances, causing death and serious injuries. This stems from the State's interest in controlling conduct which endangers the lives of its citizens, whether it be recklessly operating an automobile or an automobile plant. In some cases where State and local prosecutors have obtained convictions against company officials and managers for acts against their employees that constitute crimes under State law, the convictions have been overturned on the ground that Federal worksite health and safety laws preempt a State from pursuing criminal actions.

III. FINDINGS AND CONCLUSIONS

1. OSHA's record with respect to seeking criminal penalties for workplace safety violations and fatalities is dismal. Since its creation by Congress in 1970, OSHA has referred only 42 cases to the Justice Department for possible criminal action. Only 14 of those cases were prosecuted, resulting in 10 convictions, but no jail sentences. No one has ever spent a day in jail for violating the OSH Act. OSHA's record of criminal referrals is even bleaker when compared to the growing number of State and local prosecutions for workplace related fatalities and serious injuries. For example, since 1973 the State of California has prosecuted over 250 cases involving workplace related deaths, injuries, and illnesses, and in the past 8 years there have been 112 successful prosecutions.

2. The criminal penalty provisions of the OSH Act, as presently written and as enforced by OSHA, provides no deterrent to employers violating the statute. A company official who willfully and recklessly violates Federal OSHA laws stands a greater chance of winning a State lottery than being criminally charged by the Federal Government for workplace safety violations. The current system, which relies primarily on citations, abatement, fines, and education and training, is insufficient to ensure that every workplace is safe and healthful. The weak criminal sanctions in the Federal OSH Act are outdated and need to be strengthened and utilized more by OSHA to be a deterrent. In most areas of the law, the prospect of criminal prosecution and imprisonment has a substantial deterrent effect whereas civil fines can often be passed on as part of the cost of doing business.

3. While much of the failure by the Federal Government to seek criminal sanctions for violating Federal OSH laws stems from re-

---

[5] These figures are as of February 1, 1988.
[6] Tr. refers to the printed record of the February 4, 1988, subcommittee hearing on "Criminal Penalties for OSHA Violations."

luctance by OSHA to proceed, part of the blame rests with the Justice Department. The Justice Department has been slow to act on cases referred by OSHA. Some cases have been pending at the Justice Department, without a response, since 1985. With a few exceptions the Justice Department and the U.S. Attorney's Office have consistently declined to prosecute these types of cases. Since 1981, of the 17 cases referred by OSHA for criminal action, there has been one guilty plea, one indictment, and in two cases action is pending by the U.S. Attorney. In seven cases prosecution was declined by either the Justice Department or the U.S. Attorney. In the remaining six cases there has been no response from the Justice Department or the U.S. Attorney.

4. By "backing off" in the *Film Recovery* case because there was an ongoing criminal investigation by the State of Illinois, by inaction and silence, and by sending mixed signals, OSHA hasn't helped to resolve the issue of whether the Federal OSH Act preempts the traditional police power of the State to prosecute criminal acts that occur in the workplace. This confusion surrounding the preemption question has discouraged some State and local prosecutors from bringing criminal charges in egregious cases, and may have had the effect of shielding employers from responsibility for criminal conduct.

IV. Recommendations

1. OSHA should take an official position on the preemption question and should issue a policy statement. That position should be that the Federal OSH Act, as written, does not preempt the use of historic police powers by the States to prosecute employers for acts against their employees that constitute crimes under State law.

2. There is a need for a real partnership between the Federal Government and the States in pursuing criminal action in workplace safety cases, similar to the partnership that exists in prosecuting drug dealers. This partnership should be premised on cooperation, sharing information and coordinating investigations of workplace accidents and fatalities.

3. Congress should increase the criminal penalties provided by the OSH Act and expand the application of criminal sanctions to include violations which result in serious injuries. Criminal penalties do not now apply to willful safety violations unless there is a resulting fatality. Thus, an employer who willfully and recklessly exposes workers to mercury poisoning, causing permanent brain damage and other serious injury, is not criminally liable under the OSH Act unless a worker dies. Permanent brain damage is not enough to trigger criminal penalties.

4. There is no real program in place at OSHA to handle criminal investigations. OSHA should establish a special criminal investigation unit in its regional offices. Modeled after programs set up in some State and local prosecutors' offices, this new OSHA unit should have necessary expertise in criminal investigations and be available to respond to workplace fatalities 24 hours a day. A prompt response to a workplace fatality or serious injury and a thorough investigation are key elements in building a successful criminal case.

V. DISCUSSION

A. PYMM THERMOMETER

The *Pymm Thermometer* case dramatically illustrates the inadequacies of the present Federal regulatory scheme for dealing with workplace safety violations.

In January 1981, a worker at the Pymm Thermometer plant in Brooklyn, NY, wrote to OSHA:

> Mercury is being used, gas and ovens. Please, we don't know how to describe any more violations, but we are sure there are more. Please send an inspector down to see for himself. We only make the minimum wage, so at least we will know our health is okay.

In March 1981, OSHA inspected the Pymm plant and found serious violations. No protective gear was being used to reduce workers' exposure to mercury—no respirator masks, no aprons, and no gloves. Work surfaces were covered with mercury, and even the areas where workers ate their lunch were contaminated with mercury. OSHA issued a citation, assessed a fine of $1,400, and set a deadline of October 1981 for the company to clean up the factory. However, over the next few years, OSHA regularly extended the compliance deadline.

In 1984, the New York City Department of Health was alerted by a local doctor to elevated levels of mercury in the body of a Pymm worker. The NYC Health Department went to the Pymm factory, inspected it, conducted tests, found violations of the health code and discovered elevated levels of mercury in the workers.

In October 1985, tipped off by a former Pymm worker, an OSHA inspector discovered a hidden cellar operation at the Pymm plant—a cellar virtually without ventilation, filled with broken thermometers, with pools of mercury on the floor, and noxious vapors in the air, which produced permanent brain damage in one employee, Vidal Rodriguez, and exposed many others to serious health risks.

In 1986, OSHA issued citations against Pymm for exposing workers to dangerous levels of mercury and assessed fines of over $100,000. To date, the company has paid just $22,410 in fines, contesting the remainder.

Two months later, the Kings County (Brooklyn) district attorney, in cooperation with the New York State attorney general, brought a criminal prosecution, charging Pymm Thermometer, its owners and officers, William and Edward Pymm, with criminal assault and reckless endangerment for knowingly and continually exposing workers to a toxic substance, mercury.

Kings County District Attorney Elizabeth Holtzman explained:

> The theory of the Pymm prosecution was that assaulting a worker with a toxic substance, such as mercury, is as serious and criminal as assaulting a person with a knife or gun. (Tr. 11.)

New York State Attorney General Robert Abrams told the subcommittee:

I can assure you, however, that the injured workers in these cases are fully aware that they have been the victims of violent crimes. People sometimes recover from the most terrible of beatings. People never recover from mercury poisoning. (Tr. 18.)

In November 1987, after a 4-week trial, the jury in the New York criminal prosecution returned a guilty verdict, finding the company and two of its executives guilty of assault and reckless endangerment by exposing workers to mercury. Under New York law these crimes are punishable by up to 15 years in prison. However, the guilty verdict was subsequently overturned by the trial judge on the ground that State prosecution was preempted by the Federal OSH Act. That decision is under appeal in the New York courts.

B. THE NEED TO BEEF UP CRIMINAL PROVISIONS OF THE FEDERAL OSH ACT

The criminal provisions of the OSH Act are limited to a willful violation of a standard that results in a fatality, falsification of records, and giving advance notice of an OSHA inspection. Criminal sanctions are not applicable to cases of injuries or illnesses that do not result in a death. Thus, in the case of *Pymm Thermometer* where a worker suffered permanent brain damage from mercury, OSHA could not statutorily pursue criminal prosecution. It is simply unacceptable to have death as the trigger point. In comparison, under the Mine Safety and Health Act [MSHA], death is not required for there to be criminal action.

Even in workplace safety violation cases where there is a fatality, under the OSH Act the maximum penalty is only a $10,000 fine and 6 months' imprisonment. This "slap on the wrist" penalty is so low that motivation is taken from the Justice Department to pursue criminal prosecution in workplace safety cases. By contrast, the Resource Conservation and Recovery Act [RCRA], which deals with hazardous waste provides for a penalty of up to $250,000 or up to 15 years' imprisonment for knowingly putting a person in imminent danger of death.

While the primary emphasis of the OSH Act is on civil not criminal penalties, there is a need to beef up and strengthen the criminal provisions of the statute.

The threat or imposition of jail time can have a substantial deterrent effect not achieved through other mechanisms. Jan Chatten-Brown, a special assistant for occupational safety and health protection in the Los Angeles district attorney's office, testified:

> Civil penalties can simply be passed on as part of the cost of doing business. For a corporate officer to face even a few days of jail time is generally of greater consequence than long prison sentences for most criminals. (Tr. 43.)

Cook County State's Attorney Richard Daley told the subcommittee:

> The fines being levied against corporations have not been enough to deter serious neglect and abuse. Corporations such as Film Recovery have protected themselves from fines by filing for bankruptcy. . . . Instead, we need

to deter corporate officers, such as the Film Recovery executives, with criminal sanctions. Imprisonment is one penalty that cannot be passed on to others. (Tr. 37.)

The criminal provisions of the 18-year-old OSH statute are outdated and need to be strengthened so that the Federal Government can effectively and meaningfully prosecute cases of murder and serious injury that result from willful disregard for worker safety.

C. OSHA AND THE PREEMPTION QUESTION

While OSHA has failed to seek criminal penalties for workplace safety violations, State and local prosecutors in dozens of jurisdictions across the country have used the States' historic police powers to prosecute employers for willful conduct that has caused workers to be killed or injured on the job.

In the past 2 years the Los Angeles district attorney's office has prosecuted 18 such cases. Los Angeles Assistant District Attorney Jan Chatten-Brown describes one such prosecution:

> Our first involuntary manslaughter prosecution was against the president of a small drilling company who sent a worker down a 33-foot hole—if you can envision this—that was only 16 to 18 inches in diameter.
>
> The worker was lowered into the hole that was being drilled for an elevator shaft with his foot through a sling. He had no safety harness. The air was not tested. And the sides of the well were not encased.
>
> When the worker went into seizures and the rescue personnel responded, they were told that they could not pump oxygen into the hole because the sides of the wall might collapse. Therefore, by the time they were able to remove the victim, he was dead. (Tr. 41.)

Many of the resulting State court convictions have been challenged or appealed on the ground that the Federal OSH Act preempts State prosecution for workplace injuries and fatalities. The preemption claim has been raised in State courts by defendant employers who seek to use the Federal OSHA law as a shield against criminal prosecution. While an enormous amount of time and money is being spent on appeals on the preemption issue, OSHA and the Department of Labor have not taken an official position on the preemption question.

By its inaction and silence in some cases, and mixed signals in others, OSHA is only adding to the confusion. This confusion and uncertainty have had the effect of thwarting criminal prosecutions in some jurisdictions.

In the *Film Recovery* case, OSHA investigators backed off and deferred to the State of Illinois. As Jerry Thorn, Deputy Solicitor of Labor for National Operations, explained to the subcommittee:

> [T]he investigation that went on in *Film Recovery*, as I understand it, we were there—OSHA's inspectors were there either simultaneously with Mr. Daley's people, but when our OSHA inspectors found out that there was clear interest on the part of the State of Illinois and the county prosecutor's office with respect to the death, they some-

what backed off their investigation and simply let them, as I understand it, take over. (Tr. 89.)

By "backing off" in *Film Recovery* because the State of Illinois was involved in the case, OSHA tacitly acknowledged the State's right to act in terms of criminal prosecutions.

OSHA is not a disinterested bystander in this matter. OSHA should take an official position on the preemption question and should issue a policy statement. In addition, OSHA should not wait for a particular case to work its way up to the U.S. Supreme Court, but rather should file *amicus* briefs in various cases pending on appeal in State courts, including the *Chicago Magnet Wire Corp* case, which is pending before the Illinois Supreme Court.

OSHA should take the position that the States have clear authority under the Federal OSH Act, as it is written, to prosecute employers for acts against their employees which constitute crimes under State law.

Nothing in the OSH Act or its legislative history suggests that Congress intended to shield employers from criminal liability in the workplace or to preempt enforcement of State criminal laws of general application, such as murder, manslaugher, and assault.

Generally speaking, preemption is not read into a statute, and must be unmistakable. It would be most unusual for Congress to displace ordinary criminal laws. Further, the OSH Act is basically an antipreemption statute. Section 8(b)(4) provides:

> Nothing in this act shall be construed to supersede or in any manner affect any workmen's compensation law or to enlarge or diminish or affect in any other manner the common law or statutory rights, duties, or liabilities of employers and employees under any law with respect to injuries, disease, or death of employees arising out of, or in the course of employment.

It would have been most illogical for Congress specifically to authorize a private right to employees to pursue claims under State tort law for injuries incurred in the course of employment while at the same time prohibiting States from using their police power and criminal laws to punish the intentional acts that caused these same injuries.

Section 18 of the OSH Act, however, provides that "[n]othing in this Act shall prevent any State agency or court from asserting jurisdiction under State law over any occupational safety or health issue with respect to which no standard is in effect under section 6" and that "Any State which, at any time, desires to assume responsibility for development and enforcement therein of occupational safety and health standards . . . with respect to which a Federal standard has been promulgated . . . shall submit a State plan. . . ." In this provision forbidding States without an approved plan from setting or enforcing occupational safety and health standards, Congress' focus was on administrative regulation of standards and was not meant to apply to or to deprive employees of the longstanding protection provided by State criminal laws.

As Cook County State's Attorney Richard Daley stated:

> We are not enforcing Government standards, as OSHA does. Rather, we are enforcing our criminal code to protect the people of Cook County from gross misconduct. I submit to you that there is no conflict of jurisdiction on the issue. On the contrary, the responsibilities and objectives of OSHA and local prosecutors can complement each other quite effectively. (Tr. 32)

Daley cited an example of how absurd and detrimental to public safety it would be if Congress had intended to preempt State criminal prosecutions of employers acts against their employees:

> For example, if there was an explosion in a factory stemming from hazardous conditions and dozens of workers died, local prosecutors would be preempted from prosecution.
>
> Yet if the explosion resulted in the deaths of residents in the surrounding area, or a passerby, or a delivery person, we would not be preempted from prosecution.
>
> All these deaths would occur due to the same reckless or negligent conduct. But we could prosecute only for the deaths of those who were not employed by the factory. (Tr. 32.)

Further, the imbalance of criminal penalties would mean that in the above example, the employer, under State law, could face up to 25 years in prison for the death of a nonemployee, but under the OSH Act, could receive only 6 months in prison for a worker's death. We cannot imagine that Congress intended such a result.

The States have an interest in controlling conduct that endangers the lives of their citizens whether it be at home, at work, or on the road. State and local prosecutors should be commended and encouraged to continue their efforts to protect people in their workplaces by utilizing the historic police power of the State to prosecute workplace injuries and fatalities as criminal acts.

D. COOPERATION WITH THE STATES

OSHA's record with respect to cooperation with State and local prosecutors has been spotty. Cook County State's Attorney Richard Daley told the subcommittee:

> Unfortunately, cooperation is not the current norm. OSHA has been reluctant, at best, to cooperate in our State prosecution of these cases. (Tr. 32.)

It is absurd that local OSHA offices do not routinely fully cooperate with local law enforcement officials who seek to prosecute crimes that occur at the workplace. There is a need for a real partnership between the Federal Government and State and local prosecutors in the area of worker safety, similar to the partnership that exists in prosecuting drug dealers and environmental polluters.

At the urging of subcommittee members, on June 15, 1988, OSHA belatedly issued a memorandum to its field offices concerning cooperation with State and local prosecutors in cases where employees have been killed or injured on the job. While the memorandum encourages cooperation in State or local prosecutions "to the fullest extent appropriate" and is a step in the right direction, more coordination is needed to achieve a real partnership.

VI. Summary

OSHA's record in referring cases for criminal action is dismal. Part of the problem is that OSHA "cannot" and part is it "will not." Deficiencies in the Federal OSH statute preclude OSHA from seeking criminal sanctions in cases such as *Pymm Thermometer,* where there was no fatality. In cases such as *Film Recovery,* where there was a fatality, OSHA has regularly and consistently chosen not to seek criminal penalties. There is an institutional reluctance by OSHA, the Justice Department, and the U.S. Attorney's Office to pursue criminal prosecutions in workplace safety cases.

There is a need for OSHA to be more aggressive and timely in using available criminal sanctions. Unless the OSH statute is beefed up and vigorously enforced by OSHA to punish criminally those who show willful disregard for worker safety, some employers will continue "to get away with murder."

APPENDIX B

Letter from
Thomas M. Boyd, Assistant Attorney General
For Legislative and Intergovernmental Affairs,
U.S. Department of Justice, to Rep. Tom Lantos (D-Calif)
on DOJ's Position on Federal Preemption of State Prosecutions

Dec. 9, 1988

APPENDIX B

Office of Legislative and Intergovernmental Affairs

Office of the Assistant Attorney General   Washington, D.C. 20530

DEC 09 1988

The Honorable Tom Lantos
Chairman, Subcommittee on Employment and Housing
Committee on Government Operations
House of Representatives
Washington, D.C. 20515

Dear Mr. Chairman:

This letter is in response to your letter dated October 14, 1988, received in the Criminal Division on October 19, 1988, formally transmitting to the Department of Justice a copy of a report of the Committee on Government Operations, based on a study by the Employment and Housing Subcommittee. This report is entitled "Getting Away with Murder in the Workplace: OSHA's Nonuse of Criminal Penalties for Safety Violations." You have requested the response and comments of the Department of Justice on the issues raised in this report.

The report you have sent us concludes that inadequate use has been made of criminal penalty provisions of the Occupational Safety and Health Act. The report notes that since the establishment of OSHA in 1970, 42 cases have been referred for criminal prosecution, of which approximately one-third were ultimately prosecuted. The committee notes that some states have prosecuted many more workplace safety cases. The report expresses the committee's conclusions that the present criminal penalty provisions of the statute provide an inadequate deterrent to violations. The committee has expressed concern about the length of time some OSHA cases are pending at the Department of Justice before a prosecutorial decision is made and the number of cases in which prosecution is ultimately declined. The committee also expresses concern about factual and legal problems which are interfering with cooperation between the States and the Federal Government in the enforcement of workplace safety laws. In particular, the committee has found that in some cases in which state and local law enforcement officials have used the historical police powers of the State to prosecute employers for acts which are crimes under State laws, the convictions have been overturned on the ground the Federal workplace safety laws preempt a State from pursuing criminal actions.

The committee has concluded its report with recommendations. On the issue of Federal enforcement, the committee has recommended that Congress increase the penalties for criminal violations of the Act and expand the application of criminal sanctions to include violations which result in serious injuries. The committee has also recommended the establishment of special criminal investigative units in the OSHA regional offices. On the issue of the relationship between Federal and State enforcement, the committee has recommended cooperation and coordination similar to that which exists in prosecuting drug cases. The committee has also recommended that OSHA take an official position that the Federal OSH Act does not preempt the use of the historic police powers by the States to prosecute employers for acts against their employees that constitute crimes under State law.

The Department shares the committee's concern about the adequacy of the penalties provided by statute for criminal violations of OSHA safety standards. Prior to the passage of the Comprehensive Crime Control Act in 1984, the maximum penalty for a defendant's first OSHA conviction, no matter how egregious the conduct, was a $10,000 fine and six months imprisonment. See 29 U.S.C. § 666(e). For offenses committed after December 31, 1984, however, the maximum fine for a misdemeanor resulting in loss of human life has now been raised to $250,000 for an individual defendant, and $500,000 for an organization which is a defendant. See 18 U.S.C. §§3623 (now repealed) and 3571. The Department of Justice welcomes these increased penalties and would be happy to see an increase in the period of imprisonment authorized for a criminal violation of OSHA safety standards. We would also be inclined to give serious consideration to proposals to expand the application of criminal sanctions to include violations which lead to serious injuries, in addition to those which lead to the death of an employee.

The Department also shares the committee's view that cooperation with State law enforcement efforts is desirable. As for the legal issue of preemption, we express no views as to the relationship between Federal and State laws regulating the workplace in the context of civil enforcement, or of the circumstances under which a State may assume responsibility for the development and enforcement of occupational safety and health standards under the OSH Act. As for the narrower issue as to whether the criminal penalty provisions of the OSH Act were intended to preempt criminal law enforcement in the workplace and preclude the States from enforcing against employers the criminal laws of general application, such as murder, manslaughter, and assault, it is our view that no such general preemption was intended by Congress. As a general matter, we see nothing in the OSH Act or its legislative history which indicates that Congress intended for the relatively limited criminal penalties provided by the Act to deprive employees of the protection provided by State criminal laws of general applicability.

The committee has concluded that the Department of Justice and the United States Attorney's offices have an "institutional reluctance" to pursue criminal prosecutions in workplace safety cases. In fact, the Department of Justice considers each OSHA referral individually, on its own merits, in light of the same prosecutorial considerations applicable to other violations of federal criminal law. Since federal judicial resources are not sufficient to permit prosecution of every potential federal offense, an exercise of prosecutorial discretion is required in each case.

Among the factors considered in evaluating a case for potential prosecution are the threshold issues of whether there is probable cause to believe the potential defendant's conduct constitutes a federal offense and whether the admissible evidence is likely to be sufficient to obtain and sustain a conviction. Other potentially pertinent factors include federal law enforcement priorities, the nature and seriousness of the offense, the deterrent effect of prosecution, the potential defendant's culpability in connection with the offense, the potential defendant's history of criminal activity, the person's willingness to cooperate in the investigation or prosecution of others, and the probable sentence or other consequences if the person is convicted. Also potentially pertinent are whether the person is subject to effective prosecution in another jurisdiction and whether there is an adequate non-criminal alternative to prosecution. The United States Attorneys must also consider the immediate practical problems of allocating prosecutorial resources within their districts, as well as local federal law enforcement priorities.

The way in which these factors interact in consideration of an OSHA case may vary considerably from case to case. As a general matter, the deterrent value of prosecution is often a strong factor in favor of prosecution. Countervailing factors often include whether the evidence is sufficient to meet the government's burden of proof beyond a reasonable doubt, particularly on the issue of the necessary criminal intent. Another frequent countervailing factor is whether a sentence sufficient to the seriousness of the offense and to the necessary investment of prosecutorial resources could be obtained in the event of a conviction. In this regard, we think it likely that the increased fines available will increase the prosecutive appeal of OSHA cases.

We understand the committee's concern about the amount of time during which some OSHA cases are under consideration by the Department of Justice or the U.S. Attorneys' offices without a response. It should be noted that OSHA referrals, like the results of other investigations of federal offenses, vary considerably in their complexity. In some cases, additional investigation is required, sometimes by a grand jury. During this period, and until a prosecutorial decision is made, a public response concerning the status of the investigation would not be appropriate. Nevertheless, while we see no sound reason to afford OSHA referrals less thorough consideration than other alleged violations of federal criminal law, we agree with the committee that more expeditious treatment is desirable whenever possible.

We share the committee's concern that criminal enforcement of OSHA be effective, and that Federal and State law enforcement in the area of workplace safety be cooperative and complementary. We appreciate your providing us with a copy of the committee's report and look forward to future cooperation in the enforcement of the occupational safety and health laws.

Sincerely,

Thomas M. Boyd
Assistant Attorney General

APPENDIX C

State of Illinois v. Chicago Magnet Wire Corp.

Feb. 2, 1989

Docket No. 65588—Agenda 12—March 1988.
THE PEOPLE OF THE STATE OF ILLINOIS, Appellant, v. CHICAGO MAGNET WIRE CORPORATION et al., Appellees.

JUSTICE WARD delivered the opinion of the court:

The issue we consider on this appeal is whether the Occupational Safety and Health Act of 1970 (OSHA) (29 U.S.C. §651 et seq. (1982)) preempts the State from prosecuting the defendants, in the absence of approval from OSHA officials, for conduct which is regulated by OSHA occupational health and safety standards.

Indictments returned in the circuit court of Cook County charged the defendants, Chicago Magnet Wire Corporation, and five of its officers and agents, Anthony Jordan, Kevin Keane, Frank Asta, Gerald Colby and Allan Simon, with aggravated battery (Ill. Rev. Stat. 1985, ch. 38, pars. 12—4(a), (c)) and reckless conduct (Ill. Rev. Stat. 1985, ch. 38, par. 12—5). The individual defendants were also charged with conspiracy to commit aggravated battery (Ill. Rev. Stat. 1985, ch. 38, par. 8—2(a)). In substance, the indictments alleged that the defendants knowingly and recklessly caused the injury of 42 employees by failing to provide for them necessary safety precautions in the workplace to avoid harmful exposure to "poisonous and stupifying substances" used by the company in its manufacturing processes. On the defendants' motion, the trial court dismissed the charges, holding that OSHA has preempted the State from prosecuting the defendants for the conduct alleged in the indictments. The appellate court affirmed (157 Ill. App. 3d 797), and we granted the State's petition for leave to appeal under Supreme Court Rule 315 (107 Ill. 2d R. 315(a)).

Defendant Chicago Magnet Wire Corporation is an Illinois corporation whose principal business is the coating of wire with various substances and chemical compounds. Anthony Jordan, Kevin Keane, Allan Simon, Frank Asta and Gerald Colby are officers or managerial agents of the corporation.

The indictments charged that the defendants unreasonably exposed 42 employees to "poisonous and stupifying substances" in the workplace and prevented the employees from protecting themselves by "failing to provide necessary safety instructions and necessary safety equipment and sundry health monitoring systems." The indictments also alleged that the defendants improperly stored the substances, provided inadequate ventilation and maintained dangerously overheated working conditions.

Counts of the indictments charging the defendants with aggravated battery alleged that the defendants exposed the employees to the toxic substances with "the conscious awareness that a substantial probability existed that their acts would cause great bodily harm" in violation of section 12—4(a) of the Criminal Code of 1961 (Ill. Rev. Stat. 1985, ch. 38, par. 12—4(a)). Other counts charging the defendants with aggravated battery alleged that the defendants knowingly committed acts "with the awareness that a substantial probability existed" that their conduct would cause the employees to "take by deception [of the employer], for other than medical purposes, poisonous and stupifying substances" in violation of section 12—4(c) (Ill. Rev. Stat. 1985, ch. 38, par. 12—4(c)). The defendants were charged with reckless conduct in that they exposed the employees to substances "in a reckless manner, which caused harm to and endangered *** [their] bodily safety *** by consciously disregarding a substantial and unjustifiable risk of harm *** which constitute[s] a gross deviation from the standard of care which a reasonable person would exercise in this situation." (Ill. Rev. Stat. 1985, ch. 38, par. 12—5.) The conspiracy counts alleged that the individual defendants conspired with the intent to commit aggravated battery and charged that in furtherance of the conspiracy, the defendants committed certain overt acts. Ill. Rev. Stat. 1985, ch. 38, par. 8—2(a).

The circuit court dismissed the indictments, holding that OSHA preempts the States from prosecuting employers for conduct which is governed by Federal occupational health and safety standards, unless the State has received approval from OSHA officials to administer its own occupational safety and health plan. The court stated that because the conduct of the defendants set out in the indictments was governed by OSHA occupational health and safety standards, and the State had not received approval from OSHA officials to administer its own plan, it could not prosecute the defendants for such conduct.

The extent to which State law is preempted by Federal legislation under the supremacy clause of the Constitution of the United States is essentially a question of congressional intendment. (See *Malone v. White Motor Corp.* (1978), 435 U.S. 497, 504, 55 L. Ed. 2d 443, 450, 98 S. Ct. 1185, 1190; *Retail Clerks International Association, Local 1625 v. Schermerhorn* (1963), 375 U.S. 96, 11 L. Ed. 2d 179, 84 S. Ct. 219.) Thus, if Congress, when acting within constitutional limits, explicitly mandates the preemption of State law within a stated situation, we need not proceed beyond the statutory language to determine that State law is preempted. (*Pacific Gas & Electric Co. v. State Energy Resources Conservation & Development Comm'n* (1983), 461 U.S. 190, 203, 75 L. Ed. 2d 752, 765, 103 S. Ct. 1713, 1722.) Even absent an express command by Congress to preempt State law in a particular area, preemptive intent may be inferred where "the scheme of federal regulation is sufficiently comprehensive to make reasonable the inference that Congress 'left no room' for supplementary state regulation" (*Hillsborough County v. Automated Medical Laboratories, Inc.* (1985), 471 U.S. 707, 713, 85 L. Ed. 2d 714, 721, 105 S. Ct. 2371, 2375; *Rice v. Santa Fe Elevator Corp.* (1947), 331 U.S. 218, 230, 91 L. Ed. 1447, 1459, 67 S. Ct. 1146, 1152), or where the regulated field is one in which "the federal interest is so dominant that the federal system will be assumed to preclude enforcement of state laws on the same subject" (*Rice v. Santa Fe Elevator Corp.* (1947), 331 U.S. 218, 230, 91 L. Ed. 1447, 1459, 67 S. Ct. 1146, 1152; *Hines v. Davidowitz* (1941), 312 U.S. 52, 85 L. Ed. 581, 61 S. Ct. 399). Congressional intent to preempt State law may also be inferred where " 'the object sought to be obtained by the federal law and the character of obligations imposed by it may reveal the same purpose.' " *Fidelity Federal Savings & Loan Association v. de la Cuesta* (1982), 458 U.S. 141, 153, 73 L. Ed. 2d 664, 675, 102 S. Ct. 3014, 3022, quoting *Rice v. Santa Fe Elevator Corp.* (1947), 331 U.S. 218, 230, 91 L. Ed. 1447, 1459, 67 S. Ct. 1146, 1152.

The declared purpose of OSHA is "to assure so far as possible every working man and woman in the Nation safe and healthful working conditions and to preserve our human resources." (29 U.S.C. §651(b) (1982).) To this end, Congress gave the Secretary of Labor the authority "to set mandatory occupational safety and health standards" for the workplace and to secure compliance with those standards by imposing civil and criminal sanctions for their violation. (See 29 U.S.C. §651(b)(3) (1982).) An "occupational health and safety standard" is defined as "a standard which requires conditions, or the adoption or use of one or more practices, means, methods, operations, or processes, reasonably necessary or appropriate to provide safe or healthful employment and places of employment." (29 U.S.C. §652(8) (1982).) OSHA also imposes a duty on employers, separate and independent from specific standards set by the Secretary, to provide a workplace "free from recognized hazards that are causing or are likely to cause death or serious physical harm to his employees." 29 U.S.C. §654(a) (1982).

Congress also authorized the Secretary to conduct investigations and on-site inspections of workplaces and to institute enforcement proceedings for violations of OSHA standards. (See 29 U.S.C. §§657 through 659, 662 (1982).) For violations of specific OSHA standards or section 654(a), OSHA authorizes the imposition of civil fines ranging from $1,000 to $10,000 (29 U.S.C. §§666(a) through (c) (1982)). Criminal fines of $10,000 may be imposed for giving unauthorized advanced notification of an OSHA inspection or knowingly making false statements on any OSHA filing. (See 29 U.S.C. §§666(e), (f) (1982).) OSHA also provides for prison terms of up to six months for wilful violations of OSHA standards that result in an employee's death. 29 U.S.C. §666(e) (1982).

The defendants read section 18(a) of OSHA (29 U.S.C. §667(a) (1982)) to mean that under it Congress explicitly provided that the States are preempted from asserting jurisdiction over any occupational health and safety issue that is governed by OSHA occupational health and safety standards unless the State obtains approval from OSHA officials to administer its own occupational health and safety plan under section 18(b). Section 18 provides:

"(a) Nothing in this chapter shall prevent any State agency or court from asserting jurisdiction under State law over any occupational safety or health issue with respect to which no standard is in effect under section 655 of this title.

(b) Any State which, at any time, desires to assume responsibility for development and enforcement therein of occupational safety and health standards relating to any occupational safety or health issue with respect to which a Federal standard has been promulgated under section 655 of this title shall submit a State plan for the development of such standards and their enforcement." 29 U.S.C. §667 (1982).

The defendants state that the conduct alleged in the indictments is governed by OSHA occupational health and safety standards. Specifically, they claim that OSHA standards define permissible exposure limits for the toxic substances which allegedly injured their employees and that OSHA also regulates the conduct that the prosecution says rendered the company's workplace unsafe. The defendants contend that therefore the trial court correctly held that, because the State had not received approval from OSHA officials pursuant to section 18(b) to prosecute the conduct set out in the indictments, the charges must be dismissed. We disagree.

Contrary to this argument, we cannot say that the language of section 18 of OSHA can reasonably be construed as explicitly preempting the enforcement of the criminal law of the States as to conduct governed by OSHA occupational health and safety standards. The language of section 18 refers only to a State's development and enforcement of "occupational health and safety standards." (29 U.S.C. §667(a) (1982).) Nowhere in section 18 is there a statement or suggestion that the enforcement of State criminal law as to federally regulated workplace matters is preempted unless approval is obtained from OSHA officials.

The defendants argue, however, that because the charges set out in the indictments are based on conduct related to an alleged failure to maintain a safe work environment for their employees, in practical effect, the State is attempting to enforce occupational health and safety standards. They contend that the primary purpose of punishing conduct under criminal law is to deter conduct that society deems harmful and to secure conformity with acceptable norms of behavior. In that way, the criminal law establishes standards of care in society. When applied to conduct in the workplace, the defendants argue, criminal law serves the same purpose as OSHA, *i.e.*, to compel adherence to a particular standard of safety that will minimize the risk of injury.

It is the defendants' contention that in enacting OSHA, Congress intended to preempt all State laws to the extent that they regulate workplace safety. They cite regulations promulgated by the Secretary which they say so interpret OSHA. Specifically, they note that section 1901.2 provides:

> "Section 18(a) of [OSHA] is read as preventing any State agency or court from asserting jurisdiction under State law over any occupational safety or health issue to which a Federal standard has been issued." 29 C.F.R. §1901.2 (1986).

They cite too this regulation promulgated by the Secretary:

> "[OSHA's preemption provisions] apply to all state or local laws which relate to an issue covered by a Federal standard, without regard to whether the state law would conflict with, complement, or supplement the Federal standard, and without regard to whether the state law appears to be 'at least as effective as' the Federal standard." Hazard Communication Standard, 52 Fed. Reg. 31,852, 31,860 (1987).

We cannot accept the defendants' contention that it must be concluded that Congress intended to preempt the enforcement of State criminal laws in regard to conduct of employers in the workplace because the State criminal laws implicitly enforce occupational health and safety standards.

Although the imposition of sanctions under State penal law may effect a regulation of behavior as OSHA safety standards do, regulation through deterrence, however, is not the sole purpose of criminal law. For example, it also serves to punish as a matter of retributive justice. Too, whereas OSHA standards apply only to specific hazards in the workplace, criminal law reaches to regulate conduct in society in general. In contrast, occupational health and safety standards are promulgated under OSHA primarily as a means of regulating conduct to prevent injuries in the workplace.

It is to be observed also that for the most part OSHA imposes strict liability for violation of its standards, and that the criminal charges here allege that the defendants knowingly or recklessly injured several of their employees by unreasonably exposing them to toxic substances in the workplace. In order to be convicted of the charges, the State must establish that the defendants not only committed acts causing injury but that they also had the charged mental state, *i.e.*, that they recognized the risk of injury and nevertheless wilfully failed to take precautions to prevent injury. Thus, the criminal charges here do not set any new or other standards for workplace safety but rather seek to impose an additional sanction for an employer's conduct that, if proved, would certainly violate the duty set out in section 654(a) of OSHA (29 U.S.C. §654(a) (1982)).

There is nothing in the structure of OSHA or its legislative history which indicates that Congress intended to preempt the enforcement of State criminal law prohibiting conduct of employers that is also governed by OSHA safety standards. We would observe that the Supreme Court declared in *Jones v. Rath Packing Co.* (1977), 430 U.S. 519, 525, 51 L. Ed. 2d 604, 614, 97 S. Ct. 1305, 1309, that "[w]here *** the field which Congress is said to have pre-empted has been traditionally occupied by the States, *** 'we start with the assumption that the historic police powers of the States were not to be superseded by the Federal Act unless that was the clear and manifest purpose of Congress.'" See also *Hillsborough County v. Automated Medical Laboratories, Inc.* (1985), 471 U.S. 707, 715, 85 L. Ed. 2d 714, 722-23, 105 S. Ct. 2371, 2376; *San Diego Building Trades Council v. Garmon* (1959), 359 U.S. 236, 244, 3 L. Ed. 2d 775, 782, 79 S. Ct. 773, 779.

Certainly, the power to prosecute criminal conduct has traditionally been regarded as properly within the scope of State superintendence. (See *Knapp v. Schweitzer* (1958), 357 U.S. 371, 375, 2 L. Ed. 2d 1393, 1398, 78 S. Ct. 1302, 1305; *Patterson v. New York* (1977), 432 U.S. 197, 201, 53 L. Ed. 2d 281, 287, 97 S. Ct. 2319, 2322.) The regulation of health and safety has also been considered as "primarily, and historically" a matter of local concern. (See *Hillsborough County v. Automated Medical Laboratories, Inc.* (1985), 471 U.S. 707, 719, 85 L. Ed. 2d 714, 725, 105 S. Ct. 2371, 2378.) It cannot be said that it was the "clear and manifest" purpose of Congress to preempt the application of State criminal laws for culpable conduct of employers simply because the same conduct is also governed by OSHA occupational health and safety standards.

Although the provisions of OSHA are comprehensive, that Congress, in section 18, invited the States to administer their own occupational health and safety plans demonstrates that it did not intend to preclude supplementary State regulation. Indeed, section 651 of OSHA (29 U.S.C. §651 (1982)) provides that the States are "to assume the fullest responsibility for the administration and enforcement of their occupational safety and health laws." It seems clear that the Federal interest in occupational health and safety was not to be exclusive.

Too, considering that until the recently increased interest in environmental safety charges were rarely brought under State law for conduct relating to an employer's failure to maintain a safe workplace, it would be unreasonable to say that Congress considered the preemption of State criminal law when enacting OSHA. (See *Pratico v. Portland Terminal Co.* (1st Cir. 1985), 783 F.2d 255, 266 ("Our review of the legislative history of OSHA suggests that it is highly unlikely that Congress considered the interaction of OSHA regulations with other common law and statutory schemes other than worker's compensation").) Indeed, OSHA provides principally civil sanctions and only a few minor criminal sanctions for violations of its standards. Even for wilful violations of OSHA standards which result in an employee's death an employer can be sentenced only to a maximum of six months' imprisonment. There is no penalty provided for conduct which causes serious injury to workers. It seems clear that providing for appropriate criminal sanctions in cases of egregious conduct causing serious or fatal injuries to employees was not considered. Under these circumstances, it is totally unreasonable to conclude that Congress intended that OSHA's penalties would be the only sanctions available for wrongful conduct which threatens or results in serious physical injury or death to workers.

We judge that the purpose underlying section 18 was to ensure that OSHA would create a nationwide floor of effective safety and health standards and provide for the enforcement of those standards. (See *United Airlines, Inc. v. Occupational Safety & Health Appeals Board* (1982), 32 Cal. 3d 762, 654 P.2d 157, 187 Cal. Rptr. 387.) It was not fear that the States would apply more stringent standards or penalties than OSHA that concerned Congress but that the States would apply lesser ones which would not provide the necessary level of safety. The comment has been made: "Congress *** sought uniform national standards not to facilitate commerce but to prevent the 'race for the bottom' that occurred when each state set its own standards. Congress favored a federal law 'so that those states providing vigorous protection would not be disadvantaged by those that did not.'" (See Note, *Getting Away With Murder: Federal OSHA Preemption of State Criminal Prosecutions for Industrial Accidents*, 101 Harv. L. Rev. 535, 550 (1987).) While additional sanctions imposed through State criminal law enforcement for conduct also governed by OSHA safety standards may incidentally serve as a regulation for workplace safety, there is nothing in OSHA or its legislative history to indicate that Congress intended to preempt the enforcement of State criminal law simply because of its incidental regulatory effect.

A question with resemblance to the one here was before the Supreme Court in *Silkwood v. Kerr-McGee Corp.* (1984), 464 U.S. 238, 78 L. Ed. 2d 443, 104 S. Ct. 615. There, the Court addressed the issue of whether State courts are preempted under the Atomic Energy Act from assessing punitive damages against defendants that cause injuries by excessive radiation. In *Pacific Gas & Electric Co. v. State Energy Resources Conservation & Development Comm'n* (1983), 461 U.S. 190, 205, 75 L. Ed. 2d 752, 765, 103 S. Ct. 1713, 1722, the Court had held that under the Atomic Energy Act the States are precluded from regulating the safety aspects of nuclear energy. The defendant argued that a "State-sanctioned award of punitive damages *** punishes and deters conduct relating to radiation hazards" and therefore should be preempted by the Atomic Energy Act. The Court upheld the award notwithstanding the fact that it would have an incidental regulatory effect, stating:

> "In sum, it is clear that in enacting and amending the Price-Anderson Act, Congress assumed that state-law remedies, in whatever form they might take, were available to those injured by nuclear incidents. This was so even though it was well aware of the NRC's exclusive authority to regulate safety matters. No doubt there is tension between the conclusion that safety regulation is the exclusive concern of the federal law and the conclusion that a State may nevertheless award damages based on its own law of liability. But as we understand what was done over the years in the legislation concerning nuclear energy, Congress intended to stand by both concepts and to tolerate whatever tension there was between them. We can do no less. It may be that the award of damages based on the state law of negligence or strict liability is regulatory in the sense that a nuclear plant will be threatened with damages liability if it does not conform to state standards, but that regulatory consequence was

something that Congress was quite willing to accept." *Silkwood*, 464 U.S. 238, 256, 78 L. Ed. 2d 443, 457, 104 S. Ct. 615, 625.

We note, too, that Congress expressly stated that OSHA was not intended to preempt two bases of liability that, like criminal law, operate to regulate workplace conduct and implicitly set safety standards—State workers' compensation and tort law. Section 4(b)(4) of OSHA provides:

> "Nothing in this chapter shall be construed to supersede or in any manner affect any workmen's compensation law or to enlarge or diminish or affect in any other manner the common law or statutory rights, duties, or liabilities of employers and employees under any law with respect to injuries, diseases, or death of employees arising out of, or in the course of, employment." 29 U.S.C. §653(b)(4) (1982).

There is little if any difference in the regulatory effect of punitive damages in tort and criminal penalties under the criminal law. (See Restatement (Second) of Torts §908, comment *b* (1970).) We see no reason, therefore, why what the Court declared in *Silkwood* should not be applied to the preemptive effect of OSHA. Also, if Congress, in OSHA, explicitly declared it was willing to accept the incidental regulation imposed by compensatory damages awards under State tort law, it cannot plausibly be argued that it also intended to preempt State criminal law because of its incidental regulatory effect on workplace safety.

It is a contention of the defendants that it is irrelevant that the State is invoking criminal law jurisdiction as long as the conduct charged in an indictment or information is conduct subject to regulation by OSHA. The defendants argue that the test of preemption is whether the conduct for which the State seeks to prosecute is in any way regulated by Federal legislation. The defendants assert that because the conduct charged in the indictments is conduct regulated under OSHA, a State prosecution for that conduct is preempted by OSHA. The contention is not convincing.

Simply because the conduct sought to be regulated in a sense under State criminal law is identical to that conduct made subject to Federal regulation does not result in State law being preempted. When there is no intent shown on the part of Congress to preempt the operation of State law, the "inquiry is whether 'there exists an irreconcilable conflict between the federal and state regulatory schemes.'" (*Rice v. Norman Williams Co.* (1982), 458 U.S. 654, 659, 73 L. Ed. 2d 1042, 1049, 102 S. Ct. 3294, 3298-99; *Huron Portland Cement Co. v. City of Detroit* (1960), 362 U.S. 440, 443, 4 L. Ed. 2d 852, 856, 80 S. Ct. 813, 815; *Amalgamated Association of Street, Electric Ry. & Motor Coach Employees of America v. Lockridge* (1971), 403 U.S. 274, 285-86, 29 L. Ed. 2d 473, 482, 91 S. Ct. 1909, 1917.) A conflict arises where "compliance with both federal and state regulations is a physical impossibility" (*Florida Lime & Avocado Growers, Inc. v. Paul* (1963), 373 U.S. 132, 142-43, 10 L. Ed. 2d 248, 257, 83 S. Ct. 1210, 1217), or when State law "stands as an obstacle to the accomplishment and execution of the full purposes and objectives of Congress" (*Hines v. Davidowitz* (1941), 312 U.S. 52, 67, 85 L. Ed. 581, 587, 61 S. Ct. 399, 404).

The defendants argue that the prosecutions here would conflict with the purposes of OSHA. They say that Congress intended that under OSHA the Federal government was to have exclusive authority to set occupational health and safety standards. The standards were to be set only after extensive research to assure that the standards would minimize injuries in the workplace but at the same time not be so stringent that compliance would not be economically feasible. (See 29 U.S.C. §655(b)(5) (1982).) The defendants correctly point out that although the States are given the opportunity to enforce their own occupational health and safety standards, the plan submitted must contain assurances that the State will develop and enforce standards "at least as effective" as OSHA's. (See 29 U.S.C. §667 (1982).) Even after a State plan is approved, the Occupational Safety and Health Administration retains jurisdiction to enforce its own standards until it determines, based on three years of experience, that the State's administration of the plan is "at least as effective" as OSHA's. 29 U.S.C. §667 (1982).

The defendants maintain that Federal supervision over State efforts to enforce their own workplace health and safety programs would be thwarted if a State, without prior approval from OSHA officials, could enforce its criminal laws for workplace conduct of employers which is also subject to OSHA standards. They say that the States would thus be permitted to impose standards so burdensome as to exceed the bounds of feasibility or so vague as not to provide clear guidance to employers.

We believe the concern of the defendants is unfounded. We cannot see that State prosecutions of employers for conduct which is regulated by OSHA standards would conflict with the administration of OSHA or be at odds with its goals or purposes. On the contrary, prosecutions of employers who violate State criminal law by failing to maintain safe working conditions for their employees will surely further OSHA's stated goal of "assur[ing] so far as possible every working man and woman in the Nation safe and healthful working conditions." (29 U.S.C. §651(b) (1982).) State criminal law can provide valuable and forceful supplement to insure that workers are more adequately protected and that particularly egregious conduct receives appropriate punishment.

The defendants' statements that the State will now have the ability to enforce more stringent standards than OSHA's does not persuade. As stated, the charges here are based on the defendants' alleged wilfull failure to remove workplace hazards which create a substantial probability that they will cause injuries to their employees. Thus, employers are not left without guidance as to what standard of care they must meet. Too, in practical terms, if a defendant were in compliance with OSHA standards it is unlikely that the State would bring prosecutive action. Enforcement of State criminal law in the workplace will not "stand as an obstacle to the accomplishment and execution of the full purposes and objectives of Congress." *Hines v. Davidowitz* (1941), 312 U.S. 52, 67, 85 L. Ed. 581, 587, 61 S. Ct. 399, 404.

To adopt the defendants' interpretation of OSHA would, in effect, convert the statute, which was enacted to create a safe work environment for the nation's workers, into a grant of immunity for employers responsible for serious injuries or deaths of employees. We are sure that that would be a consequence unforeseen by Congress.

The question here has been considered by a few courts. The appellate court of Wisconsin in *State ex rel. Cornellier v. Black* (Wis. App. 1988), 144 Wis. 2d 745, 425 N.W.2d 21, held that the State's authority to enforce its criminal laws in the workplace has not been preempted by OSHA. The court stated:

"There is nothing in OSHA which we believe indicates a

compelling congressional direction that Wisconsin, or any other state, may not enforce its homicide laws in the workplace. Nor do we see any conflict between the act and sec. 940.06, Stats. To the contrary, compliance with federal safety and health regulations is consistent, we believe, with the discharge of the state's duty to protect the lives of employees, and all other citizens, through enforcement of its criminal laws. Wisconsin is not attempting to impose a penalty for violation of any safety regulations. It is only attempting to impose the sanctions of the criminal code upon one who allegedly caused the death of another person by reckless conduct. And the fact that that conduct may in some respects violate OSHA safety regulations does not abridge the state's historic power to prosecute crimes." (144 Wis. 2d at 755, 425 N.W.2d at 25.)

(A divided court held to the contrary in *People v. Hegedus* (1988), 169 Mich. App. 62, 425 N.W.2d 729, *leave to appeal granted in part* (1988), 431 Mich. 870, 429 N.W.2d 593; *Sabine Consolidated, Inc. v. State* (Tex. App. 1988), 756 S.W.2d 865, citing the opinion of our appellate court in this case (157 Ill. App. 3d 797), also held *contra*.)

We would note that on September 27, 1988, the congressional committee on government operations approved and adopted a report on the question of whether State criminal prosecutions for workplace safety violations are preempted by OSHA. The committee concluded that inadequate use has been made of the criminal penalty provisions of the Act and recommended to Congress that "OSHA should take the position that the States have clear authority under the Federal OSH Act, as it is written, to prosecute employers for acts against their employees which constitute crimes under State law." Report of House Comm. on Government Operations, Getting Away with Murder in the Workplace: OSHA's Nonuse of Criminal Penalties for Safety Violations, H.R. Rep. No. 1051, 100th Cong., 2d Sess. 9 (1988).

The People as supplemental authority cite a letter from the Department of Justice to the chairman of the committee. The letter of the Department of Justice, responding to the report, states in part that the Department shared the concerns of the committee as to the adequacy of the statutory criminal penalties provided for violations of OSHA and also observes:

"As for the narrower issue as to whether the criminal penalty provisions of the OSH Act were intended to preempt criminal law enforcement in the workplace and preclude the States from enforcing against employers the criminal laws of general application, such as murder, manslaughter, and assault, it is our view that no such general preemption was intended by Congress. As a general matter, we see nothing in the OSH Act or its legislative history which indicates that Congress intended for the relatively limited criminal penalties provided by the Act to deprive employees of the protection provided by State criminal laws of general applicability."

The defendants offered supplemental authorities also, arguing that the Department's view was not entitled to deference and was not binding on a court. It, of course, does not bind a court and, whether entitled to deference or not, it is certainly not inappropriate to note that the view of the governmental department charged with the enforcement of OSHA is also the view of this court.

In view of our holding that the State is not preempted from conducting prosecutions, we need not address the defendants' motion to strike portions of the State's brief and certain grand jury testimony.

For the reasons given, the judgments of the appellate court and circuit court are reversed and the cause is remanded to the circuit court of Cook County for further proceedings.

Judgments reversed;
cause remanded.

APPENDIX D

State Prosecution Contacts

ARIZONA:	Marty Woelfle Assistant Attorney General for Environmental Crimes Organized Crime and Racketeering Division Arizona Attorney General's Office 1275 W. Washington Phoenix, Ariz. 85007 (602) 542-3881
CALIFORNIA:	Ira Reiner, District Attorney for the County of Los Angeles Jan Chatten-Brown, Assistant District Attorney 210 W. Temple St. Los Angeles, Calif. 90012 (213) 974-5903
CONNECTICUT:	John A. Connelly Waterbury State's Attorney P.O. Box 2157 Waterbury, Conn. 06702 (203) 756-4431
ILLINOIS:	Frank J. Parkerson, Assistant State's Attorney Cook County (Ill.) District Attorney's Office 500 Richard J. Daley Center Chicago, Ill. 60602 (312) 443-5365
INDIANA:	David H. Coleman Hendricks County Prosecuting Attorney P.O. Box 59 Danville, Ind. 46122 (317) 745-9283
MARYLAND:	Elizabeth Volz, Assistant Attorney General Environmental Crimes Unit Maryland Attorney General's Office 2500 Broening Highway Baltimore, Md. 21224 (301) 631-3025

MASSACHUSETTS: Scott Harshbarger
Middlesex County District Attorney
40 Thorndike St.
Cambridge, Mass. 02141
(617) 494-4050

MICHIGAN: Theodore Klimaszewski
Assistant Attorney General for the State of Michigan
6520 Mercantile Way, Suite 1
Lansing, Mich. 48913
(517) 334-6013

Michael Modelski
Oakland County Assistant Prosecuting Attorney
Court House Tower
Pontiac, Mich. 48053
(313) 858-5230

MINNESOTA: Nancy Leppink, Special Assistant Attorney General
Minnesota Attorney General's Office
520 Lafayette Road
Minneapolis, Minn.
(612) 296-0695

NEW YORK: Elizabeth Holtzman
Brooklyn (N.Y.) District Attorney
Municipal Building
Brooklyn, N.Y. 12201
(718) 802-2995

NORTH CAROLINA: Ronald Stephens
Durham County District Attorney
Judicial Building
Durham, N.C. 27701
(919) 560-6840

OREGON: Stephanie Smythe
Assistant Attorney General
100 Justice Building
Salem, Ore. 97310
(503) 378-4732

TENNESSEE: James R. White
State Labor Commissioner
501 Union Building, Second Floor
Nashville, Tenn. 37219
(615) 741-1991

TEXAS:	Kenneth Oden Travis County Prosecuting Attorney 314 W. 11th, Room 300 Austin, Texas 78701 (512) 473-9415
VIRGINIA:	Cathleen M. Pritchard, Assistant Commonwealth's Attorney City of Virginia Beach Municipal Center Virginia Beach, Va. 23456 (804) 427-4401
WASHINGTON:	Norm Maleng Prosecuting Attorney W554 King County Courthouse Seattle, Wash. 98104 (206) 583-2200
WISCONSIN:	E. Michael McCann Milwaukee County District Attorney 821 W. State St. Milwaukee, Wisc. 53223 (414) 278-4646 Perry Foltz Rock County (Wis.) District Attorney 51 S. Main St. Janesville, Wis. 53545 (608) 755-2115

APPENDIX D

TEXAS:
Kenneth Oden
Travis County Prosecuting Attorney
515 W. 11th, Room 300
Austin, Texas 78701
(512) 475-9415

VIRGINIA:
Cathleen M. Friedland, Assistant Commonwealth's Attorney
City of Virginia Beach
Municipal Center
Virginia Beach, Va. 23456
(804) 427-4401

WASHINGTON:
Norm Maleng
Prosecuting Attorney
W554 King County Courthouse
Seattle, Wash. 98104
(206) 583-2200

WISCONSIN:
E. Michael McCann
Milwaukee County District Attorney
821 W. State St.
Milwaukee, Wisc. 53223
(414) 278-4646

Perry Folts
Rock County (Wis.) District Attorney
51 S. Main St.
Janesville, Wis. 53545
(608) 755-2115

Copyright © 1985 The Bureau of National Affairs, Inc.

APPENDIX E

**Memorandum from John A. Pendergrass,
Assistant Secretary, Occupational Safety & Health Administration,
Regarding Cooperation in State or Local Criminal Prosecutions**

U.S. Department of Labor Occupational Safety and Health Administration
Washington, D.C. 20210

Reply to the Attention of:

JUN 15 1988

MEMORANDUM FOR: REGIONAL ADMINISTRATORS

FROM: JOHN A. PENDERGRASS
Assistant Secretary

SUBJECT: Cooperation in State or Local Criminal Prosecutions

This memorandum provides guidance to the field on OSHA's cooperation with State and local government entities undertaking criminal prosecutions in cases where employees have been killed or injured on the job. Please share this policy with the State plan States in your Region.

OSHA supports all efforts to encourage occupational safety and health and to this end will cooperate in State or local prosecutions to the fullest extent appropriate. This policy of cooperation shall not be construed, however, as a statement of OSHA policy regarding the legal question of preemption which remains a complex issue to be resolved by the courts when raised by an affected party.

When you receive a request for records from State or local officials, you shall consult with the Regional Solicitor regarding the possible effect of a State or local prosecution on any pending or potential Federal case, criminal or civil. This consultation is also necessary to protect the integrity of the files, whether open or closed, from disclosure to other persons.

Further, as part of this cooperation, the following procedural principles shall be followed:

   a. Notify the Director of Field Programs of such State and local prosecutions, who will in turn coordinate with the Directors of Compliance Programs and of Federal State Operations and the National Office Solicitor.

   b. Coordinate any response with the Director of Field Programs and the Regional Solicitor in those cases where the testimony or written opinion of an OSHA official is requested. (Department of Labor guidelines set forth in 29 CFR Part 2, Subpart C, Employees Served with Subpoenas must be followed).

Please provide a copy of this memorandum to the State designees in your Region. If you have any questions, please contact Sandy Taylor on 523-8111.

APPENDIX F

**High Risk Occupational Disease
Notification and Prevention Act (S 582)**

APPENDIX F

101ST CONGRESS
1ST SESSION

S. 582

To notify workers who are at risk of occupational disease in order to establish a system for identifying and preventing illness and death of such workers, and for other purposes.

IN THE SENATE OF THE UNITED STATES

MARCH 15 (legislative day, JANUARY 3), 1989

Mr. METZENBAUM (for himself, Mr. KENNEDY, Mr. PELL, Mr. DODD, Mr. SIMON, Mr. HARKIN, Mr. ADAMS, Ms. MIKULSKI, Mr. MITCHELL, Mr. DASCHLE, Mr. GORE, Mr. INOUYE, Mr. MOYNIHAN, Mr. RIEGLE, Mr. CRANSTON, Mr. WIRTH, Mr. BURDICK, Mr. KERRY, Mr. LAUTENBERG, Mr. SARBANES, and Mr. CONRAD) introduced the following bill; which was read twice and referred to the Committee on Labor and Human Resources

---

A BILL

To notify workers who are at risk of occupational disease in order to establish a system for identifying and preventing illness and death of such workers, and for other purposes.

1 *Be it enacted by the Senate and House of Representa-*
2 *tives of the United States of America in Congress assembled,*
3 **SECTION 1. SHORT TITLE; TABLE OF CONTENTS.**
4   (a) SHORT TITLE.—This Act may be cited as the
5 "High Risk Occupational Disease Notification and Preven-
6 tion Act".
7   (b) TABLE OF CONTENTS.—

Sec. 1. Short title; table of contents.
Sec. 2. Findings and purpose.
Sec. 3. Definitions.
Sec. 4. Risk Assessment Board.
Sec. 5. Employee notification and counseling.
Sec. 6. Means of employee notification.
Sec. 7. Occupational and environmental health centers.
Sec. 8. Research, training, and education.
Sec. 9. Employee medical monitoring; discrimination against employees; confidentiality.
Sec. 10. Enforcement authority.
Sec. 11. Reports to Congress.
Sec. 12. Subjects of Federal agency studies.
Sec. 13. Regulations.
Sec. 14. Authorization of appropriations.
Sec. 15. Effective date.

SEC. 2. FINDINGS AND PURPOSE.

(a) FINDINGS.—Congress finds that—

    (1) during the past 2 decades, considerable scientific progress has been made in—

        (A) the identification of hazardous substances, agents, and processes;

        (B) the identification of medical problems associated with exposure to such substances, agents, and processes; and

        (C) the diagnosis and treatment of diseases related to such exposure;

    (2) progress also has been made in controlling the exposure of individuals to such substances, agents, and processes;

    (3) despite the progress described in paragraphs (1) and (2), there are significant gaps in efforts to promote the health and safety of individuals exposed to such substances, agents, and processes;

(4) potentially harmful substances, physical agents, and processes are in wide industrial and commercial use in the United States;

(5) a significant number of workers suffer disability or death or both wholly or partially as a result of being exposed to occupational health hazards;

(6) diseases caused by exposure to occupational health hazards constitute a substantial burden on interstate commerce and have an adverse effect on the public welfare;

(7) workers have a basic and fundamental right to know that they have been exposed to an occupational health hazard and are at risk of contracting an occupational disease;

(8) there is a period of time between exposure and the onset of disease when it often is possible to intervene medically in the biological process of disease either to prevent or, by early detection, successfully treat many disease conditions;

(9) social and family services that reinforce health-promoting behavior can reduce the risk of contracting an occupational disease;

(10) by means of established epidemiological, clinical, and toxicological studies, it is possible to define

and identify specific worker populations at risk of contracting occupational diseases;

(11) there is no established national program for identifying, notifying, counseling, and medically monitoring worker populations at risk of occupational diseases;

(12) there is a lack of adequately trained professionals, as well as appropriately staffed and equipped health facilities, to recognize and diagnose occupational diseases;

(13) there is a need for increased research to identify and monitor worker populations at risk of occupational diseases; and

(14) through prevention and early detection of occupational disease the staggering costs of medical treatment and care in the United States can be substantially reduced.

(b) PURPOSE.—It is the purpose of this Act—

(1) to establish a Federal program to notify individual employees within populations at risk of occupationally induced disease that they are at risk because of exposure to an occupational health hazard, and to counsel them appropriately;

(2) to authorize and direct the certification of health facilities that have a primary purpose of educat-

ing, training, and advising physicians, nurses, and other professionals in local communities throughout the United States to recognize, diagnose, and treat occupational disease;

(3) to expand Federal research and education efforts to improve means of identifying and monitoring worker populations at risk of occupational disease; and

(4) to establish a set of protections prohibiting discrimination against employees on the basis of identification and notification of occupational disease risk.

SEC. 3. DEFINITIONS.

Except as otherwise provided in this Act, for the purpose of this Act:

(1) BOARD.—The term "Board" means the Risk Assessment Board established under section 4(a).

(2) COMMERCE.—The term "commerce" means trade, traffic, commerce, transportation, or communication among the several States, or between a State and any place outside thereof, or within the District of Columbia, or a possession of the United States (other than the Trust Territory of the Pacific Islands), or between points in the same State but through a point outside thereof.

(3) EMPLOYEE.—The term "employee" means—

>   (A) an employee of an employer who is employed in a business of the employer that affects commerce; or
>
>   (B) a former employee who—
>
>   (i) was formerly employed by an employer in a business of the employer that at the time of employment affected commerce; and
>
>   (ii) as to whom any Federal agency maintains records pertaining to work history, or the employer maintains personnel records, medical records, or exposure records.
>
>   (4) EMPLOYER.—The term "employer" means a person engaged in a business affecting commerce who has employees, including the United States or any State or political subdivision of a State.
>
>   (5) HAZARD COMMUNICATION STANDARD.—The term "hazard communication standard" means the standard contained in section 1910.1200 of title 29 of the Code of Federal Regulations in effect on January 1, 1987.
>
>   (6) INSTITUTE.—The term "Institute" means the National Institute for Occupational Safety and Health.
>
>   (7) MEDICAL MONITORING.—The term "medical monitoring" means periodic examinations or laboratory

tests to diagnose or aid in the diagnosis of a disease that has been the subject of a notice.

(8) OCCUPATIONAL HEALTH HAZARD.—The term "occupational health hazard" means a chemical, physical, or biological agent, generated by or integral to the work process and found in the workplace, or an industrial or commercial process found in the workplace, for which there is statistically significant evidence (based on clinical or epidemiologic study conducted in accordance with established scientific principles) that chronic health effects have occurred in persons exposed to such agent or process. The term includes a chemical that is a carcinogen, toxic or highly toxic agent, reproductive toxin (including an agent that may cause a miscarriage or birth defect), irritant, corrosive, sensitizer, hepatotoxin, nephrotoxin, neurotoxin, an agent that acts on the hematopoietic system, and an agent that damages a lung, skin, eye, or mucous membrane.

(9) PERSON.—The term "person" means one or more individuals, partnerships, associations, corporations, business trusts, legal representatives, or any organized group of persons.

(10) POPULATION AT RISK OF DISEASE.—The term "population at risk of disease" means a class or category of employees—

    (A) exposed to an occupational health hazard under working conditions (such as concentrations of exposure, or durations of exposure, or both) comparable to the clinical or epidemiologic data referred to in paragraph (8); and

    (B) identified and designated as a population at risk of disease by the Board pursuant to section 4(c).

(11) SECRETARY.—The term "Secretary" means the Secretary of Health and Human Services.

SEC. 4. RISK ASSESSMENT BOARD.

(a) ESTABLISHMENT.—

(1) IN GENERAL.—There is established within the Department of Health and Human Services, the Risk Assessment Board.

(2) MEMBERSHIP.—

    (A) IN GENERAL.—The Board shall consist of seven members. Each member shall be appointed by the Secretary from a list of three nominees provided by the National Academy of Sciences. In making appointments under this paragraph, the Secretary may request additional lists.

(B) PUBLIC HEALTH SERVICE EMPLOYEES.—Four members of the Board shall be career or commissioned Public Health Service employees.

(C) NON-PUBLIC HEALTH SERVICE EMPLOYEES.—Three members of the Board shall be appointed from among individuals who are not career or commissioned Public Health Service employees.

(D) SPECIALISTS.—The Board shall include two physicians specializing in occupational medicine, an epidemiologist, a toxicologist, an industrial hygienist, an occupational health nurse, and an occupational biostatistician.

(3) TERM OF OFFICE.—

(A) PUBLIC HEALTH SERVICE MEMBERS.—The terms of members appointed under paragraph (2)(B) shall be 5 years, except that of the members first appointed—

(i) 1 member shall be appointed for 2 years;

(ii) 1 member shall be appointed for 3 years;

(iii) 1 member shall be appointed for 4 years; and

(iv) 1 member shall be appointed for 5 years.

(B) OTHER MEMBERS.—The terms of members appointed under paragraph (2)(C) shall be 5 years, except that of the members first appointed—

(i) 1 member shall be appointed for 1 year;

(ii) 1 member shall be appointed for 3 years; and

(iii) 1 member shall be appointed for 5 years.

(4) CHAIRMAN.—The Secretary shall designate 1 member to serve as Chairman of the Board.

(5) VACANCIES.—Any member appointed to fill a vacancy in the Board that occurs prior to the expiration of a term shall be appointed to serve for the remainder of that term.

(6) REPORTING.—The Board shall report to the Secretary through the Director of the Institute.

(7) STAFF.—The Secretary shall provide full-time staff personnel necessary to carry out the functions of the Board.

(8) COMPENSATION.—Section 5316 of title 5, United States Code, is amended by adding at the end thereof the following new paragraph:

"Members, Risk Assessment Board, Department of Health and Human Services (7).".

(b) INDEPENDENCE OF BOARD.—In the exercise of its functions, powers, and duties, the Board shall be independent of the Secretary and the other offices and officers of the Department unless otherwise specifically provided in this Act.

(c) FUNCTIONS OF BOARD.—

(1) IN GENERAL.—

(A) DUTIES.—The Board shall—

(i) review pertinent medical and other scientific studies and reports concerning the incidence of disease associated with exposure to occupational health hazards;

(ii) identify and designate from the review, and from field assessments where appropriate, those populations at risk of disease that should receive notification pursuant to this Act, including the size, nature, and composition of the populations to be notified;

(iii) develop an appropriate form and method of notification that will be used by the Secretary, or agents of the Secretary described under section 6, to notify the designated populations at risk of disease; and

(iv) determine the appropriate type (if any) of medical monitoring or beneficial health counseling, or both, for the disease associated with the risk, which shall be described in the notification under section 5(b)(6).

(B) PANEL OF EXPERTS.—The Board may appoint an expert or a panel of experts on the particular disease that is the subject of the notice. The report of such expert or panel on the Board's recommendation shall be included in the hearing record.

(C) INFORMATION REQUESTS.—The Board, consistent with section 552a of title 5, United States Code, may request information from any Federal agency or other government or private organization for the purpose of obtaining studies and reports conducted or initiated with respect to actual or potential occupational health hazards. The information shall be furnished consistent with provisions for Federal access set forth under the Occupational Safety and Health Act of 1970 (29 U.S.C. 651 et seq.) and the Federal Mine Safety and Health Act of 1977 (30 U.S.C. 801 et seq.), and regulations promulgated pursuant to such Acts.

(2) IDENTIFICATION OF POPULATIONS AT RISK OF DISEASE.—In identifying populations at risk of disease, the Board shall consider the following factors based on the best available scientific evidence—

>(A) the extent of clinical and epidemiologic evidence that specific substances, agents, or processes may be a causal factor in the etiology of chronic illnesses or long-latency diseases among employees exposed to such substances, agents, or processes in specific working conditions (such as concentration of exposure, or durations of exposure, or both);

>(B) the extent of supporting evidence from clinical, epidemiologic, or toxicologic studies that specific substances, agents, or processes may be a causal factor in the etiology of chronic illnesses or long-latency diseases among individuals exposed to such substances, agents, or processes;

>(C) the employees involved in particular industrial classifications and job categories who are or have been exposed to such substances, agents, or processes under working conditions (such as

concentrations, or durations, or both) that may be a causal factor in the etiology of the illnesses or diseases;

(D) the extent of the increased risk of illness or disease created by occupational health hazards alone or in combination with such factors as smoking and diet; and

(E) other medical, health, and epidemiological factors, including consistency of association, specificity of association, strength of association, dose-response relationships, biological plausibility, temporal relationships, statistical significance, and the health consequences of notifying or failing to notify a population at risk.

(3) DESIGNATION OF IDENTIFIED POPULATIONS FOR NOTIFICATION.—

(A) DESIGNATION.—In designating populations at risk of disease for notification, the Board shall consider the extent to which particular populations may derive health benefits from receipt of notification. The Board shall undertake as its first priority to designate populations likely to benefit from medical monitoring or health counseling.

(B) POSSIBLE FACTORS.—In making the designation required by this paragraph, the Board may consider—

 (i) exposures for which there exists a permanent standard promulgated under section 6(b)(5) of the Occupational Safety and Health Act of 1970 (29 U.S.C. 655(b)(5));

 (ii) the extent of medical monitoring already available to employee populations covered by the permanent standards;

 (iii) the need to notify former employees as well as current employees; and

 (iv) the extent to which notification may prevent miscarriages and birth defects.

(C) NOTIFICATION.—The Board, in making determinations, and the Institute, in giving or coordinating notification, shall notify as many employees at risk of disease as appropriations and the best available scientific evidence permit. The Secretary shall include a detailed explanation of the reasons for the notification determinations in the report submitted pursuant to section 11(b).

(4) DETERMINATION.—If the Board determines that a class or category of employees is a population at

risk of disease to be notified pursuant to this Act, the Board shall—

 (A) make such a determination pursuant to subsection (d); and

 (B) within 10 days of making such a determination, transmit to the Secretary the classes or categories of employees to be notified under section 5.

(d) PROCEDURES.—

 (1) NOTICE OF PROPOSED DETERMINATION.—For each population designated for notification, the Board shall issue a notice of proposed determination.

 (2) CONTENTS OF NOTICE.—The notice required by paragraph (1) shall—

  (A) be published in the Federal Register,

  (B) set forth which classes or categories of employees are being considered for inclusion as an employee population to be notified, and a concise statement of the basis for their inclusion and the contents of the proposed notice as specified in section 5(b) (other than subparagraphs (C) and (F) of section 5(b)(6));

  (C) provide for the public to submit written views on the proposed determination within 60 days of the notice; and

(D) provide for a hearing within 45 days of the notice at which the public may express views on the proposed determination of the Board.

(3) FINAL DETERMINATION.—The Board shall issue a final determination within 60 days after the hearing based on the record developed pursuant to paragraph (2). The final determination shall be deemed to be a final agency action.

(4) EXTENSION.—The Board may, in exceptional circumstances and for good cause shown, extend the time between the issuance of the notice described in paragraph (2), and the issuance of a final determination under paragraph (3), except that the extension may not exceed 150 days for the total period of time beginning with the issuance of the notice.

(5) ACTION.—Any aggrieved person may bring a civil action for mandamus in the appropriate United States district court if the final agency action is not completed within 105 days or 150 days, as the case may be.

(e) BOARD AGENDA.—Within 6 months after the Board is appointed and every 6 months thereafter, the Board shall publish in the Federal Register an agenda listing the chemical, physical, or biological agents and industrial or commercial processes that are under review by the Board or that the

Board anticipates may, within the ensuing 6 months, be reviewed by the Board to decide whether to issue a notice of proposed determination. For each item on the agenda, the Board shall, if available, identify (1) the population to be evaluated with respect to the agent or process, and (2) the name and telephone number of a knowledgeable agency official. The Board may at any time publish a supplement to an agenda adding agents or processes that the Board anticipates will be subject to review prior to the next regularly scheduled publication of an agenda.

(f) BOARD REVIEW.—With respect to a final determination by the Board, not later than 5 years after the initial issuance of notification and not later than each 5 years thereafter, the Board shall review—

    (1) new scientific data relevant to the determination in order to assess the appropriateness and accuracy of the notice; and

    (2) the appropriateness of medical monitoring practices under section 9.

SEC. 5. EMPLOYEE NOTIFICATION AND COUNSELING.

(a) NOTIFICATION OF POPULATION AT RISK.—On a determination by the Board that a given class or category of employees is a population at risk of disease to be notified pursuant to this Act, the Secretary shall make every reasonable effort to ensure that each individual within such population is notified of the risk. The Secretary, through the Institute, shall direct the notification required by this section.

(b) CONTENTS OF NOTIFICATION.—The notification shall include:

   (1) HAZARD.—An identification of the occupational health hazard, including the name, composition, and properties of known chemical agents.

   (2) DISEASES.—The disease or diseases associated with exposure to the occupational health hazard, and the fact that such association pertains to classes or categories of employees.

   (3) EXTENT OF THE RISK.—The extent of the risk of such disease or diseases for the population at risk compared to the population at large.

   (4) LATENCY PERIODS.—Any known latency periods from the time of exposure to the time of clinical manifestation of a disease.

   (5) POSSIBLE CONTRIBUTING FACTORS.—Any known information concerning the extent of increased risk of illness or disease associated with exposure to the occupational health hazard in combination with exposure to nonoccupational factors.

   (6) COUNSELING.—Counseling information appropriate to the nature of the risk, including—

(A) the advisability of initiating a personal medical monitoring program;

(B) the most appropriate type or types of medical monitoring or beneficial health counseling or both for the disease associated with the risk;

(C) the name and address of the nearest occupational and environmental health center certified under this Act;

(D) the protections for notified employees, as established under section 9;

(E) employer responsibilities with respect to medical monitoring for notified employees, as established under section 9; and

(F) the telephone number of the hot line established under subsection (c).

(c) TELEPHONE INFORMATION.—The Institute shall establish a toll-free long distance telephone "hot line" for employees notified under this section or their personal physicians, for the purpose of providing additional medical and scientific information concerning the nature of the risk and its associated disease.

(d) DISSEMINATION OF INFORMATION.—The Institute, after consultation with the Board, shall prepare and distribute other medical and health promotion material and information on any risk subject to notification under this section and its associated disease as the Institute and the Board consider appropriate.

(e) ACCESS TO INFORMATION.—In carrying out the notification responsibilities under this section, the Secretary, consistent with section 552a of title 5, United States Code, may request information from—

    (1) any Federal agency, or State or political subdivision of a State, solely for the purpose of obtaining names, addresses, and work histories of employees subject to notification under this section;

    (2) any employer insofar as Federal access already is provided for under the Occupational Safety and Health Act of 1970 (29 U.S.C. 651 et seq.) and the Federal Mine Safety and Health Act of 1977 (30 U.S.C. 801 et seq.), and regulations promulgated pursuant to such Acts; and

    (3) any employer insofar as such information is maintained by such employer under a Federal or State law concerning occupational safety and health matters.

(f) LIABILITY.—The United States or any agency or employee thereof (including any employer or government acting pursuant to section 6) shall not be subjected to suit or judicial or nonjudicial proceedings of any kind that seek monetary damages with respect to or arising out of any act or omission performed pursuant to this Act, including the failure to perform any act or omission pursuant to this Act. This subsection shall not apply to—

>(1) an employee of the United States for any act or omission that is a knowing and deliberate violation of a provision of the Act to the extent that Federal law otherwise authorizes a suit against that individual for monetary damages; and

>(2) an employer or government acting pursuant to section 6, for any act or omission that is a knowing or reckless violation of a provision of the Act.

(g) JUDICIAL REVIEW.—

>(1) PETITION.—Any person adversely affected or aggrieved by a determination of the Board under this Act is entitled to judicial review of the determination in the United States Court of Appeals wherein such person resides or has the principal place of business of the person or in the United States Court of Appeals for the District of Columbia circuit on a petition filed in such court. A person may be adversely affected or aggrieved by one or more of the following Board determinations:

>>(A) The determination that an agent or process is or is not an occupational health hazard.

    (B) The determination of the class or category of employees that is a population at risk of disease.

    (C) The determination as to what constitutes appropriate medical monitoring or beneficial counseling for the designated population at risk.

Any petition filed pursuant to this section shall be filed within 30 days after such determination by the Board. On the filing of a petition, the Secretary shall certify the hearing record.

  (2) REVIEW.—The court shall review the determination of the Board based on the hearing record.

  (3) JUDICIAL ACTION.—The court shall set aside the determination of the Board if the determination is found to be—

    (A) arbitrary, capricious, or an abuse of discretion;

    (B) contrary to constitutional right, power, privilege, or immunity;

    (C) in excess of statutory jurisdiction, authority, or limitations;

    (D) without observance of procedure required by law; or

    (E) unsupported by substantial evidence on the record.

(4) STAY.—The commencement of proceedings under this subsection shall not operate as a stay of the requirement on the Secretary to notify employees unless the court specifically orders a stay based on a determination by the court that the complaining party is highly likely to succeed on the merits.

SEC. 6. MEANS OF EMPLOYEE NOTIFICATION.

(a) RESPONSIBILITY OF SECRETARY.—Except as otherwise provided in this section, the Secretary shall be responsible for notifying employees at risk of disease, as determined by the Board.

(b) COOPERATION WITH PRIVATE EMPLOYERS AND STATE AND LOCAL GOVERNMENTS.—

(1) IN GENERAL.—In carrying out notification responsibilities under subsection (a), the Secretary is encouraged to cooperate to the extent practicable with private employers and State and local governments.

(2) CERTIFICATION OF PRIVATE EMPLOYERS OR STATE OR LOCAL GOVERNMENTS.—

(A) IN GENERAL.—Upon request by a private employer or a State or local government, the Secretary may certify the private employer or State or local government to conduct notification of its current or former employees, or both, who are members of populations determined to be at risk. Such certification shall require inclusion in the notification of the information described in section 5(b) and shall be in accordance with regulations issued by the Secretary.

(B) ADMINISTRATION.—No private employer or State or local government certified under this paragraph may receive payment for the cost of such notification from the United States, or have a right of access to Federal records for the purposes of carrying out the notification.

(C) FORM OF NOTIFICATION.—The form of notification adopted by a private employer or State or local government shall conform, to the maximum extent practicable, to a model notification form issued by the Board under section 4(c).

(c) EMPLOYEES NOT CURRENTLY EXPOSED.—

(1) IN GENERAL.—In the case of former employees and employees for whom no exposure to the occupational health hazard occurred in the course of employment with their current employer as of the time the notice was issued, the notification shall be transmitted to each employee in the designated population at risk of disease who was exposed to the occupational health hazard within 30 years prior to the date of notification.

(2) INDIVIDUAL NOTIFICATION.—Notification shall be on an individual basis, except that if individual notification is not reasonably possible, the notifying entity shall make use of public service announcements and other means of notification appropriate to reach the population at risk.

(d) EMPLOYEES CURRENTLY EXPOSED.—

(1) IN GENERAL.—In the case of employees for whom any exposure to the occupational health hazard occurred in the course of current employment, notification shall be transmitted to individual employees wherever reasonably possible.

(2) LIMITING RULE.—If individual notification is not reasonably possible, the notifying entity shall make use of public service announcements and other means of notification appropriate to reach the population at risk. Such means may include working with employers to post prominently notices as specified in section 5(b).

(e) VARIANCES.—

(1) APPLICATION FOR VARIANCE.—Within 30 days after the Board issues a final determination identifying a population at risk of disease for notification, an employer who employs or has employed employees within that population may apply to the Institute for a determination exempting the employer's employees

from the population at risk, if the employer believes that, as a result of significant mitigating factors, the employer's employees are not members of a population at risk of disease. The application of the employer shall describe in detail the basis for the application.

(2) HEARING.—If the Institute concludes that any application raises an issue of material fact that is subject to reasonable dispute, the Institute, within 30 days after the receipt of an application, shall—

    (A) publish a notice so stating in the Federal Register; and

    (B) schedule a hearing on the disputed issues.

All applications for a variance with respect to any one population at risk shall be consolidated in a single hearing.

(3) EXEMPTION.—During the pendency of any application before the Institute, the Secretary shall be exempted from the notification requirements of the final determination adopted by the Board with respect to any parties seeking a variance. Each such hearing before the Institute shall be completed within 60 days of the notice of hearing.

(4) DECISION ON VARIANCE.—Within 30 days after the close of the hearing, or, where no hearing is

held, within 30 days of the receipt of an application, the Institute shall issue a decision granting a variance to any employer who has demonstrated by a preponderance of the evidence that the employees of the employer should not be included within the population at risk of disease. The Institute shall deny a variance to all other employers. In determining whether a variance should be granted as to any specific employee or group of employees to be notified, the Institute shall evaluate whether there are significant mitigating factors for such employee or employees, including work practices, health and safety programs, engineering controls, or other factors that are fundamentally different from the factors evidenced by the data relied on by the Board, that substantially eliminate the risk of developing the occupational disease under examination.

(5) LIMITATION.—No employer who has not applied for a variance may avail itself of any decision by the Institute granting a variance to some other employer. Determinations by the Board may not be challenged in any action brought pursuant to this subsection.

SEC. 7. OCCUPATIONAL AND ENVIRONMENTAL HEALTH CENTERS.

(a) SELECTION FROM AMONG EXISTING FACILITIES.—

(1) ESTABLISHMENT AND CERTIFICATION.—Within 90 days after the effective date of this Act, the Secretary shall establish and certify 10 health centers. The Secretary shall select the 10 health centers from among the educational resource centers of the National Institute for Occupational Safety and Health and similar facilities of the National Institute for Environmental Health Sciences, the National Cancer Institute, and other private or governmental organizations designated by the Secretary. At a later date, the Secretary may establish and certify additional health centers from among the health care facilities described in this paragraph.

(2) BASIS FOR SELECTION.—In carrying out paragraph (1), the Secretary shall base selection on ability and experience in the recognition, diagnosis, and treatment of occupationally related diseases, capacity to offer training to physicians, nurses, and other professionals, and geographical proximity for designated populations.

(b) FUNCTIONS OF CENTERS.—A center shall—

(1) provide education, training, and technical assistance to personal physicians and other professionals who serve employees notified under section 5; and

(2) be capable, in the event that adequate facilities are not otherwise reasonably available, of providing diagnosis, treatment, and medical monitoring for employees notified under section 5.

SEC. 8. RESEARCH, TRAINING, AND EDUCATION.

(a) IN GENERAL.—

(1) IMPROVED METHODS OF MONITORING AND IDENTIFICATION.—The Institute shall conduct or provide for research, training, and education designed to improve the means of identifying employees exposed to occupational health hazards and improve medical assistance to such employees. The research, training, and education shall include—

(A) studying the etiology, and development of occupationally related diseases, and the development of disabilities resulting from such diseases;

(B) developing means of medical monitoring of employees exposed to occupational health hazards;

(C) examining the types of medical treatment of workers exposed to occupational health hazards, and means of medical intervention to prevent the deterioration of the health and functional capacity of employees disabled by occupational diseases;

 (D) studying and developing medical treatment and allied health services to be made available to employees exposed to occupational health hazards; and

 (E) sponsoring epidemiological, clinical, and laboratory research to identify and define additional employee populations at risk of disease.

(2) AUTHORITY TO EMPLOY EXPERTS AND CONSULTANTS.—In carrying out activities under this section, the Institute is authorized to engage the services of experts and consultants, as the Institute considers necessary.

(b) EDUCATION.—Part F of title VII of the Public Health Service Act is amended by inserting after section 788 (42 U.S.C. 295g–8) the following new section:

"SEC. 788A. GRANTS AND CONTRACTS FOR TRAINING AND CURRICULUM DEVELOPMENT IN OCCUPATIONAL MEDICINE.

"(a) EXISTING PROGRAMS.—

"(1) IN GENERAL.—The Secretary may make grants to, and enter into contracts with, schools of medicine and schools of nursing in which occupational medicine or occupational health programs exist on the date of enactment of this section to assist such programs in meeting the costs of carrying out projects to—

"(A) provide continuing education for faculty in departments of internal medicine and family medicine or in schools of nursing in order to enable such faculty to provide instruction in the diagnosis and treatment of occupational diseases;

"(B) develop, publish, and disseminate curricula and training materials concerning occupational medicine or health for use in undergraduate medical or nursing training; or

"(C) establish, for residents in graduate medical education programs in internal medicine, family medicine, and other specialties with a primary care focus, or in graduate nursing programs in schools of nursing, training programs in occupational medicine or health consisting of clinical training, for periods of between 1 and 4 months, in settings such as medical facilities, union offices, and industrial worksites.

"(2) PREFERENCES.—In making grants and entering into contracts under this subsection, the Secretary shall give preference to applicants that demonstrate—

"(A) the ability to recruit a significant number of participants to participate in the project to be carried out under the grant or contract (in the case of a project described in subparagraph (A) or (C) of paragraph (1)); and

"(B) expertise and experience in the provision of continuing education in occupational medicine or health (in the case of a project described in subparagraph (A) of such paragraph) or the provision of residency training in occupational medicine or health (in the case of a project described in subparagraph (C) of such paragraph).

"(b) NEW PROGRAMS.—

"(1) IN GENERAL.—The Secretary may make grants to, and enter into contracts with, schools of medicine and schools of nursing in which, on the date of enactment of this section, there do not exist training programs in occupational medicine or health. The purpose of grants and contracts under this subsection shall be to provide support for projects to provide training in occupational medicine or health for faculty who are certified in internal medicine or family medicine by the appropriate national medical specialty board or faculty who have similar qualifications in professional nursing.

"(2) Project requirements.—Each project for which a grant or contract is made under this subsection shall—

"(A) be based in a graduate medical education program in internal medicine or family medicine or in graduate programs in a school of nursing;

"(B) have an arrangement with an accredited training program in occupational medicine or health for the provision of training in occupational medicine or health to the faculty selected by the recipient of the grant or contract under this subsection; and

"(C) have a plan for the use of the faculty receiving training with a grant or contract under this section to provide education and training in occupational medicine or health to other individuals.

"(c) Minimum Number of Schools.—The Secretary shall, during the period October 1, 1989, through September 30, 1992, make grants to, and enter into contracts with, not less than 10 schools of medicine or schools of nursing under subsections (a) and (b).

"(d) AVAILABILITY OF CERTAIN FUNDS.—Amounts described in section 14(b)(2) of the High Risk Occupational Disease Notification and Prevention Act shall be available to carry out this section.

"(e) DEFINITIONS.—For the purpose of this section:

"(1) GRADUATE MEDICAL EDUCATION PROGRAM.—The term 'graduate medical education program' has the same meaning as in section 789(b)(4)(A).

"(2) SCHOOL OF NURSING.—The term 'school of nursing' has the same meaning as in section 853(2).".

SEC. 9. EMPLOYEE MEDICAL MONITORING; DISCRIMINATION AGAINST EMPLOYEES; CONFIDENTIALITY.

(a) EMPLOYEE MEDICAL MONITORING.—

(1) IN GENERAL.—For any employee who is a member of a population that is determined by the Board to be at risk of disease, the medical monitoring recommended by the Board as a result of exposure to the occupational health hazard shall be provided or made available by the current employer at no additional cost to the employee if any part of such exposure occurred in the course of the employee's employment by that employer. If the benefits are made available through an existing employer health plan, the employee may be required to meet deductibles or copayments generally required under the existing employer health

plan. Any such current employer shall be required to provide monitoring only for employees who—

 (A) are notified individually under section 5; or

 (B) the employer knows or has reason to know are members of the population at risk, as determined by the Board.

(2) SPECIAL RULES FOR MEDICAL MONITORING.—

 (A) MONITORING RECOMMENDED BY BOARD.—The medical monitoring required under this Act shall be limited to the monitoring recommended by the Board.

 (B) MEANS.—The means of providing such medical monitoring shall be left to the employer's judgment consistent with sound medical practices.

 (C) SPECIAL SMALL BUSINESS LIMITATION.—An employer with 50 or fewer employees may not be required to pay more than an amount of $250 for medical monitoring for any employee in any year. Such amount shall be adjusted annually after 1989 based on the Consumer Price Index for medical care services maintained by the Bureau of Labor Statistics.

(b) BENEFIT REDUCTION PROHIBITED.—

(1) IN GENERAL.—If, following a determination by the Board under this Act, the employee's physician medically determines that an employee who is a member of a population at risk shows evidence of the development of the disease described in the notice or other symptoms or conditions increasing the likelihood of incidence of such disease, the employee shall have the option of being transferred to a less hazardous or nonexposed job. If within 10 working days after the employee has exercised the option and transmitted to the employer a copy of the initial determination, the employer's medical representative has not requested independent reconsideration of such determination, the employee shall be removed to a less hazardous or nonexposed job and shall maintain earnings, seniority, and other employment rights and benefits as though the employee had not been removed from the former job. In providing such alternative job assignment, the employer shall not be required to violate the terms of any applicable collective bargaining agreement, and shall not be required to displace, lay off, or terminate any other employee.

(2) INDEPENDENT RECONSIDERATION.—If the employer's medical representative requests independent reconsideration of the initial medical determination

under paragraph (1), the employee's physician and the employer's medical representative shall, within 14 working days of the transmittal of the initial determination, submit the matter to another mutually acceptable physician for a final medical determination, which shall be made within 21 working days of the transmittal of the initial determination unless otherwise agreed by the parties. If the two medical representatives have been unable to agree on another physician within 14 working days, the Secretary or the Secretary's local designee for such purpose shall immediately, at the request of the employee or the employee's physician, appoint a qualified independent physician who shall make a final medical determination within the 21-working day period specified in this paragraph, unless otherwise agreed by the parties. The employer shall bear all costs related to the procedure set forth in this paragraph.

(3) EMPLOYEES SUBJECT TO MEDICAL REMOVAL.—An employer shall be required to provide medical removal protection only for employees who—

    (A) are notified individually under section 5; or

(B) the employer knows or has reason to know are members of the population at risk, as determined by the Board.

(4) SPECIAL RULES FOR MEDICAL REMOVAL.—

(A) COURSE OF EMPLOYMENT.—An employer shall be required to provide such protection only if any part of the employee's exposure to the occupational health hazard occurred in the course of the employee's employment by that employer.

(B) LESS HAZARDOUS JOB AVAILABLE.—The medical removal protection described in this subsection shall be provided for as long as a less hazardous or nonexposed job is available. The availability of such a job shall depend on the employee's skills, qualifications, and aptitudes and the job's requirements. Where such job is not available, the medical removal protection shall be provided for a period not to exceed 12 months.

(C) FOLLOWUP MEDICAL SURVEILLANCE.—The employer may condition the provision of medical removal protection on the employee's participation in followup medical surveillance for the occupational health effects in question based on the procedure set forth in this subsection.

(D) REDUCTION FOR COMPENSATION.—The employer's obligation to provide medical removal protection shall be reduced to the extent that the employee receives compensation for earnings lost during the period of removal, or receives income from employment with another employer made possible by virtue of the employee's removal.

(E) ALTERNATIVE EMPLOYMENT.—An employee who is receiving medical removal protection and for whom no less hazardous or nonexposed job is available must undertake reasonable good faith efforts to obtain alternative employment.

(5) SPECIAL SMALL BUSINESS LIMITATIONS.—An employer shall not be required to provide medical removal protection for employees if the employer—

(A) has—

(i) during calendar years 1989 and 1990, 100 or fewer employees at the time medical removal protection is requested; or

(ii) during subsequent calendar years, 50 or fewer employees at the time medical removal protection is requested; and

(B) made or is in the process of making a reasonable good faith effort to eliminate the occu-

1 pational health hazard that is the basis for the
2 medical removal decision.
3 (c) DISCRIMINATION PROHIBITED.—
4 (1) IN GENERAL.—No employer or other person
5 shall discharge or in any manner discriminate against
6 any employee, or applicant for employment, on the
7 basis that the employee or applicant is or has been a
8 member of a population that has been determined by
9 the Board to be at risk of disease.
10 (2) REQUIRED EXPOSURE.—The subsection shall
11 not apply if the position that the applicant seeks re-
12 quires exposure to the occupational health hazard that
13 is the subject of the notice.
14 (3) REMOVAL TO LESS HAZARDOUS JOB.—If it is
15 medically determined pursuant to subsection (b) that an
16 employee should be removed to a less hazardous or
17 nonexposed job, an employer may effect such a remov-
18 al without violating this subsection so long as the em-
19 ployee maintains the earnings, seniority, and other em-
20 ployment rights and benefits, as though the employee
21 had not been removed from the former job.
22 (4) ALTERNATIVE JOB ASSIGNMENT.—An em-
23 ployer with 100 or fewer employees for years 1989
24 through 1990, and thereafter an employer with 50 or
25 fewer employees, may transfer an employee who is or

has been a member of a population at risk to another job without violating this subsection so long as the new job has earnings, seniority, and other employment rights and benefits as comparable as possible to the job from which the employee has been removed. In providing such alternative job assignment, the employer shall not violate the terms of any applicable collective bargaining agreement.

(d) LIMITATIONS FOR AGRICULTURAL WORKERS.—Provisions of this Act relating to medical removal protection shall not apply to any seasonal agricultural worker employed by an employer for less than 6 months of continuous employment. The Secretary, using existing authorization, shall provide that in the case of seasonal agricultural workers employed by an employer for less than 6 months of continuous employment, the medical monitoring recommended by the Board is provided through the Migrant Health Program of the Bureau of Health Care Delivery and Assistance of the Department of Health and Human Services using funds appropriated under section 14. An amount not to exceed $1,000,000 for each fiscal year, from funds authorized to be appropriated by this Act, shall be set aside, if necessary, to carry out the preceding sentence.

(e) CONFIDENTIALITY.—The records of the identity, diagnosis, prognosis, or treatment of any individual employee that are maintained in connection with the performance of any function authorized by this Act shall be confidential and may not be disclosed unless—

    (1) authorized by another provision of this Act and necessary to carry out such provision; or

    (2) on the written consent of such employee or the personally designated representative of the employee.

SEC. 10. ENFORCEMENT AUTHORITY.

    (a) RECORDKEEPING.—The Secretary shall require such recordkeeping by the Institute or by employers acting pursuant to section 6 as is necessary to monitor the numbers, types, and results of notification under this Act.

    (b) ACTIONS BY THE SECRETARY.—

        (1) INJUNCTIVE RELIEF.—Whenever the Secretary determines that an employer has engaged, is engaged, or is about to engage in an act or practice constituting a violation of this Act or any rule or regulation promulgated under this Act, other than a violation of section 9, the Secretary may bring an action in the appropriate United States district court to enjoin such acts or practices. On a proper showing, an injunction or permanent or temporary restraining order shall be granted without bond.

        (2) CIVIL PENALTY.—The Secretary may bring an action in the appropriate United States District

1 Court against an employer acting pursuant to section 6
2 for any act or omission that is a knowing or reckless
3 violation of a provision of this Act or any rule or regu-
4 lation promulgated under this Act. Any employer who
5 violates this Act (or a rule or regulation promulgated
6 under this Act) as set forth in the preceding sentence
7 shall be assessed a civil penalty of not more than
8 $10,000 for each violation.
9 (c) REVIEW OF EMPLOYEE COMPLAINTS.—
10 (1) IN GENERAL.—
11 (A) APPLICATION FOR REVIEW.—Any em-
12 ployee who is aggrieved by a violation of section
13 9 may, within 6 months after such violation
14 occurs, apply to the Secretary of Labor for a
15 review of such alleged violation.
16 (B) INVESTIGATION.—On receipt of such ap-
17 plication, the Secretary of Labor shall cause such
18 investigation to be made as the Secretary of
19 Labor considers appropriate.
20 (C) ACTION.—If, after such investigation,
21 the Secretary of Labor determines that a reasona-
22 ble cause exists to believe that a violation has oc-
23 curred, the Secretary of Labor shall bring an
24 action in any appropriate United States district
25 court. In any such action, the United States dis-

1 trict court shall have jurisdiction for cause shown
2 to restrain violations of section 9, and to order all
3 appropriate relief under subsection (d) or (e). In
4 any action brought by the Secretary of Labor pur-
5 suant to this subsection, an employer shall be
6 given a reasonable opportunity to prove by a pre-
7 ponderance of the evidence that an individual who
8 received notification pursuant to section 5 is not a
9 member of a population at risk, except that a de-
10 termination by the Board that has not been set
11 aside under section 5(g) may not be challenged in
12 any such action.
13   (D) DEFENSE.—It shall be a defense to any
14 action brought to enforce rights under section 9(a)
15 that—
16   (i) an employee who received individual
17   notification failed to assert the rights of the
18   employee under section 9(a) within 1 year
19   after receiving such notification, except for
20   good cause shown; or
21   (ii) an employee who did not receive in-
22   dividual notification but had reason to know
23   that the employee was a member of a popu-
24   lation at risk who is entitled to rights under
25   section 9(a) neglected or omitted to assert

the rights based on the lapse of at least 1 year, and circumstances are sufficient to cause prejudice to the adverse party.

(2) DETERMINATION BY SECRETARY.—Within 90 days of the receipt of the application filed under this subsection, the Secretary of Labor shall notify the complainant of the determination of the Secretary of Labor under paragraph (1). If the Secretary of Labor finds that there was not reasonable cause to believe that a violation occurred, the Secretary shall issue an order denying the application and informing the applicant of the rights of the applicant under paragraph (3).

(3) APPEAL.—

(A) DENIAL OF APPLICATION.—Any person adversely affected or aggrieved by a determination of the Secretary of Labor under paragraph (2) is entitled to judicial review of the determination in the appropriate United States District Court on a petition filed in such court within 30 days after such determination by the Secretary of Labor. The court may set aside the determination by the Secretary of Labor under paragraph (2) only if the determination is found to be—

(i) arbitrary, capricious, or an abuse of discretion;

     (ii) contrary to constitutional right, power, privilege, or immunity;

     (iii) in excess of statutory jurisdiction, authority, or limitations;

     (iv) without observance of procedure required by law; or

     (v) unsupported by substantial evidence on the record.

If the court sets aside the determination of the Secretary of Labor under paragraph (2), the employee may bring an action of the type authorized by paragraph (1)(C).

  (B) MEMBER OF THE POPULATION AT RISK RULE.—In any action brought by an aggrieved person under subparagraph (A), the aggrieved person shall be given a reasonable opportunity to prove, by a preponderance of the evidence, that an individual who did not receive notification is a member of a population at risk, except that a determination by the Board that has not been set aside under section 5(g) may not be challenged in any such action.

  (C) FAILURE TO ACT WITHIN 90 DAYS.—If the Secretary of Labor has not acted within 90 days pursuant to paragraph (2), an applicant may bring a civil action for mandamus in the appropriate United States District Court.

(d) REINSTATEMENT AND OTHER RELIEF.—Any employee who is injured in violation of section 9 shall be restored to his or her employment and shall be compensated for—

    (1) any lost wages (including fringe benefits and seniority);

    (2) costs associated with medical monitoring that are incurred up to the time when the discrimination is fully remedied; and

    (3) costs associated with bringing the allegation of violation.

(e) CIVIL PENALTIES.—Any person that violates section 9 shall be liable for a civil penalty of not more than $10,000 for each violation.

(f) EXCLUSIVITY OF REMEDY.—Except as otherwise expressly provided in this Act, remedies provided in this Act shall be exclusive remedies with respect to any acts or omissions taken pursuant to or alleged to be in violation of this Act.

(g) EFFECT ON OTHER LAWS.—

    (1) ADMISSABILITY OF TESTIMONY AND EVIDENCE.—No testimony, whether oral or written, and no documents or physical evidence of any type that

would prove, tend to prove, or that is offered in an attempt to prove—

    (A) that the Board has made a finding or determination that an employee or an employee population is or is not a member of or is or is not a population at risk of disease as determined under this Act;

    (B) that an employee or employee population is or is not to receive (or has or has not received) notification under this Act; or

    (C) that medical evaluation or monitoring or removal is or is not to be initiated (or has or has not been initiated) under this Act;

shall be admissible in or considered in connection with or form the basis of a ruling in any civil, criminal, administrative, or other judicial or quasi-judicial proceeding of any type, whether brought under Federal or State law other than a claim brought pursuant to section 5(f), 5(g), 10(b), or 10(c). The admission or reliance on any such testimony, documents, or physical evidence shall automatically constitute reversible error.

  (2) MENTAL OR EMOTIONAL DISTRESS CLAIMS.—

    (A) PROHIBITION.—No person shall bring any tort or workers' compensation claim based on mental or emotional harm, fear of disease, or stress resulting directly or indirectly from any report, finding, notice, medical evaluation decision, or monitoring decision made under this Act, from any other action taken under this Act, or from any failure to take an action required by this Act.

(B) APPLICABILITY.—The prohibition set forth in subparagraph (A) applies whether the person seeking to bring such a claim—

(i) has been directly subject to such a report, finding, notice, medical evaluation decision, monitoring decision, other action, or failure to take required action; or

(ii) has learned about such a report, finding, notice, medical evaluation decision, monitoring decision, other action, or failure to take any required action that directly affected another person.

(3) CONSTRUCTION RULE.—Nothing in this Act shall preclude the admission into evidence of—

(A) the results of any medical evaluation or monitoring;

(B) any medical and other scientific studies and reports concerning the incidence of disease associated with exposure to occupational health hazards; or

(C) any data related to exposure to occupational health hazards for individual employees,

in connection with any claim for compensation, loss or damage brought under State or Federal law. Notification pursuant to this Act shall not be relevant in determining whether such a claim is timely under any applicable statute of limitations.

(h) PROHIBITION ON ACTIONS AGAINST PHYSICIANS FOR GOOD FAITH DETERMINATIONS UNDER SECTION 9(b).—No action may be brought for any claim based on a good faith determination made by a physician under section 9(b).

SEC. 11. REPORTS TO CONGRESS.

(a) HAZARD COMMUNICATION STANDARD REPORT.— The Secretary of Labor shall report to Congress annually, not later than January 15 of each year, regarding implementation and enforcement of the hazard communication standard. The report shall include detailed information on—

(1) monitoring and enforcement; significant areas of noncompliance; and penalties assessed and steps taken to correct the noncompliance;

(2) efforts to evaluate the hazard communication standard;

(3) efforts to assist employers to comply with the hazard communication standard;

(4) efforts to educate employees to their rights under the hazard communication standard; and

(5) efforts to comply with Federal court decisions requiring or encouraging an expanded scope for the hazard communication standard.

(b) OCCUPATIONAL DISEASE NOTIFICATION REPORT.—The Secretary shall report to Congress annually, not later than January 15 of each year, regarding implementation and enforcement of notification under this Act. The report shall include detailed information on—

(1) numbers, types and results of notifications carried out pursuant to sections 5 and 6;

(2) research efforts carried out pursuant to section 8;

(3) training and education efforts for employees, personal physicians, nurses, and other professionals carried out pursuant to sections 7 and 8;

(4) enforcement efforts carried out pursuant to section 10; and

(5) efforts to assist employers under this Act.

SEC. 12. SUBJECTS OF FEDERAL AGENCY STUDIES.

(a) NOTIFICATION REQUIRED.—Each Federal agency that conducts epidemiologic studies on occupational disease initiated after the effective date of this Act shall establish procedures for notifying the subjects of such studies of the findings of such study. If the findings are that the subjects are at risk of disease, the notification shall include the information specified in section 5(b), except that required by subparagraphs (D), (E), and (F) of section 5(b)(6). No notice under this section shall impose any liabilities or create any rights under section 9.

(b) METHOD OF NOTICE.—All occupational epidemiologic studies conducted by a Federal agency initiated after the effective date of this Act shall include in the study design specific methods for notifying living subjects or their immediate family members that they are part of a population at risk of disease.

SEC. 13. REGULATIONS.

The Secretary shall prescribe such regulations as may be necessary to carry out this Act.

SEC. 14. AUTHORIZATION OF APPROPRIATIONS.

(a) IN GENERAL.—There are authorized to be appropriated $25,000,000 for each of the fiscal years 1990 through 1992 to carry out this Act and section 788A of the Public Health Service Act (as added by section 8(b) of this Act).

(b) SET-ASIDE.—

(1) RESEARCH, TRAINING, AND EDUCATION.—Of the total amount appropriated under subsection (a) for

a fiscal year, at least $4,000,000 shall be available to carry out section 8 of this Act and section 788A of the Public Health Service Act.

(2) EDUCATION.—Of the total amount available under paragraph (1) for a fiscal year, at least $1,000,000 shall be available to carry out section 788A of the Public Health Service Act.

SEC. 15. EFFECTIVE DATE.

(a) IN GENERAL.—Except as may be otherwise provided in this Act, this Act and the amendments made by this Act shall become effective January 1, 1990, or 6 months after the date of enactment of this Act, whichever occurs first.

(b) BOARD.—The Board shall be appointed within 60 days after the effective date of this Act.

(c) REGULATIONS.—The Secretary shall issue regulations necessary to administer the Act within 120 days after the effective date of this Act.

APPENDIX G

**Bill to Amend the OSH Act of 1970
To Improve Regulation of the Construction Industry (S 930)**

101ST CONGRESS
1ST SESSION

S. 930

To amend the Occupational Safety and Health Act of 1970 to establish an Office of Construction, Safety, Health, and Education within OSHA, to improve inspections, investigations, reporting, and recordkeeping in the construction industry, to require certain construction contractors to establish construction safety and health programs and onsite plans and appoint construction safety specialists, and for other purposes.

IN THE SENATE OF THE UNITED STATES

MAY 4 (legislative day, JANUARY 3), 1989

Mr. DODD (for himself and Mr. LIEBERMAN) introduced the following bill; which was read twice and referred to the Committee on Labor and Human Resources

A BILL

To amend the Occupational Safety and Health Act of 1970 to establish an Office of Construction, Safety, Health, and Education within OSHA, to improve inspections, investigations, reporting, and recordkeeping in the construction industry, to require certain construction contractors to establish construction safety and health programs and onsite plans and appoint construction safety specialists, and for other purposes.

1 *Be it enacted by the Senate and House of Representa-*
2 *tives of the United States of America in Congress assembled,*

SECTION 1. SHORT TITLE; TABLE OF CONTENTS.

(a) SHORT TITLE.—This Act may be cited as the "Construction Safety, Health, and Education Improvement Act of 1989".

(b) TABLE OF CONTENTS.—The table of contents is as follows:

Sec. 1. Short title; table of contents.
Sec. 2. References to the Occupational Safety and Health Act of 1970.
Sec. 3. Definitions.
Sec. 4. Office of Construction, Safety, Health, and Education.
Sec. 5. Inspections, investigations, reporting, and recordkeeping.
Sec. 6. Construction safety and health programs.
Sec. 7. Onsite construction safety and health plans.
Sec. 8. Construction safety specialists.
Sec. 9. Construction Safety and Health Training Academy.
Sec. 10. National Institute for Occupational Safety and Health.
Sec. 11. Penalties.
Sec. 12. Advisory Committee on Construction Safety and Health.
Sec. 13. Budget contents.
Sec. 14. Effective dates.

SEC. 2. REFERENCES TO THE OCCUPATIONAL SAFETY AND HEALTH ACT OF 1970.

Except as otherwise specifically provided, whenever in this Act an amendment or repeal is expressed in terms of an amendment to, or repeal of, a section or other provision, the reference shall be considered to be made to a section or other provision of the Occupational Safety and Health Act of 1970 (29 U.S.C. 651 et seq.).

SEC. 3. DEFINITIONS.

Section 3 (29 U.S.C. 652) is amended—

(1) in paragraph (5), by inserting after "employees" the following: "(including a self-employed contractor in the construction industry)";

(2) in paragraph (6), by inserting after 'before the period at the end thereof the following: "(including a self-employed contractor in the construction industry, at such times as the contractor personally performs construction work)"; and

(3) by adding at the end thereof the following new paragraphs:

"(15) The term 'Academy' means the Construction Safety and Health Training Academy established under section 38.

"(16) The term 'Advisory Committee on Construction Safety and Health' means the Advisory Committee established under section 107(e) of the Contract Work Hours and Safety Standards Act (40 U.S.C. 333(e)).

"(17) The term 'construction contractor' means a person (including a prime contractor, general contractor, or subcontractor) who enters into a contract with a construction owner for the performance of construction work.

"(18) The term 'construction owner' means a person who—

"(A) owns, leases, or has effective control of—

                "(i) real property, with or without improvements; or

                "(ii) a structure or other improvement on real property; and

            "(B) performs, or intends to perform, construction work on such property or improvement.

        "(19) The term 'construction safety specialist' means—

            "(A) an individual who is hired and retained, and performs the duties required, under section 37; or

            "(B) in any case in which such individual is not appointed pursuant to section 34(d), an employee designated by a construction contractor to perform such duties.

        "(20) The term 'construction work' means work for construction, alteration, or repair, or any combination thereof, including painting and decorating. Such term shall include such work performed under a contract between an employer and an agency of the United States or any State or political subdivision of a State.

        "(21) The term 'construction worksite' means a site where construction work is performed.

"(22) The term 'serious injury' means any injury or illness that results in—

    "(A) the permanent removal of a part of the body;

    "(B) a part of the body being rendered functionally useless; or

    "(C) a substantial reduction of a bodily function or efficiency, on or off the job.".

SEC. 4. OFFICE OF CONSTRUCTION, SAFETY, HEALTH, AND EDUCATION.

The Act (29 U.S.C. 651 et seq.) is amended by adding at the end thereof the following new section:

"SEC. 34. OFFICE OF CONSTRUCTION, SAFETY, HEALTH, AND EDUCATION.

"(a) ESTABLISHMENT.—There is established in the Occupational Safety and Health Administration an Office of Construction, Safety, Health, and Education (hereinafter in this section referred to as the "Office") to ensure safe and healthy working conditions in the performance of construction work.

"(b) DUTIES.—The Office shall—

    "(1) carry out the powers, duties, and responsibilities of the Secretary under this Act that relate to safety and health in the performance of construction work, including powers, duties, and responsibilities prescribed under subsections (h) through (k) of section 8 and sections 35 through 38;

"(2) develop mandatory criteria for construction safety and health programs established under section 35 and onsite construction safety and health plans established under section 36;

"(3) assume control of a construction site following an accident that results in a fatality to—

"(A) prevent the destruction of any evidence that would assist in the investigation of the accident; and

"(B) oversee rescue operations conducted in response to the accident;

"(4) assist the Advisory Committee on Construction Safety and Health in the development of training courses and curriculum for certification of construction safety specialists and in the training and testing program of the Academy;

"(5) consult with and advise employers, employees, and labor organizations as to effective means of preventing occupational injuries and illnesses in construction work and ensure equal access to such consultative services;

"(6) increase awareness of construction site safety through education, training, and outreach programs;

"(7) provide technical experts who can assist construction contractors with specific technical inquiries; and

"(8) carry out such other duties as are assigned to the Office by law.

"(c) PERSONNEL.—

"(1) DIRECTOR.—The Office shall be headed by a Director, appointed by the Secretary, who shall be responsible for carrying out the duties of the Office prescribed under subsection (b).

"(2) EMPLOYEES.—The Director shall—

"(A) in accordance with the civil service laws, appoint and fix the compensation of such officers and employees as may be necessary to carry out this section; and

"(B) prescribe the powers, duties, and responsibilities of all officers and employees engaged in carrying out this section.

"(3) ADDITIONAL INSPECTORS.—Subject to appropriations, the Secretary shall employ such additional inspectors as are necessary to carry out the duties of the Office prescribed under subsection (b). Such inspectors shall be in addition to inspectors employed under this Act on the effective date of this section.

"(4) SMALL BUSINESS LIAISON.—The Director shall designate an employee of the Office to serve as a small business liaison. The liaison shall be responsible for providing assistance for complying with new and existing standards established under this Act for the performance of construction work (in the form of manuals, videos, and audiotapes and other outreach programs) to trade associations and other small businesses.

"(d) EXEMPTIONS.—

"(1) IN GENERAL.—The Secretary may issue regulations that provide an exemption from the requirements of section 35, 36, or 37, in whole or in part, for certain types of construction projects, operations, or construction contractors, if the Secretary finds that such a requirement is not feasible for such type of project, operation, or contractor and will not promote a significant increase in employee safety and health.

"(2) PROCEDURE.—In issuing such regulations, the Secretary shall—

"(A) condition the exemption on compliance with alternative requirements that are feasible for such project, operation, or contractor and are capable of promoting employee safety and health;

"(B) base the exemption on written information submitted to the Secretary;

"(C) publish notice of the exemption in the Federal Register; and

"(D) provide an opportunity for public comment and response to such submitted information.".

SEC. 5. INSPECTIONS, INVESTIGATIONS, REPORTING, AND RECORDKEEPING.

Section 8 (29 U.S.C. 657) is amended by adding at the end thereof the following new subsections:

"(h)(1) The Secretary shall establish an effective and fair system for construction worksite inspections.

"(2) In establishing such system, the Secretary shall establish construction worksite inspection priorities intended to ensure that resources for enforcement are concentrated on construction worksites and operations having a high potential for fatalities or serious injuries and illnesses.

"(3) In establishing such priorities, the Secretary shall give due weight to the record of the compliance of—

"(A) an employer with standards established under this Act and the Contract Work Hours and Safety Standards Act (40 U.S.C. 327 et seq.); and

"(B) a construction owner and an employer with this Act, including compliance with recordkeeping and reporting requirements of this Act.

"(4) The Secretary shall inspect on a priority basis construction projects of construction owners and construction worksites of employers having a higher than average incident frequency or severity rate of injuries or illnesses for the specific type of construction operation involved.

"(5) The Secretary shall use reports and notices filed under subsections (j) and (k) to develop the system of prioritized inspections required under this subsection.

"(6) The inspection priority system required under this subsection shall not have the effect of limiting—

"(A) inspections conducted by the Secretary on the basis of complaints by employees or employee representatives or complaints of imminent dangers; or

"(B) inspections intended as a followup to prior enforcement actions or proceedings.

"(7) No construction owner or employer engaged in construction work shall be—

"(A) excluded by the Secretary from inspections conducted under this subsection; or

"(B) advised directly or indirectly as to the status of the owner or employer or placement on the inspection priority system required under this subsection.

"(8) No advance notice of an inspection conducted under this subsection shall be provided to any person.

"(i)(1) Each construction contractor shall maintain accurate records concerning accidents and injuries at a construction worksite.

"(2) The records required under paragraph (1) shall specify—

"(A) the name, business address, and telephone number of the construction owner;

"(B) the location of the construction worksite;

"(C) the name, business address, and telephone number of the employer whose employee or employees were killed or injured or could have been killed or injured by the incident;

"(D) the name and business address of the project contractor or pertinent general contractor at the construction worksite;

"(E) the date and time of the incident;

"(F) the type of incident (such as whether the incident involved a fire, explosion, or building collapse);

"(G) the number and nature of fatalities or injuries resulting from the incident;

"(H) the number of persons hospitalized as a result of the incident;

"(I) the number of persons unaccounted for at the time the report is made; and

"(J) the identity and mailing address of the construction safety specialist responsible for investigating the incident.

"(3) Such information shall be available for any inspection conducted under this Act.

"(4) Any construction contractor who fails to maintain the information required by this subsection shall be assessed a civil penalty in an amount to be determined by the Secretary.

"(j)(1) Except as otherwise provided in this paragraph, an employer shall report to the appropriate regional office of the Occupational Safety and Health Administration by telephone or telegraph (not later than 24 hours after the occurrence of the incident) any incident involving construction work that results in—

"(A) a serious injury;

"(B) a fatality;

"(C) a structural failure that leads to the collapse of a place of employment; or

"(D) a potential collapse of a place of employment.

"(2) The employer shall submit a followup written report to the regional office as soon as practicable after a report is made under paragraph (1).

"(3) Each report required under this subsection shall specify—

"(A) the name, business address, and telephone number of the construction owner of the project;

"(B) the location of the project;

"(C) the name, business address, and telephone number of the employer or employers whose employee or employees were killed or seriously injured;

"(D) the date and time of the occurrence;

"(E) the nature of the occurrence;

"(F) the number of fatalities and serious injuries;

"(G) the number of persons hospitalized;

"(H) the number of persons unaccounted for at the time the report is made; and

"(I) the name, business address, and telephone number of the construction safety specialist submitting the report.

"(4)(A) The Secretary shall conduct an inspection of the construction worksite, or the pertinent area of the site, to investigate all reports of incidents described in paragraph (1), as well as reports of those categories of serious injuries as the Secretary, by regulation, prescribes.

"(B) The Secretary shall be granted access to the site.

"(C) The Secretary shall conduct the inspection as soon as practicable, but no later than 24 hours, after receipt of a construction safety report under paragraph (1), unless the

Secretary determines that conditions at the site would make an inspection dangerous.

"(5) The construction contractor shall take appropriate measures (as defined in regulations issued by the Secretary) to prevent the destruction of any evidence that would assist the Secretary in the investigation of the occurrence.

"(6)(A) As soon as practicable after the investigation, the Secretary shall prepare and make public a narrative description of the occurrence.

"(B) Such description shall include a statement of all of the items referred to in paragraph (3) and shall contain appropriate recommendations for the prevention of a similar occurrence in the future.

"(C) A copy of such description shall be provided by the Secretary to the construction contractor or the construction safety specialist and shall be made available to employees and their representatives.

"(k)(1) Except as provided in paragraph (2), prior to the commencement of construction work on a construction project, the construction safety specialist shall provide to the Secretary notice in writing that provides—

"(A) a description of the project;

"(B) the name, principal business address, and telephone number of—

"(i) the construction owner;

"(ii) the project construction safety specialist; and

"(iii) the prime and general contractors.

"(C) the total number of other employers who the specialist estimates will be engaged on the project;

"(D) the municipal address of the project or its location with respect to the nearest public highway;

"(E) the date construction work will commence and the anticipated duration of such work;

"(F) the estimated total costs of the project for labor and material, including labor and material for work performed by subcontractors; and

"(G) the expected number of employees to be employed on such project at—

"(i) the commencement of construction work;

"(ii) the time when the peak number of employees is expected to be employed; and

"(iii) the conclusion of construction work.

"(2) If it is necessary to perform construction work on a project immediately in order to prevent injury to persons, such work may be commenced without complying with paragraph (1), except that, in any such case, the notice shall be given to the Secretary as soon as practicable after the construction work on the project commences.

"(3)(A) At the completion of a construction project or at intervals of 1-year duration of a project, whichever occurs earlier, the construction safety specialist for the project shall submit to the Secretary a report of all fatalities and serious injuries and illnesses suffered by employees, and of all structural failures that occurred, on the project. If there have been no fatalities or serious injuries or illnesses, or structural failures, on the project, the report shall so state.

"(B) The report shall provide—

"(i) the name, business address, and telephone number of the construction owner, construction safety specialist, and each employer who was engaged on the project;

"(ii) the name of each employer whose employee or employees suffered death, serious injury, or illness;

"(iii) a description of the nature of the work performed by such employer on the project;

"(iv) a description of the nature of the work performed by each employee at the time the employee suffered death or serious injury or illness; and

"(v) the date of each such occurrence.

"(4) The Secretary shall—

"(A) issue such regulations as are necessary to carry out this subsection; and

"(B) prepare and make available standard forms to be used for compliance with the requirements of this subsection.".

SEC. 6. CONSTRUCTION SAFETY AND HEALTH PROGRAMS.

The Act (29 U.S.C. 651 et seq.) (as amended by section 4 of this Act) is further amended by adding at the end thereof the following new section:

"SEC. 35. CONSTRUCTION SAFETY AND HEALTH PROGRAMS.

"(a) REQUIREMENT.—Except as provided in section 34(d), a construction contractor shall establish a written safety and health program (hereinafter in this section referred to as the 'program') in accordance with this section.

"(b) MANAGER.—

"(1) IN GENERAL.—The program shall provide for the assignment of a construction safety specialist or other employee of the contractor who is responsible for general management of the program.

"(2) DUTIES.—The responsibilities and authorities of such employee shall be specified in writing as part of the safety program.

"(c) GENERAL SAFETY AND HEALTH TRAINING.—

"(1) IN GENERAL.—The program shall provide for ensuring that all supervisory personnel and all employees of the contractor engaged in construction work on the project shall receive, or have received within the immediately preceding 12-month period, general safety and health training in a manner prescribed by the Secretary, with the advice of the Advisory Committee on Construction Safety and Health.

"(2) CONSTRUCTION WORKSITE INSTRUCTION.—The construction worksite instruction required under subsection (g) may be counted toward fulfillment of the minimum training requirements prescribed under paragraph (1).

"(3) DEMONSTRATION OF COMPLIANCE.—The Secretary shall establish a procedure under which a contractor will be able to demonstrate to the Secretary that the employees of the contractor have fulfilled the training requirements prescribed under paragraph (1).

"(4) RECORDKEEPING.—The Secretary shall require a contractor to maintain current records showing the amount of training and instruction received by each employee and supervisor, and the subjects in which such training and instruction have been received, pursuant to paragraph (1) and subsection (g).

"(d) INSPECTIONS.—The program shall provide for regular safety inspections of the construction worksite by the contractor, and for emergency and special inspections as may be appropriate.

"(e) REPORTING.—The program shall provide for the reporting of fatalities, serious injuries, and other injuries and illnesses, in accordance with the requirements of this Act and other applicable laws, standards, and regulations.

"(f) EMERGENCY EVACUATION PLANS.—The program shall provide for an emergency evacuation plan, including the location of first-aid facilities and the routes of exiting from areas during emergencies.

"(g) CONSTRUCTION WORKSITE INSTRUCTION.—The program shall provide for the instruction of employees in—

"(1) the recognition and avoidance of unsafe and unhealthy conditions;

"(2) the standards and regulations applicable to the work activities of each employee;

"(3) use of construction worksite equipment and required personal protective equipment; and

"(4) the handling and use of poisons, caustics, flammable liquids, gases, and other toxic or harmful substances.

"(h) MATERIAL SAFETY DATA SHEETS.—The program shall provide for—

"(1) obtaining material safety data sheets from manufacturers, distributors, and other suppliers of chemicals and other hazardous materials that the contractor shall use on the project; and

"(2) making available those sheets to their own employees and to employers of other employees who may be exposed.

"(i) COPIES.—The program shall provide for making available to all employees and employee representatives, prior to the commencement of construction work for that contractor, copies of the program.

"(j) MEETINGS.—The program shall provide for regular safety and health construction worksite meetings to be conducted with employees to review and update the program.

"(k) NOTICES.—The program shall provide for the placement of a notice, to be furnished by the Secretary, in a conspicuous place or places where notices to employees are customarily posted—

"(1) informing employees that a program exists on the construction worksite; and

"(2) identifying the name and location of the employee of the contractor who is responsible for general management of the program under subsection (b).

"(l) EVALUATIONS AND ANALYSES.—

"(1) IN GENERAL.—The program shall provide for a procedure under which an employee or employee representative may request the contractor to conduct an evaluation of a perceived hazardous condition or an analysis of a perceived harmful material or substance on a construction worksite.

"(2) CONSTRUCTION SAFETY PROFESSIONAL.—Hazardous condition evaluations and substance analyses shall be conducted for the contractor under paragraph (1) by an individual who, by possession of a recognized degree, certificate, or professional standing, and who by extensive knowledge, training, and experience, has successfully demonstrated the ability of the individual to solve or resolve problems relating to worker safety and health involving construction work.

"(3) WRITTEN REPORT.—A written report of such an evaluation and analysis shall be provided immediately to the construction safety specialist and the employee who requested the evaluation or analysis.

"(4) DENIAL.—If the contractor denies a request to conduct an evaluation or analysis under paragraph (1), the denial and the specific reasons for the denial shall be reduced to writing and provided immediately to the construction safety specialist, the employee who requested the evaluation or analysis, and any employee representative of the employee.".

SEC. 7. ONSITE CONSTRUCTION SAFETY AND HEALTH PLANS.

The Act (29 U.S.C. 651 et seq.) (as amended by section 6 of this Act) is further amended by adding at the end thereof the following new section:

"SEC. 36. ONSITE CONSTRUCTION SAFETY AND HEALTH PLANS.

"(a) REQUIREMENT.—Except as provided in section 34(d), a construction contractor shall develop and maintain an onsite project safety and health plan for each construction project that—

"(1) includes a construction process plan that meets the requirements of subsection (b);

"(2) includes a hazard analysis that meets the requirements of subsection (c); and

"(3) meets the other requirements prescribed in subsection (d).

"(b) CONSTRUCTION PROCESS PLAN.—The construction process plan required under subsection (a) shall—

"(1) describe the construction process to be used, including specific references to critical points and conditions in the process that require special attention;

"(2) identify the means that will be used to ensure the structural stability of all buildings, structures, and excavations during the process;

"(3) contain a list of all inspections and tests required (including a schedule for such inspections and

        1   tests) and the criteria established for continuation of
        2   construction based on the inspection and test results;
        3   and

        4   "(4) make appropriate references to the hazard
        5   analysis prepared in accordance with subsection (c).

        6   "(c) HAZARD ANALYSIS.—The hazard analysis re-
        7   quired under subsection (a) shall—

        8   "(1) identify any possibilities for major safety fail-
        9   ures of the project that could occur throughout the
       10   construction process; and

       11   "(2) provide instructions for the prevention of haz-
       12   ards throughout the construction process.

       13   "(d) OTHER REQUIREMENTS.—The plan required
       14   under subsection (a) shall—

       15   "(1) protect workers against the hazards antici-
       16   pated on the project and described in the hazard
       17   analysis;

       18   "(2) contain provisions for the construction work-
       19   site specific education and training of supervisory per-
       20   sonnel and employees in the recognition, avoidance,
       21   and prevention of unsafe and unhealthy conditions;

       22   "(3) establish construction worksite specific bench-
       23   marks and procedures for monitoring by the construc-
       24   tion safety specialist of compliance with such plan by
       25   all contractors on the project;

"(4) provide for the posting of a notice, to be furnished by the Secretary, in a conspicuous place or places where notices to employees are customarily posted, informing employees that such plan exists for the project, and identifying the project construction safety specialist, the means for contacting the specialist on the project, and the address and telephone number of the principal place of business of the specialist;

"(5) provide for maintaining on the construction worksite and making available to all employers and employees and their representatives, on request, a copy of such plan;

"(6) provide that a construction safety specialist shall, on behalf of the contractor, stop work at, or remove affected employees from, an area in which danger exists if the specialist believes that an imminent danger exists that cannot be eliminated immediately through actions not requiring the stoppage of such work or the removal of affected employees; and

"(7) comply with all other requirements of this Act.

"(e) APPROVAL.—The construction safety specialist shall evidence the approval of the plan by affixing the signature of the specialist and the certification number on a copy of the plan required under subsection (a)(1).

"(f) COPY.—The construction safety specialist shall provide to each employer on the project, prior to commencement of work by the contractor, a copy of the plan.

"(g) NOTIFICATION OF HAZARDS AND VIOLATIONS.—

"(1) IN GENERAL.—If the construction safety specialist determines that there exists—

"(A) a condition or work practice that violates any Federal, State, or local occupational safety or health law, standard, regulation, or order;

"(B) a failure to comply with the program established under this section or the safety and health program of a contractor; or

"(C) an occupational safety or health hazard, the specialist shall notify the responsible contractor or contractors and direct such contractor or contractors to correct the condition, work practice, or hazard.

"(2) WRITING.—Such notification shall be reduced to writing by the construction safety specialist as soon as practicable.

"(3) COPIES.—A copy of the notification shall be—

"(A) provided by the construction safety specialist to the responsible contractor or contractors; and

"(B) be made available, on request, to all affected employers, employees, and employee representatives on the project and to the Secretary.".

SEC. 8. CONSTRUCTION SAFETY SPECIALISTS.

The Act (29 U.S.C. 651 et seq.) (as amended by section 7 of this Act) is further amended by adding at the end thereof the following new section:

"SEC. 37. CONSTRUCTION SAFETY SPECIALISTS.

"(a) REQUIREMENT.—Except as provided in section 34(d), a construction contractor shall ensure that a construction safety specialist—

"(1) is hired and retained to perform the duties prescribed by this Act at a construction worksite; and

"(2) performs such duties.

"(b) OVERALL RESPONSIBILITY.—Notwithstanding subsection (a), the construction contractor shall have overall responsibility for the safety of a construction worksite with respect to construction safety, rules, and practices.

"(c) SUBCONTRACTORS AND SUPERVISORS.—The construction contractor shall—

"(1) ensure that each subcontractor and supervisor at a construction worksite is informed of the identity and duties of the construction safety specialist; and

"(2) require such subcontractor and supervisor to comply with the policies of the construction safety specialist regarding health and safety at the construction worksite.

"(d) COMPLIANCE WITH ONSITE CONSTRUCTION SAFETY AND HEALTH PLAN.—

"(1) IN GENERAL.—A construction safety specialist shall ensure compliance at the construction worksite with the onsite construction safety and health plan required under section 36 and with guidelines established by the Secretary and by the Advisory Committee on Construction Safety and Health.

"(2) SAFETY REPORT.—A construction safety specialist shall maintain a detailed safety report with respect to the construction worksite, including information concerning unsafe practices, actions, and conditions, and any accidents, injuries, or fatalities at such worksite.

"(3) HAZARDS.—If a construction safety specialist discovers a hazard at a construction worksite, the construction safety specialist shall—

"(A) order the construction contractor to eliminate such hazard; and

"(B) inform the construction contractor, the construction owner of the construction worksite, and the Secretary of any imminent dangers at the worksite.

"(e) QUALIFICATIONS.—To be considered a construction safety specialist under this Act, an individual shall—

"(1)(A)(i) have obtained written certification of completion of a minimum 40-hour course of study in construction health and safety, the curriculum of which shall be established jointly by the Secretary, the Advisory Committee on Construction Safety and Health, and the Office of Construction, Safety, Health, and Education established under section 34; and

"(ii) continue to receive certification for successful completion of a refresher course at least once each 3 years;

"(B) have completed an on-the-job training program conducted by a certified construction safety specialist, with such training conducted for no less than 18 months and covering each aspect of a project, from the initial stages through project completion; or

"(C) have met satisfactory equivalency education and experience criteria established by the Secretary;

"(2) be capable of identifying existing and predictable hazards or conditions at a construction worksite that are unsanitary, hazardous, or dangerous to construction workers.

"(f) RESPONSIBILITIES.—On behalf of the construction contractor, the construction safety specialist shall—

"(1) prior to the commencement of construction work on the project, prepare or approve an onsite project safety and health program applicable to the entire project in accordance with section 36, except that if it is not practicable to prepare or approve the entire program prior to the commencement of construction work—

"(A) that part of the program that is feasible to prepare or approve at such time shall be prepared or approved; and

"(B) the program as it relates to the specific major phases of construction work to be performed on the subject shall be prepared or approved prior to the commencement of each such phase;

"(2) prior to the commencement of construction work by each contractor on the project, review and approve the safety and health program of such contractor to ensure its consistency with the onsite project safety and health program and compliance with this Act;

"(3) monitor the implementation by contractors of their safety and health programs;

"(4) ensure compliance by contractors with reporting requirements established by this Act;

"(5) obtain, or require a contractor to obtain, the assistance of an individual or organization professionally qualified and duly licensed to perform architectural or engineering services if the construction safety specialist determines that—

"(A) such assistance is required for compliance with the onsite safety and health program, a safety and health program of the contractor, or this Act; or

"(B) a particular aspect of the work involves such safety hazards, is so highly technical, or requires such special expertise for safe construction that the contractor could not reasonably be expected to be aware of the risks; and

"(6) perform all other responsibilities assigned to a construction safety specialist under this Act.".

SEC. 9. CONSTRUCTION SAFETY AND HEALTH TRAINING ACADEMY.

The Act (29 U.S.C. 651 et seq.) (as amended by section 8 of this Act) is further amended by adding at the end thereof the following new section:

"SEC. 38. CONSTRUCTION SAFETY AND HEALTH TRAINING ACADEMY.

"(a) ESTABLISHMENT.—There is established in the Occupational Safety and Health Administration a Construction

Safety and Health Training Academy (hereinafter in this section referred to as the 'Academy'), to be headed by a Director appointed by the Secretary.

"(b) FUNCTIONS.—The Academy shall be responsible for—

"(1) the training of—

"(A) employees of the Office of Construction Safety, Health, and Education who conduct inspections of construction worksites; and

"(B) other persons as the Secretary, with the advice of the Advisory Committee on Construction Safety and Health, shall consider appropriate; and

"(2) the training and certification of construction safety specialists who have fulfilled the requirements of a standardized training course and testing program developed or approved by the Academy, with the advice of the Advisory Committee on Construction Safety and Health.

"(c) COOPERATIVE AGREEMENTS.—In carrying out this section, the Academy may enter into cooperative educational and training agreements with educational institutions, State governments, labor organizations, and construction industry employers.

"(d) EQUAL ACCESS.—The Secretary and the Academy shall ensure that employees, employee representatives, and employers have equal access to the services made available pursuant to this section.".

SEC. 10. NATIONAL INSTITUTE FOR OCCUPATIONAL SAFETY AND HEALTH.

Section 22 (29 U.S.C. 671) is amended—

(1) by striking out "in the Department of Health, Education and Welfare" each place it appears in subsections (a) and (b) and inserting in lieu thereof "as an agency of the Public Health Service"; and

(2) by striking out "Secretary of Health, Education and Welfare" each place it appears and inserting in lieu thereof "Secretary of Health and Human Services".

SEC. 11. PENALTIES.

Subsection (e) of section 17 (29 U.S.C. 666(e)) is amended to read as follows:

"(e)(1)(A) Except as provided in subparagraph (B), any employer who—

"(i) willfully violates any standard, rule, or order promulgated pursuant to section 6, or any regulation prescribed pursuant to this Act, and thereby causes serious injury to any employee; or "(ii) under circumstances evincing an extreme indifference to human life, recklessly engages in conduct that violates such standard, rule, order, or regulation and creates a grave risk of serious injury to any employee, and thereby causes serious injury to such employee,

shall, on conviction, be fined in accordance with title 18, United States Code, or imprisoned not more than 5 years, or both.

"(B) If a conviction under subparagraph (A) is for a violation committed after a first conviction of such employer, such employer shall be fined in accordance with title 18, United States Code, or imprisoned not more than 10 years, or both.

"(2)(A) Except as provided in subparagraph (B), any employer who—

"(i) willfully violates any standard, rule, or order promulgated pursuant to section 6, or any regulation prescribed pursuant to this Act, and thereby causes the death of any employee; or

"(ii) under circumstances evincing an extreme indifference to human life, recklessly engages in conduct violates such standard, rule, order, or regulation and that creates a grave risk of death to any employee, and thereby causes the death of such employee, shall, on conviction, be fined in accordance with title 18, United States Code, or imprisoned not more than 10 years, or both.

"(B) If a conviction under subparagraph (A) is for a violation committed after a first conviction of such employer, such employer shall be fined in accordance with title 18, United States Code, or imprisoned not more than 20 years, or both.

"(3) For purposes of this subsection, an employer acts recklessly with respect to a result or circumstance described by this subsection if the employer is aware of and consciously disregards a substantial and unjustifiable risk that such result will occur or that such circumstance exists. The risk must be of such a nature and degree that disregarding it constitutes a gross deviation from the standard of conduct that a reasonable person would observe in the situation.".

SEC. 12. ADVISORY COMMITTEE ON CONSTRUCTION SAFETY AND HEALTH.

Section 107(e) of the Contract Work Hours and Safety Standards Act (40 U.S.C. 333(e)) is amended by adding at the end thereof the following new paragraph:

"(4) The Advisory Committee shall have the authority to schedule meetings and set meeting agendas, establish subcommittees, conduct research and issue reports, retain experts and consultants, employ secretarial and clerical personnel, and purchase office equipment and research material as may be necessary to carry out its functions.".

SEC. 13. BUDGET CONTENTS.

In the preparation of the budget message required under section 1105 of title 31, United States Code, the President shall set forth as separate appropriation accounts—

    (1) amounts for appropriation for construction industry safety and health activities conducted pursuant to the Occupational Safety and Health Act of 1970 (29 U.S.C. 651 et seq.) and the Contract Work Hours and Safety Standards Act (40 U.S.C. 327 et seq.); and

    (2) amounts required for appropriation for nonconstruction occupational safety and health activities conducted pursuant to the Occupational Safety and Health Act of 1970 and the Contract Work Hours and Safety Standards Act.

SEC. 14. EFFECTIVE DATES.

(a) IN GENERAL.—Except as provided in subsection (b), this Act and the amendments made by this Act shall become effective on October 1, 1991.

(b) CONSTRUCTION SAFETY SPECIALISTS.—Section 37 of the Occupational Safety and Health Act of 1970 (as added by section 8 of this Act) shall become effective on October 1, 1993.

APPENDIX H

Proposed OSHA Rules:

**Generic Standard for Exposure Monitoring
[53 FR 37591, Sept. 27, 1988]**

**Medical Surveillance Programs for Employees
[53 FR 37595 Sept. 27, 1988]**

OSHA ADVANCE NOTICE OF PROPOSED RULEMAKING ON A GENERIC STANDARD FOR EXPOSURE MONITORING
[53 FR 37591, Sept. 27, 1988]

DEPARTMENT OF LABOR

Occupational Safety and Health Administration

29 CFR Part 1910

[Docket No. H-029]

Generic Standard for Exposure Monitoring

AGENCY: Occupational Safety and Health Administration (OSHA), Labor.

ACTION: Advance notice of proposed rulemaking.

SUMMARY: Notice is given that OSHA is undertaking, through rulemaking procedures under section 6(b) of the Occupational Safety and Health Act of 1970 (the Act), 29 U.S.C. 655(b), an evaluation of the feasibility and usefulness of adoption of a generic standard on exposure monitoring for employees exposed to toxic substances. A generic standard is one that addresses a health related issue rather than a substance. The Agency is interested in determining if generic exposure monitoring requirements could be used, for example, to simplify development of future rules that where necessary, would contain exposure monitoring provisions, or could be used to provide for exposure measurement of employees who are not now entitled to such monitoring. Though OSHA has adopted exposure limits for the several hundred substances listed in the Z-tables contained in 29 CFR 1910.1000, there is no provision requiring that exposure monitoring be performed on employees exposed in excess of those limits. Thus, adoption of exposure monitoring requirements applicable to § 1910.1000 may be warranted.

DATES: Comments and information must be received on or before December 27, 1988.

ADDRESSES: Written submissions in response to this notice should be submitted in quadruplicate to the Docket Officer, Docket No. H-029, Room N2625, Occupational Safety and Health Administration, U.S. Department of Labor, 200 Constitution Avenue NW., Washington, DC 20210, telephone 202-523-7894. All written submissions and documents mentioned in this notice will be available for inspection and copying in Room N2625 at the above address.

FOR FURTHER INFORMATION CONTACT: Mr. James Foster, Director, Office of Information and Consumer Affairs, Room N3647, Occupational Safety and Health Administration, 200 Constitution Avenue NW., Washington, DC 20210, telephone (202) 523-8148.

SUPPLEMENTARY INFORMATION: OSHA's 6(b) health standards (with the exception of the 13 work practice standards governing carcinogens i.e. 29 CFR 1910.1003–1910.1016) contain employee exposure monitoring provisions as stipulated by section 6(b)(7) of the Occupational Safety and Health Act of 1970, 29 U.S.C. 655 et seq. ("the Act".) These provisions, for the most part, address similar issues from standard to standard. These issues cover such topics as initial monitoring, frequency of monitoring, whether to use personal or area sampling, whether to use full shift or grab sampling, the notification of monitoring results to employees or employee representatives, the least acceptable accuracy of measurement, the provision and procedures for employees to observe monitoring, and the recordkeeping of employee exposure monitoring results.

The Agency is interested in determining if generic exposure monitoring requirements could be used to simplify the development of future rules that would necessarily contain exposure monitoring provisions or could be used where such requirements are not already in place. A generic standard would not be substance or industry specific but could apply to a broad number of chemicals and industries. By "where requirements are not already in place," the Agency is considering the application to some or all of the several hundred substances listed in the Z-tables contained in 29 CFR 1910.1000. Although OSHA has adopted exposure limits for these substances, there are currently no provisions in 29 CFR 1910.1000 requiring monitoring of employees' exposure. OSHA anticipates that the generic standard for exposure monitoring would establish broad performance criteria for an acceptable exposure monitoring program which could include such provisions which have been adopted in most 6(b) standards and which are not specific to the chemical being regulated. However, OSHA is willing to consider other suggested regulatory and nonregulatory options which would accomplish our objectives of simplifying rulemaking and offering increased protection through exposure monitoring programs.

As an adjunct to the development of criteria for workplace exposure monitoring OSHA has initiated an evaluation of the effectiveness of exposure monitoring requirements in existing OSHA standards, in order to determine what has worked and what has not worked in practice, and if these requirements could be improved.

Comments and Information Requested

In order to assist the Agency in gathering as much information as possible for use in determining the usefulness of a generic exposure monitoring standard, OSHA has prepared a list of questions soliciting comment on issues pertinent to this rulemaking. OSHA requests that interested persons provide as much detail as possible in answer to the questions. Please explain the reasons for your responses and discuss why a particular action is advisable. The Agency also requests that interested persons submit additional comments and information on other issues deemed relevant that are not addressed by the questions in this notice. The information submitted in response to this notice will aid the Agency in determining whether to proceed with development of a notice of proposed rulemaking. All comments will become part of the record of any resulting rulemaking and will be carefully considered in the development of any proposed regulation.

In accordance with the provisions of Executive Order No. 12291 and the Regulatory Flexibility Act (Pub. L. 96-354, 94 Stat. 1164, 5 U.S.C. 601 et seq., OSHA will determine whether this is a major action and if so, will prepare a Regulatory Impact Analysis. OSHA has prepared some questions to obtain information on technical feasibility, and economic and environmental impact of regulatory actions on affected industries in general and, on small businesses in particular. OSHA requests cost data related to the issues raised in the ANPR. To provide an accurate assessment and to best assist the Agency in estimating costs of regulatory alternatives, information should be as detailed as possible and should, if possible, discuss the effectiveness of various strategies in complying with a generic standard. (By cost data on strategies, OSHA is referring to the costs of any necessary monitoring, calibration, and analytical equipment incurred in complying with a generic standard. OSHA is also referring to additional costs incurred by such activities as taking samples, maintaining monitoring equipment, and keeping necessary records as required by a generic exposure monitoring standard.)

Issues

Data, views, and arguments are solicited on all of the issues described below as well as on other relevant issues.

1. Value of Exposure Monitoring

Exposure monitoring programs are used primarily to detect instances of employee overexposure and to indicate the effectiveness of an employer's exposure control program.

(a) Is exposure monitoring an effective means of determining overexposure and control method adequacy to enable the implementation of appropriate corrective actions to reduce exposure? If not, in what situations is it not?

(b) What exposure information is useful to determine that remedial action is necessary to reduce workplace heath risks?

(c) With respect to the objectives of exposure monitoring that were mentioned in the proceeding discussion, what data or information do you have that suggest that exposure monitoring programs meet these objectives? For example, are exposure monitoring programs effective tools for protection against overexposure or conditions that may lead to overexposure? For what specific objectives is exposure monitoring most effective? Least effective? Would it be appropriate for OSHA to adopt mandatory generic exposure monitoring provisions? Why? Why not?

(d) How should OSHA assess the cost and effectiveness of exposure monitoring as well as the balance between the two?

2. Criteria

Would it be possible or beneficial to develop standardized criteria to determine when and what workplace exposure monitoring is "necessary for the protection of employees" under section 6(b)(7) of the Act? Such criteria, for example, could be used to decide when mandatory exposure monitoring is appropriate, the frequency of monitoring, and the adequacy of certain monitoring methods. Should such criteria be developed through rulemaking or should they be developed as guidelines, perhaps by NIOSH or an expert advisory group? Do criteria or guidelines exist, perhaps in the public arena, which would be suitable for adoption by OSHA?

3. Effectiveness of Existing OSHA Requirements

(a) How effective have exposure monitoring requirements in OSHA standards been in ensuring compliance with the PELs? Is there information available about the effectiveness of specific standards? Have these requirements achieved the specific objectives which they were intended to achieve? What have been the costs?

(b) Could the exposure monitoring requirements be more effective? Could they be written so as to better complement and reinforce the other provisions of the standards? How should standards be written so as to optimize the value of exposure monitoring?

4. Scope and Application

OSHA is presently considering two options with respect to the applicability of a generic standard on exposure monitoring:

(a) OSHA could develop a standard that would be available for incorporation by reference into newly-developed 6(b) standards or revisions of existing standards (inluding revisions of the PELs found in the § 1910.1000 (Z tables.) Please comment on the appropriateness of this option, the impact that the implementation of this option would have on your industry, the technical and economic problems you foresee with implementation, and the benefits and advantages of implementing this option.

(b) Alternatively, adoption of generic exposure monitoring requirements only applicable to the substance in the Z tables found in § 1910.1000 is being considered. Monitoring could be required when the employer suspects that exposure levels could exceed a Z-table PEL or to ensure that levels do not exceed the PEL (by the use of initial monitoring.)

(1) Please comment on the appropriateness of this option, the impact that the implementation of this option would have on your industry, the problems you foresee with the implementation, and the benefits and advantages in implementing this option.

(2) If OSHA incorporates a generic exposure monitoring standard into § 1910.1000, are sampling and analytical methods available to the employer to determine compliance with the PELs found in the Z-tables? If not, which substances cannot be feasibly monitored by employers and why?

(3) Are there other reasons why some of the § 1910.1000 substances would be more appropriate for coverage by a generic exposure monitoring standard than others?

(c) Technological feasibility of exposure measurement has been an issue during some of OSHA's 6(b) rulemaking efforts (e.g., asbestos (51 FR 22612) and ethylene oxide (49 FR 25734). These issues in some cases have included the availability of monitoring equipment for measuring exposures at the proposed permissible exposure limit, action level, short term exposure limit, or excursion limit. Has there been new information on the technological feasibility of exposure measurement which should be considered that was not considered in the previous rulemakings for 6(b) standards? Do recent evaluations of exposure monitoring programs exist that result in a better understanding of the role of exposure monitoring and reveal a more efficient way of targeting technical resources. If so, could this information be used in determining the applicability of the generic rule?

(d) Please identify any special processes or situations where a generic exposure monitoring standard would not be applicable.

5. Initial Monitoring and the Discontinuation of Monitoring

OSHA's specific 6(b) standards require that each affected employer undertake a program of initial monitoring and measurement. Subsequent monitoring depends on the results of initial measurement and is triggered by the action level. (The action level is usually defined to be half the permissible exposure limit.) Initial monitoring results not exceeding the action level normally do not require further monitoring. Results exceeding the action level but not exceeding the PEL require some program of subsequent monitoring on a regular basis. Results exceeding the PEL require more frequent subsequent monitoring on a regular basis.

(a) Please comment on the appropriateness of adopting an initial monitoring requirement in a generic standard.

(b) Should some mechanism other than an action level of PEL be used to trigger subsequent periodic monitoring? If so, what would that mechanism be and why would it be more appropriate?

(c) Are there situations where employers could estimate exposures without having to sample (such as estimating the exposure by way of the volume or mass of the chemical and the size of the room). Please describe where exposure estimation would be appropriate in lieu of exposure monitoring. What are advantages and disadvantages of estimating exposures over actual sampling? In particular, information is sought on the relative costs of estimating and monitoring exposure, and on the confidence that can be placed on exposure level determinations made by means other than sampling and analysis.

(1) Please explain how in your situation the estimation of exposures would be advantageous to actual sampling. If cost savings is the only advantage, please describe in as much detail as possible the costs that would be saved.

(2) What kinds of exposure estimation procedures are known in your industry? Please describe them in detail. In addition, if any comparisons to actual sampling have been done, please describe these comparisons and their results.

(3) OSHA's ethylene oxide (29 CFR 1910.1047(a)(2)) and asbestos (29 CFR 1910.1001) and (29 CFR 1926.58) standards contain exemptions where objective data can be relied upon to show that the form of the substance or the conditions under which it will be used make it impossible for an exposure greater than the regulated level to occur. Please specify if such a provision would be appropriate for a generic exposure monitoring standard. How should the provision be stated? What substances, designs, or processes can be given as examples to support the provision?

(4) OSHA believes that actual sampling may be more accurate than exposure estimation in most instances. If you agree or disagree with this belief, please give reasons and whatever factual data exist to support your answer.

(d) In OSHA's standards for Hazard Communication (1910.1200) and Benzene (1910.1028), exemptions are granted for mixtures in which the hazardous substance is less than a certain percentage. Should a generic standard on exposure monitoring contain similar exemptions and if so, how should the determination of such percentages be made?

(e) When exposure levels exceed the action level in the specific 6(b) standards, the employer may only discontinue a periodic monitoring program when two consecutive measurements taken several days apart show exposures to be below the action level. The employer needs to perform additional monitoring if processes, controls, or personnel have changed or some other situation has occurred which causes the employer to suspect that increased exposure has occurred.

(1) Is it reasonable to adopt this same policy of discontinuing and reestablishing monitoring in a generic standard?

(2) If you feel that OSHA should adopt another policy of discontinuing and reestablishing monitoring, please explain why and what that policy should be.

(3) Several of the 6(b) standards have set 7 days as the established number of days between the two consecutive measurements required to document reduction in exposure level below the action level. Do you agree or disagree with 7 days between two consecutive measurements as an established number of days between measurements in a generic standard? Please give reasons for you answer. If you disagree with 7 days as the established number of days, what number of days do you recommend and why?

(4) The 6(b) standards require that the employer, in addition to performing scheduled periodic monitoring, monitor whenever spills, leask, ruptures, or other breakdowns occur. Is it reasonable for OSHA to make this same additional monitoring requirement in a generic standard? If it is not, please explain why not. What other monitoring procedure would be appropriate for the situations listed above?

6. Frequency of Monitoring

OSHA's specific standards require sampling semiannually to quarterly (depending whether the initial sample and subsequent samples exceed the action level or the permissible exposure limit), while other 6(b) standards require sampling quarterly to monthly. The monitoring frequencies for the 6(b) standards were established as a result of administrative decisions in response to professional judgments as stated in the 6(b) rulemaking records. Some of the considerations for setting monitoring frequencies have been the length of time to get samples to and back from the laboratory and to properly notify employees of the results and the steps to mitigate any overexposures.

(a) Please state whether or not you agree with a seminannual to quarterly monitoring frequency and explain why.

(b) If you believe that OSHA should adopt another monitoring frequency, what frequency do you recommend and why?

(c) It may be appropriate to allow the employer to determine the monitoring frequency for his/her particular situation. This may be especially true in a situation where the exposures are consistent and predictable such as in an automated process. The employer may determine that quarterly or semiannual monitoring is not necessary because it may be possible to predict with confidence that exposures over long periods will not exceed targeted control levels. In order to implement such a policy, OSHA would have to require some form of documentation of level of exposure. In what situations would a policy of employer-determined frequency monitoring be appropriate? What form should documentation take?

(d) Would it be appropriate to allow employers to sample exposures at some times and estimate exposures at other times. In what situations would estimation be appropriate? In what situations should sampling be used to verify exposure estimation?

7. Full Shift Personal Sampling or Area and Grab Sampling

In most instances, OSHA requires full shift personal sampling rather than area sampling for its specific standards. The Agency believes that full shift personal sampling gives a truer picture of an employee's exposure. If area sampling adequately represents actual employee exposure, its use might reduce the cost burden associated with sampling. OSHA does realize that if an employee is exposed to a substance for a short period of time, methods other than personal sampling may suffice. Also, for some substances the equipment to take continuous full shift samples may not be available. Experience suggests, however, that full shift sampling is the most appropriate means of adequately monitoring employees exposed to a substance on a continuous basis.

In answering the following questions please consider the analysis of samples as well as the sampling itself:

(a) Other than potential cost savings, what other advantages may be associated with area and grab sampling over personal full shift sampling in your industry?

(b) Are there situations where area and grab sampling would be more appropriate than personal and full shift sampling? Please describe those situations. Are there situations where it might be appropriate to alternate personal and area sampling?

(c) Please describe the best available methods for conducting area and grab sampling that are known in your industry, and in what situations these methods are used and why. Also, please discuss the costs associated with equipment and procedures used to perform area and grab sampling in your industry.

(d) If the cost burden is the essential disadvantage to performing personal and full shift sampling in lieu of other methods, please describe the cost burden known in your industry.

8. Appropriate Criteria for Representative Sampling

Some 6(b) standards such as asbestos and coke ovens define the criteria for representative monitoring. For example, asbestos requires samples representing full shift exposures for each employee in each job category in each work area for each shift (29 CFR 1910.1001 (d)(1)(ii).) Other 6(b) standards developed at an earlier time, such as 1,2-dibromo-3-chloropropane (29 CFR 1910.1044) (DBCP) do not. Ethylene oxide, in addition, contains a provision where if an employer can document equivalent exposure levels for different shifts for the same operation, the employer does not have to sample for different shifts (29 CFR 1910.1047(d)(1)(iii). OSHA is inviting comment on the appropriate criteria for representative sampling in a generic standard.

(a) In a generic exposure monitoring standard should OSHA use the ethylene oxide approach where employer does not have to sample several shifts for the same operation if proper documentation exists? Please provide reasons for your answer.

(b) Could you provide data for your industry demonstrating that exposures at different shifts for the same operation do not differ?

(c) Do you have data from your industry demonstrating that exposures do not differ from one operation to the next? Otherwise, it would seem reasonable to assume representative sampling means sampling each operation (i.e. industrial process which seemingly produces a different exposure scenario from another type of industrial process) where appropriate.

(d) The Agency requests any other information or comments concerning the criteria for representative sampling.

9. Accuracy of Sampling

For the most part, OSHA does not require specific methods for sampling and analysis in its substance specific standards. OSHA's specific standards do have requirements for the accuracy of sampling and analysis, but these requirements vary, depending upon the degree of achievable accuracy of the sampling device and analytical method at the required PEL and the action level.

(a) Should OSHA require a minimally acceptable accuracy in a generic standard? If not, why? If so, what minimally acceptable accuracy would you recommend that may appropriately apply to monitoring a large number of substance? What are the reasons for your recommendation?

(b) The National Institute for Occupational Safety and Health (NIOSH) and OSHA's Salt Lake Analytical Laboratory have recommended sampling and analytical methods for OSHA's own sampling activities in determining compliance with the Agency's standards. Should OSHA require that employers use these methods or methods documented as being equivalent with respect to accuracy in a generic standard? Please provide reasons supporting your response.

(c) Should OSHA require or recommend any particular methods of calibration in a generic standard? Which methods if any? Please explain why these methods should or should not be used.

10. Notification to Employees

In accordance with section 8(c) of the Act, OSHA's specific standards require that employees be notified in writing after the receipt of the results of any monitoring performed. In addition, the employer must include a description of the corrective action taken which will reduce the exposures below the PEL, whenever monitoring results indicate that the PEL has been exceeded.

(a) What requirements for employee notification would be appropriate in a generic standard?

(b) Should the language of the requirements be similar to the language in the 6(b) standards? In other words, what form should the notification take? Should employees be notified in writing as in the 6(b) standards? Should they be notified individually or should they be notified by the posting of results in an appropriate location such as a bulletin board accessible to affected employees?

11. Employee Observation of Monitoring

OSHA's specific 6(b) standards require the employer to provide affected employees or their designated representatives the opportunity to observe monitoring. Employees are to be provided with protective equipment and whatever other protection is required for the area in which sampling is occurring. They are entitled to receive explanations of the monitoring procedures and to receive copies of any results of measurements taken.

(a) Would it be appropriate for OSHA to make this same requirement in a generic standard?

(b) If OSHA's requirement for employee observation is inappropriate for a generic standard, what type of requirements would you recommend, and why would they be more appropriate?

12. Recordkeeping

The specific 6(b) standards require that employers establish and maintain records of personal or environmental monitoring results (29 CFR 1910.20).

(a) How should a generic standard address the establishment of exposure monitoring records?

13. General Considerations

An issue during this rulemaking is whether an approach other than adoption of a generic approach might be more appropriate to effectuate the purpose of simplifying future rulemakings and requiring monitoring of employees not now being monitored.

(a) If you feel that OSHA should not develop a generic standard, please explain why.

(b) What alternative to a generic standard would be appropriate, and what would be the advantages of the alternative over the generic approach?

Public Participation

Interested persons are invited to submit comments on these and other pertinent issues relating to a generic standard for exposure monitoring by December 27, 1988. Comments should be sent in quadruplicate to the Docket Officer, Room N2625, Occupational Safety and Health Administration, U.S. Department of Labor, 200 Constitution Avenue NW., Washington, DC 20210, telephone 202-523-7894. All written comments in response to this notice will be available for inspection and copying in the Docket Office at the above address between the hours of 8:15 a.m. and 4:15 p.m., Monday through Friday. The data received in response to this Advance Notice will be carefully reviewed and will be used by OSHA to determine whether it is necessary and appropriate to pursue further regulatory activity (and the nature of that activity) regarding a generic standard for exposure monitoring.

Authority and Signature

This document was prepared under the direction of John a. Pendergrass, Assistant Secretary of Labor for Occupational Safety and Health, 200 Constitution Avenue NW., Washington, DC 20210. It is issued pursuant to section 6(b) of the Occupational Safety and Health Act (84 State. 1593; 29 U.S.C. 655).

Signed at Washington, DC, this 20th day of September 1988.

John A. Pendergrass,
Assistant Secretary of Labor.
[FR Doc. 88-22017 Filed 9-26-88; 8:45 am]
BILLING CODE 4510-26-M

OSHA ADVANCE NOTICE OF PROPOSED RULEMAKING ON MEDICAL SURVEILLANCE PROGRAMS
[53 FR 37595, Sept. 27, 1988]

29 CFR Part 1910

[Docket No. H-031]

Medical Surveillance Programs for Employees

AGENCY: Occupational Safety and Health Administration (OSHA), Labor.

ACTION: Advance notice of proposed rulemaking.

SUMMARY: Notice is given that OSHA is undertaking, through rulemaking procedures under section 6(b) of the Occupational Safety and Health Act of 1970 (the Act), 29 U.S.C. 655(b), and evaluation of the feasibility and usefulness of adoption of a generic standard on medical surveillance programs for employees exposed to toxic substances or hazardous physical agents. A generic standard is one that addresses a health related issue rather than a substance. The Agency is interested in determining if generic medical surveillance requirements could be used, for example, to simplify development of future rules that where necessary, would contain medical surveillance provisions, or could be used to provide medical protection to exposed employees who are not now entitled to medical surveillance. Though OSHA has adopted exposure limits for the several hundred substances listed in the Z-tables contained in 29 CFR 1910.1000, there is no provision requiring that medical surveillance be made available to employees exposed in excess of those limits. Thus, adoption of medical surveillance requirements applicable to § 1910.1000 may be warranted.

DATE: Comments in response to this Advance Notice should be submitted by December 27, 1988.

ADDRESS: Comments should be submitted to the Docket Officer, Occupational Safety and Health Administration, Docket No. H-031, Room N-2439, U.S. Department of Labor, 200 Constitution Avenue, NW., Washington, DC. 20210. Telephone 202-523-7894.

FOR FURTHER INFORMATION CONTACT: Mr. James F. Foster, Director, Office of Information and Consumer Affairs, OSHA, U.S. Department of Labor, Room N-3647, 200 Constitution Avenue, NW., Washington, DC. 20210, Telephone: (202) 523-8148.

SUPPLEMENTARY INFORMATION:

1. Background.

The concept of generic standards is not novel. OSHA has promulgated generic standards for Hazard Communication (29 CFR 1910.1200), and Employee Access to Exposure and Medical Records (29 CFR 1910.20), and has proposed a generic rule for Employee Exposure to Toxic Substances in Laboratories (52 FR 1212). Additionally, a draft proposed revision of the existing Respiratory Protection Standards (29 CFR 1910.134) is being reviewed. The above standards activities establish precedent for exploring the possibility of adopting generic standards.

Section 6(b)(7) of the Occupational Safety and Health Act of 1970, 29 U.S.C. 655 et. seq. ("the Act") requires the Agency to adopt, where appropriate in new standards, provisions dealing with medical surveillance. Existence of a generic standard on medical surveillance would satisfy the Act's mandate with respect to adoption of such requirements in new standards and free OSHA to concentrate on more substance specific issues in future rulemakings.

OSHA's existing 6(b) standards contain medical surveillance provisions as stipulated by the Act. Experience suggests that certain requirements, described later, have potential for generic treatment as evidenced by their consistent treatment during previous rulemakings.

The specific content of the medical examinations, frequency of testing, and other provisions triggered by the results of the medical examination, however, vary for each 6(b) standard depending on the health effect associated with exposure to the specific substance. It may be necessary, therefore, to include specific recommended medical tests and procedures in an appendix to the generic standard that would be identified as being appropriate for groups of substances with similar chemical composition, or for substances leading to similar health effects.

As an adjunct to the development of criteria for workplace medical testing, OSHA has initiated an evaluation of the effectiveness of medical monitoring and surveillance requirements in existing OSHA standards, in order to determine what has worked and what has not worked in practice, and if these requirements could be improved.

2. Key Issues to be Addressed

(a) The Value of Workplace Medical Testing

The success of workplace medical surveillance and monitoring depends on the degree to which these activities contribute in practice to the detection of occupational disease or existing medical conditions that would progress or lead to disability without appropriate medical intervention and treatment. The value of medical surveillance in the prevention of occupational disease can logically be determined by reference to the specific surveillance strategy, which may be initial screening, detection of adverse effects, predicting future health effects, collecting data for health research, or other aims. OSHA intends to examine workplace medical testing carefully, including its experience with past standards, to determine its value for each of the specific objectives.

(b) Criteria

OSHA seeks to determine if criteria could be developed to decide when medical surveillance is appropriate, the frequency of examinations and the suitability of particular tests.

(c) Scope

OSHA must determine whether this generic standard will establish mandatory medical surveillance provisions applicable only to the substances with PELs found in section 1910.1000, to all toxic substance found in the workplace, or whether it will only be available for incorporation by reference in existing or future rules.

If it is determined that this standard should only apply to the substances in the Z-tables, OSHA believes that significant benefit may still be derived from a generic rule that provides medical protection to employees who are not being medically monitored even though their exposures may be above the permissible exposure limits ("PELs") in § 1910.1000. Each PEL was established based on scientific data demonstrating the occurrence of adverse health effects. Thus, requiring medical surveillance for overexposed employees may be warranted. In addition, OSHA is developing procedures to modify the Z-table PELs of 29 CFR 1910.1000 in an expeditious manner in response to current scientific data. However, section 6(b)(7) of the Act requires incorporation of appropriate provisions for labels and other forms of warning, personal protective equipment, medical surveillance, and exposure monitoring for each substance undergoing 6(b) rulemaking. Having generic standards for each of these topics as requirements in addition to the PELs of section 1910.1000 would satisfy the Act's requirement to address these topics and permit narrower specific rulemakings dealing with PELs. A generic standard dealing with warning labels (Hazard Communication) is already in place and generic standards for personal protective equipment and exposure monitoring are currently being considered. The addition of a generic standard on medical surveillance may provide an additional mechanism to accomplish the goal of more timely rulemakings for revision of the PELs in section 1910.1000

(d) Content of the Standard

Experience suggests that certain requirements have greatest potential for generic treatment as evidenced by their consistent treatment during previous rulemakings. OSHA 6(b) standards typically require that:

(1) Employees exposed above the action level [e.g., usually one-half the PEL] are to be included in the medical surveillance program;

(2) Physicians must perform or supervise exams;

(3) Employers must pay for exams;

(4) The employer must provide certain information about his employees and their jobs to the examining physician; and

(5) The examining physician must provide certain information to the employer with respect to the state of the employee's health;

Specification for routine work histories, routine pre-employment and periodic exams, and specific testing requirements for x-ray and pulmonary function tests have been treated similarly in past standards.

It may not be feasible in a generic medical surveillance standard to adopt more extensive or more specific mandatory provisions beyond those items listed above, for broad application. Additional data must be obtained to further address this question.

(e) Categorization

Since there are over 400 substances in tables Z-1, Z-2, and Z-3, OSHA will have to determine whether it is possible to group the substances by some common feature. One common feature may be by chemical type. That is, if we categorize the substances as acids, alkalies, gases, dusts, metals and metalloids, plastics and solvents, and determine that the major health effects within each group of substances are similar, it may be possible to assign appropriate medical surveillance for each chemical group. To illustrate, exposure to phosphoric acid may cause skin burns, eye irritation and eye burns, nose and throat irritation or skin irritation. Likewise sulfuric acid may cause eye, nose, and throat irritation, eye and skin burns. Additionally, prolonged exposure to these acids may cause dental erosion and inflammation of the bronchial tubes. As shown above, these two acids have similar major health effects. Medical surveillance for both may include examination of the respiratory system, and examination of the nose, throat, teeth and skin.

Another possibility may be grouping by health effect. If we find that many substances have similar major health effects, it may be possible to prescribe specific medical surveillance for those groups of health effects. For example, for the substances that produce eye, nose, throat, and skin irritation (acetaldehyde, acetic acid, ammonia, antimony, benzyl chloride, magnesium, phosphoric acid, sulfuric acid, xylene (etc)), medical surveillance may include an examination of the respiratory system, eyes, nose, throat, and skin.

OSHA is interested in receiving data and comment on the concept of categorizing and generically apply medical requirements to substances by health effects or chemical type. Specific data is solicited later in this notice.

(f) Feasibility

During this rulemaking, OSHA will develop an analysis of the feasibility of any proposed generic medical surveillance standard, as required by the Act. In addition, pursuant to Executive Order 12291, OSHA will determine whether this is a major action and if so, will prepare a Regulatory Impact Analysis. Under the Regulatory Flexibility Act (5 U.S.C. 601 *et seq.*), if a significant impact on small entities is anticipated, a Regulatory Flexibility Analysis will also be performed.

As noted below, OSHA solicits all available information on current medical surveillance programs and costs of compliance. OSHA requests that industry-wide feasibility studies and the collection of data relevant to assisting the Agency in complying with E.O. 12291, the Regulatory Flexibility Act, and the OSH Act be initiated by interested persons as soon as possible.

Requested Public Submissions

Public comment on the discussions in this Advance Notice and other relevant issues is requested for the purpose of assisting OSHA in its evaluation of the appropriateness and feasibility of developing a generic standard on medical surveillance programs for employees.

Comment is requested, as set forth below, on issues relating to: (1) The value of workplace medical monitoring and surveillance; (2) Criteria that OSHA may use to determine when and what medical intervention is appropriate; (3) The effectiveness of medical surveillance and monitoring requirements in existing OSHA standards; (4) The scope and application of the standard; (5) Categorization and generic application by health effects or by chemical type; (6) Economic feasibility; and (7) Provisions which should be considered for inclusion in a generic medical surveillance program. OSHA also requests that interested persons submit comments and information on other issues deemed relevant that are not specifically addressed by the questions below. The data submitted in response to this notice will aid the Agency in determining whether to proceed with development of a notice of proposed rulemaking.

1. Value of Medical Surveillance and Monitoring

Medical surveillance programs can be used to achieve many objectives including (1) the prevention, detection and treatment of disease, (2) an indication of the effectiveness of an employer's hazard control program, and, (3) a measure of the effectiveness of OSHA's PELs.

(a) In what situations are medical surveillance and monitoring effective means of detecting occupational illnesses to enable the implementation of appropriate medical treatment designed to arrest progression of the disease? How should "effectiveness" be defined with regard to medical testing? Is information available on the health benefits of workplace medical testing?

(b) What information is useful to employers to determine that remedial action is necessary to reduce workplace health risks? What medical information is unnecessary? Are there certain remedial actions which should or should not be triggered by medical information?

(c) Several objectives of medical surveillance were mentioned in the preceding discussion. What data or information do you have that suggest that medical surveillance programs meet these objectives? For example, are medical surveillance programs effective tools for the protection against disease or conditions that may lead to disease? What objectives should OSHA consider incorporating into mandatory standards? For what specific objectives is medical surveillance, monitoring or screening most effective? Least effective? Would it be appropriate for OSHA to adopt mandatory generic medical surveillance provisions? Why? Why not?

(d) How should OSHA assess the costs and effectiveness of medical surveillance and monitoring as well as the balance between the two?

2. Criteria

(a) Would it be possible or beneficial to develop standardized criteria to determine when and what workplace medical testing is "appropriate" under section 6(b)(7) of the Act? Such criteria, for example, could be used to decide when mandatory medical surveillance is appropriate, the frequency of medical examinations, the suitability of particular tests, and other questions common to rulemaking. Should such criteria be developed through rulemaking or should they be developed as guidelines, perhaps by NIOSH/CDC or an expert advisory group? Do criteria or guidelines exist, perhaps in the public health arena, which would be suitable for adoption of OSHA? How should the risks inherent in the medical tests and procedures be evaluated?

(b) Which medical testing requirements in existing OSHA standards would be appropriate for the purpose of a generic standard? Why? What information is available to show that an existing substance specific medical provision would have value and be effective if adopted as a provision in the generic standard? Could the medical surveillance and monitoring requirements as set forth in existing OSHA's 6(b) standards be modified so that they would be more effective for the purposes of a generic rule?

3. Effectiveness of Existing OSHA Requirements

(a) How effective have medical testing requirements in OSHA standards been in identifying and preventing occupational disease? Is there information available about the effectiveness of specific standards? Have these requirements achieved the specific objectives which they were intended to achieve? What have been the costs?

(b) Could the medical surveillance and monitoring requirements be more effective? Could they be written so as to better complement and reinforce the other provisions of the standards? How should standards be written so as to optimize the preventive value of medical testing?

4. Scope and Application

(a) Are there groups of employees in industry that should be, but currently are not undergoing medical surveillance monitoring that is related to their activities in the workplace?

(i) How would medical surveillance benefit these employees?

(ii) What is the basis for defining those employees needing medical surveillance and those who do not, and what factors (i.e. nature of the hazard, nature of the operation, control efficacy, etc.) enter into this determination?

(iii) Do recent evaluations of general or disease-specific periodic health examinations exist that result in a better understanding of the role of medical intervention and reveal a more efficient way of targeting of medical resources? If so, how could this information be used in determining the applicability of the generic rule?

(b) Should generic medical surveillance provisions be considered only for employees exposed to substances for which OSHA has adopted PELs? If so, should medical surveillance be considered only for employees for which exposure is shown to be above either an "action level" or above the PEL? What other mechanisms or levels could be established to trigger generic medical surveillance requirements?

(c) Some of OSHA's standards (e.g., benzene and ethylene oxide) do not require medical surveillance unless workers are exposed or expected to be exposed above either the action level or PEL for a specific number of days per year. For example, the benzene standard requires that "The employer shall make available a medical surveillance program for employers who are or may be exposed to benzene at or above the action level 30 or more days per year; [and] for employees who are or may be exposed to benzene at or above the PELs 10 days or more per year * * *." These allowances are provided for in recognition of the need for a practical cut-off for who is to be included in the medical surveillance program. Should similar provisions be incorporated into the generic standard and why? If so, what should they be?

(d) If medical surveillance should be provided to employees exposed to substances for which OSHA has no PEL. what mechanism could be set forth other than exposure above a specific level such as a PEL, to trigger implementation of the medical program?

(e) If a generic standard is adopted,

should it only be available for reference in future standards before being obligatory? Why or why not? If adopted, how should the generic standard relate to existing 6(b) standards?

(f) If a generic standard is adopted should it be incorporated as a paragraph in § 1910.1000 and be mandatory where overexposure to Z-table substances occurs? Why or why not?

(g) To what extent have employers voluntarily implemented occupationally related medical surveillance programs? Please provide in detail a description of what those programs consist of and the basis for their implementation.

(h) For what kinds of workplaces would a generic medical surveillance standard be appropriate and which employees should be covered in those workplaces?

5. Categorization

(a) Are there similar biological outcomes produced by large groups of substances that can be detected by specific procedures such as x-rays, pulmonary function tests, blood count, urinalysis, etc?

(b) What groups of substances cause these similar biological outcomes?

(c) Are sufficient medical data available to demonstrate that a number of substances cause similar biological changes over similar intervals of time so that generic periodicity of medical testing can be established?

(d) Are there groups of substances for which medical tests can be justified on a periodic basis as being necessary for early detection of adverse effects that may cause to progress or may be reversed upon removal from exposure?

(e) Is there a consensus in the medical community with respect to the utility of specific medical procedures in occupational medical surveillance programs?

(f) What criteria should OSHA use in determining the appropriateness of requiring specific medical tests for certain hazards and potential hazards?

(g) How can a generic standard for medical surveillance be designed so that advances in medical surveillance procedures and technology will not render OSHA's generic provisions obsolete?

6. Economic Feasibility

In order to perform an economic feasibility analysis, it is helpful to have a financial and economic profile of the industries that may be required to implement medical surveillance programs. Affected industries may include all workplaces using toxic substances, or may only include workplaces using a substance regulated by OSHA in 1910.1000. The following information is requested to aid in preparation of that profile.

(a) What are the number of employees that could conceivably be required to be provided with medical surveillance who are not now undergoing monitoring?

(b) Though OSHA has not proposed adoption of specific components to be included in a generic medical surveillance program. What costs in your industry and in your workplace can be estimated that may be incurred in conducting a typical work-related medical surveillance program? Give the costs according to the following categories: (1) The medical examination (list the components such as history, physical, and tests); (2) lost work time (include average time lost per worker); (3) transportation, and (4) recordkeeping. Were these costs based upon your company's current medical surveillance program? If so, how many employees are included in the program and what is the size of your firm?

(c) What will be the financial impact on firms/industries if OSHA required a periodic medical exam for all workers exposed above any of the PELs listed on the Z Tables? What if the exams were required for workers exposed above one-half of any of the PELs? Be as specific as possible.

7. Provisions of the Standard

(a) What mandatory provisions could be adopted that would be common to and appropriate for all occupational medical surveillance programs?

(b) Provisions typically found in 6(b) standards require that: Employees exposed above a certain level are to be provided medical surveillance, physicians must perform or supervise medical exams, employers must pay for exams and provide certain information about his employees and their jobs to the examining physician, physicians must provide information to the employer with respect to the state of an employee's health, and provision is made for routine work histories, and routine pre-employment and periodic exams. (i) Which of the generally applicable medical surveillance provisions described above should be mandated by a generic standard and why? (ii) Which should not and why? (iii) What other general provisions exist that the Agency should consider for adoption in a generic medical surveillance standard? (iv) With respect to conducting medical exams, should health professionals other than physicians, such as occupational health nurses, be permitted to supervise or conduct such exams?

(c) For each of OSHA's existing 6(b) standards, the specific content of the medical examinations (e.g., specific tests and procedures), frequency of testing, and other provisions triggered by the results of medical examination vary with each standard depending on the health effect associated with exposure to the specific substance. (i) How can a generic rule be designed to set-forth and mandate inclusion of specific medical tests and procedures as being appropriate in individual workplaces based on the chemical composition or health effects of the substances in use? (ii) What specific tests and procedures can be equated with what common chemical compositions or health effects for inclusion in medical surveillance programs? (iii) What minimum frequency of testing is appropriate for the recommended medical procedure?

(d) If a generic medical surveillance standard is adopted, should it only mandate implementation of general administrative medical provisions while recommending specific tests and examinations in an appendix that would be required to be provided if determined relevant by the examining physician?

Public Participation

Interested persons are invited to submit comment on these and other pertinent issues relating to generic medical surveillance programs for employees by December 27, 1988. Comments should be sent in quadruplicate to the Docket Officer, at the address noted above where they will be available for inspection and copying from 8:15 a.m. to 4:45 p.m., Monday through Friday. The data received in response to this Advance Notice will be carefully reviewed and will be used by OSHA to determine whether it is necessary and appropriate to pursue further regulatory activity regarding this generic standard.

Authority and Signature

This Advance Notice of Proposed Rulemaking was prepared under the direction of John A. Pendergrass, Assistant Secretary of Labor for Occupational Safety and Health, 200 Constitution Avenue NW., Washington, DC 20210. It is issued pursuant to section 6(b) of the Occupational Safety and Health Act of 1970 (84 Stat. 1593; 29 U.S.C. 655).

Signed at Washington, DC, this 20th day of September 1988.

John A. Pendergrass,
Assistant Secretary.
[FR Doc. 88–22016 Filed 9–26–88; 8:45 am]
BILLING CODE 4510-26-M

APPENDIX I

Bureau of Labor Statistics
Guide to Recordkeeping Requirements for
Occupational Injuries and Illnesses, June 1986

Published by The Bureau of National Affairs, Inc.

A Brief Guide to Recordkeeping Requirements for Occupational Injuries and Illnesses

U.S. Department of Labor
Bureau of Labor Statistics
June 1986

The Occupational Safety and Health Act
of 1970 and 29 CFR 1904

O.M.B. No. 1220-0029
Effective April 1986

ATTENTION: OSHA RECORDKEEPER

IMPORTANT: DO NOT DISCARD. This booklet contains guidelines for keeping the occupational injury and illness records necessary to fulfill your recordkeeping obligation under the Occupational Safety and Health Act of 1970 (29 USC 651) and 29 CFR Part 1904, or equivalent State law.

Published by The Bureau of National Affairs, Inc.

U.S. Department of Labor
William E. Brock, Secretary

Bureau of Labor Statistics
Janet L. Norwood, Commissioner
June 1986

Regional Offices of the Bureau of Labor Statistics

Region I: Boston
John F. Kennedy Federal Bldg.
Government Center, Suite 1603
Boston, Mass. 02203
Phone: (617) 565-2302
- Connecticut
- Maine
- Massachusetts
- New Hampshire
- Rhode Island
- Vermont

Region II: New York
1515 Broadway, Suite 3400
New York, N.Y. 10036
Phone: (212) 944-3114
- New Jersey
- New York
- Puerto Rico
- Virgin Islands

Region III: Philadelphia
3535 Market Street
Post Office Box 13309
Philadelphia, Penn. 19101
Phone: (215) 596-1162
- Delaware
- District of Columbia
- Maryland
- Pennsylvania
- Virginia
- West Virginia

Region IV: Atlanta
1371 Peachtree St., N.E., Suite 540
Atlanta, Ga. 30367
Phone: (404) 347-3660
- Alabama
- Florida
- Georgia
- Kentucky
- Mississippi
- North Carolina
- South Carolina
- Tennessee

Region V: Chicago
Federal Office Building, 9th Floor
230 S. Dearborn Street
Chicago, Ill. 60604
Phone: (312) 353-6911
- Illinois
- Indiana
- Michigan
- Minnesota
- Ohio
- Wisconsin

Region VI: Dallas
Federal Building
525 Griffin St., Room 221
Dallas, Tex. 75202
Phone: (214) 767-6956
- Arkansas
- Louisiana
- New Mexico
- Oklahoma
- Texas

Regions VII and VIII:
Kansas City and Denver
911 Walnut Street
Kansas City, Mo. 64106
Phone: (816) 374-2830
- Colorado
- Iowa
- Kansas
- Missouri
- Montana
- Nebraska
- North Dakota
- South Dakota
- Utah
- Wyoming

Regions IX and X:
San Francisco and Seattle
71 Stevenson Street, Box 3766
San Francisco, Calif. 94119
Phone: (415) 995-5618
- Alaska
- American Samoa
- Arizona
- California
- Guam
- Hawaii
- Idaho
- Nevada
- Oregon
- Washington

Published by The Bureau of National Affairs, Inc.

Preface

The information in this pamphlet explains the requirements of the Occupational Safety and Health Act of 1970 and Title 29 of the *Code of Federal Regulations*, Part 1904 (29 CFR Part 1904) for recording and reporting occupational injuries and illnesses. The Occupational Safety and Health Act of 1970 and 29 CFR Part 1904 require employers to prepare and maintain records of occupational injuries and illnesses. The act made the Secretary of Labor responsible for the collection, compilation, and analysis of statistics of work-related injuries and illnesses. The Bureau of Labor Statistics (BLS) administers this recordkeeping and reporting system. In most States, a State agency cooperates with BLS in administering these programs.

Records of injuries and illnesses are necessary for carrying out the purposes of the act. They provide a basis for a statistical program which produces injury and illness data which are used by OSHA in measuring and directing the agency's efforts. The records are also helpful to employers and employees in identifying many of the factors which cause injuries or illnesses in the workplace. In addition, OSHA records are designed to assist safety and health compliance officers in making OSHA inspections.

This pamphlet summarizes the OSHA recordkeeping requirements of 29 CFR Part 1904, and provides basic instructions and guidelines to assist employers in fulfilling their recordkeeping and reporting obligations. Many specific standards and regulations of the Occupational Safety and Health Administration (OSHA) have additional requirements for the maintenance and retention of records of medical surveillance, exposure monitoring, inspections, accidents and other activities and incidents relevant to occupational safety and health, and for the reporting of certain information to employees and to OSHA. These additional requirements are not covered in this pamphlet. For information on these requirements, employers should refer directly to the OSHA standards or regulations or contact their OSHA Area Office.

Further information on the requirements outlined in this pamphlet is available in the free detailed report, *Recordkeeping Guidelines for Occupational Injuries and Illnesses*, which may be obtained by using the order form on page 18. Assistance can also be obtained by contacting the participating State agency or the BLS regional office for your area. The BLS regional offices are listed on the inside front cover. State agencies are listed at the end of this publication.

The following government agencies are involved in OSHA recordkeeping:

A. *The Occupational Safety and Health Administration, U.S. Department of Labor.* The Occupational Safety and Health Administration is responsible for developing, implementing, and enforcing safety and health standards and regulations. OSHA works with employers and employees to foster effective safety and health programs which reduce workplace hazards.

B. *Bureau of Labor Statistics, U.S. Department of Labor.* The Bureau of Labor Statistics is responsible for administering and maintaining the OSHA recordkeeping system, and for collecting, compiling, and analyzing work injury and illness statistics.

C. *State Agencies.* Many States cooperate with BLS in administering the OSHA recordkeeping and reporting programs. Some States have their own safety and health laws which may impose additional obligations. Employers should consult their State safety and health laws concerning these requirements.

These guidelines were prepared in the BLS Office of Occupational Safety and Health Statistics, by Stephen Newell, under the general direction of William M. Eisenberg, Associate Commissioner.

Contents

Chapters:	Page
I. Employers subject to the recordkeeping requirements of the Occupational Safety and Health Act of 1970	1
II. OSHA recordkeeping forms	3
III. Location, retention, and maintenance of records	4
IV. Deciding whether a case should be recorded and how to classify it	6
V. Categories for evaluating the extent of recordable cases	12
VI. Employer obligations for reporting occupational injuries and illnesses	13
VII. Access to OSHA records and penalties for failure to comply with recordkeeping obligations	14
Glossary of terms	15
Order form	18
List of participating State agencies	18
BLS regional offices	inside front cover

Published by The Bureau of National Affairs, Inc.

Chapter I. Employers Subject to the Recordkeeping Requirements of the Occupational Safety and Health Act of 1970

The recordkeeping requirements of the Occupational Safety and Health Act of 1970 apply to private sector employers in all States, the District of Columbia, Puerto Rico, the Virgin Islands, American Samoa, Guam, and the Trust Territories of the Pacific Islands.

Employers who must keep OSHA records

Employers with 11 or more employees (at any one time in the previous calendar year) in the following industries must keep OSHA records. The industries are identified by name and by the appropriate Standard Industrial Classification (SIC) code:

Agriculture, forestry, and fishing (SIC's 01-02 and 07-09)
Oil and gas extraction (SIC 13 and 1477)
Construction (SIC's 15-17)
Manufacturing (SIC's 20-39)
Transportation and public utilities (SIC's 41-42 and 44-49)
Wholesale trade (SIC's 50-51)
Building materials and garden supplies (SIC 52)
General merchandise and food stores (SIC's 53 and 54)
Hotels and other lodging places (SIC 70)
Repair services (SIC's 75 and 76)
Amusement and recreation services (SIC 79), and
Health services (SIC 80).

If employers in any of the industries listed above have more than one establishment with combined employment of 11 or more employees, records must be kept for *each* individual establishment.

3. **Employers who infrequently must keep OSHA records**

Employers in the industries listed below are normally exempt from OSHA recordkeeping. However, each year a small rotating sample of these employers is required to keep records and participate in a mandatory statistical survey of occupational injuries and illnesses. Their participation is necessary to produce national estimates of occupational injuries and illnesses for *all* employers (both exempt and nonexempt) in the private sector. If an employer who is regularly exempt is selected to maintain records and participate in the Annual Survey of Occupational Injuries and Illnesses, he or she will be notified in advance and supplied with the necessary forms and instructions. Employers who normally do not have to keep OSHA records include:

1. All employers with no more than 10 full- or part-time employees *at any one time* in the previous calendar year.

2. Employers in the following retail trade, finance, insurance and real estate, and services industries (identified by SIC codes):

- Automotive dealers and gasoline service stations (SIC 55)
- Apparel and accessory stores (SIC 56)
- Furniture, home furnishings, and equipment stores (SIC 57)
- Eating and drinking places (SIC 58)
- Miscellaneous retail (SIC 59)
- Banking (SIC 60)
- Credit agencies other than banks (SIC 61)
- Security, commodity brokers, and services (SIC 62)
- Insurance (SIC 63)
- Insurance agents, brokers, and services (SIC 64)
- Real estate (SIC 65)
- Combined real estate, insurance, etc. (SIC 66)
- Holding and other investment offices (SIC 67)
- Personal services (SIC 72)
- Business services (SIC 73)
- Motion pictures (SIC 78)
- Legal services (SIC 81)
- Educational services (SIC 82)
- Social services (SIC 83)
- Museums, botanical, zoological gardens (SIC 84)
- Membership organizations (SIC 86)
- Private households (SIC 88), and
- Miscellaneous services (SIC 89).

Even though recordkeeping requirements are reduced for employers in these industries, they, like nonexempt employers, must comply with OSHA standards, display the OSHA poster, and report to OSHA within 48 hours any accident which results in one or more fatalities or the hospitalization of five of more employees. Also, some State safety and health laws may require regularly exempt employers to keep injury and illness records, and some States have more stringent catastrophic reporting requirements.

C. Employers and individuals who never keep OSHA records

The following employers and individuals do not have to keep OSHA injury and illness records:

- *Self-employed individuals;*
- *Partners with no employees;*
- *Employers of domestics* in the employers' private residence for the purposes of housekeeping or child care, or both; and
- *Employers engaged in religious activities* concerning the conduct of religious services or rites. Employees engaged in such activities include clergy, choir members, organists and other musicians, ushers, and the like. However, records of injuries and illnesses occurring to employees while performing secular activities must be kept. Recordkeeping is also required for employees of private hospitals and certain commercial establishments owned or operated by religious organizations.

State and local government agencies are usually exempt from OSHA recordkeeping. However, in certain States, agencies of State and local governments are required to keep injury and illness records in accordance with State regulations.

D. Employers subject to other Federal safety and health regulations

Employers subject to injury and illness recordkeeping requirements of other Federal safety and health regulations are not exempt from OSHA recordkeeping. However, records used to comply with other Federal recordkeeping obligations may also be used to satisfy the OSHA recordkeeping requirements. The forms used must be equivalent to the log and summary (OSHA No. 200) and the supplementary record (OSHA No. 101).

Chapter II. OSHA Recordkeeping Forms

Only two forms are used for OSHA recordkeeping. One form, the OSHA No. 200, serves as both the Log of Occupational Injuries and Illnesses, on which the occurrence and extent of cases are recorded during the year; and as the Summary of Occupational Injuries and Illnesses, which is used to summarize the log at the end of the year to satisfy employer posting obligations. The other form, the Supplementary Record of Occupational Injuries and Illnesses, OSHA No. 101, provides additional information on each of the cases that have been recorded on the log.

A. The Log and Summary of Occupational Injuries and Illnesses, OSHA No. 200

The log is used for recording and classifying occupational injuries and illnesses, and for noting the extent of each case. The log shows when the occupational injury or illness occurred, to whom, the regular job of the injured or ill person at the time of the injury or illness exposure, the department in which the person was employed, the kind of injury or illness, how much time was lost, whether the case resulted in a fatality, etc. The log consists of three parts: A descriptive section which identifies the employee and briefly describes the injury or illness; a section covering the extent of the injuries recorded; and a section on the type and extent of illnesses.

Usually, the OSHA No. 200 form is used by employers as their record of occupational injuries and illnesses. However, a private form equivalent to the log, such as a computer printout, may be used if it contains the same detail as the OSHA No. 200 and is as readable and comprehensible as the OSHA No. 200 to a person not familiar with the equivalent form. It is important that the columns of the equivalent form have the same identifying number as the corresponding columns of the OSHA No. 200 because the instructions for completing the survey of occupational injuries and illnesses refer to log columns by number. It is advisable that employers have private equivalents of the log form reviewed by BLS to insure compliance with the regulations.

The portion of the OSHA No. 200 to the right of the dotted vertical line is used to summarize injuries and illnesses in an establishment for the calendar year. Every nonexempt employer who is required to keep OSHA records must prepare an annual summary for each establishment based on the information contained in the log for each establishment. The summary is prepared by totaling the column entries on the log (or its equivalent) and signing and dating the certification portion of the form at the bottom of the page.

B. The Supplementary Record of Occupational Injuries and Illnesses, OSHA No. 101

For every injury or illness entered on the log, it is necessary to record additional information on the supplementary record, OSHA No. 101. The supplementary record describes how the accident or illness exposure occurred, lists the objects or substances involved, and indicates the nature of the injury or illness and the part(s) of the body affected.

The OSHA No. 101 is not the only form that can be used to satisfy this requirement. To eliminate duplicate recording, workers' compensation, insurance, or other reports may be used as supplementary records if they contain all of the items on the OSHA No. 101. If they do not, the missing items must be added to the substitute or included on a separate attachment.

Completed supplementary records must be present in the establishment within 6 workdays after the employer has received information that an injury or illness has occurred.

Chapter III. Location, Retention, and Maintenance of Records

Ordinarily, injury and illness records must be kept by employers for *each* of their establishments. This chapter describes what is considered to be an establishment for recordkeeping purposes, where the records must be located, how long they must be kept, and how they should be updated.

A. Establishments

If an employer has more than one establishment, a separate set of records must be maintained for *each* one. The recordkeeping regulations define an establishment as "a single physical location where business is conducted or where services or industrial operations are performed." Examples include a factory, mill, store, hotel, restaurant, movie theater, farm, ranch, sales office, warehouse, or central administrative office.

The regulations specify that distinctly separate activities performed at the same physical location (for example, contract construction activities operated from the same physical location as a lumber yard) shall each be treated as a separate establishment for recordkeeping purposes. Production of dissimilar products; different kinds of operational procedures; different facilities; and separate management, personnel, payroll, or support staff are all indicative of separate activities and separate establishments.

B. Location of records

Injury and illness records (the log, OSHA No. 200, and the supplementary record, OSHA No. 101) must be kept for every physical location where operations are performed. Under the regulations, the location of these records depends upon whether or not the employees are associated with a fixed establishment. The distinction between fixed and nonfixed establishments generally rests on the nature and duration of the operation and not on the type of structure in which the business is located. A nonfixed establishment usually operates at a single location for a relatively short period of time. A fixed establishment remains at a given location on a long-term or permanent basis. Generally, any operation at a given site for more than 1 year is considered a fixed establishment. Also, fixed establishments are generally places where clerical, administrative, or other business records are kept.

1. *Employees associated with fixed establishments.* Records for these employees should be located as follows:

a. Records for employees working at fixed locations, such as factories, stores, restaurants, warehouses, etc., should be kept at the work location.
b. Records for employees who report to a fixed location but work elsewhere should be kept at the place where the employees report each day. These employees are generally engaged in activities such as agriculture, construction, transportation, etc.
c. Records for employees whose payroll or personnel records are maintained at a fixed location, but who do not report or work at a single establishment, should be maintained at the base from which they are paid or the base of their firm's personnel operations. This category includes generally unsupervised employees such as traveling salespeople, technicians, or engineers.

2. *Employees not associated with fixed establishments.* Some employees are subject to common supervision, but do not report or work at a fixed establishment on a regular basis. These employees are engaged in physically dispersed activities that occur in construction, installation, repair, or service operations. Records for these employees should be located as follows:

a. Records may be kept at the field office or mobile base of operations.
b. Records may also be kept at an established central location. If the records are maintained centrally: (1) The address and telephone number of the place where records are kept must be available at the worksite; and (2) there must be someone available at the central location during normal business hours to provide information from the records.

C. Location exception for the log (OSHA No. 200)

Although the supplementary record and the annual summary must be located as outlined in the previous section, it is possible to prepare and *maintain the log* at an alternate location or by means of data processing equipment, or both. Two requirements must be met: (1) Sufficient information must be available at the alternate location to complete the log within 6 workdays after receipt of information that a recordable case has occurred; and (2) a copy of the log updated to within 45 calendar days must be present at all times in the establishment. This location exception applies only to the log, and not to the other OSHA records. Also, it does not affect the employer's posting obligations.

D. Retention of OSHA records

The log and summary, OSHA No. 200, and the supplementary record, OSHA No. 101, must be retained in each establishment for 5 calendar years following the end of the year to which they relate. If an establishment changes ownership, the new employer must preserve the records for the remainder of the 5-year period. However, the new employer is not responsible for updating the records of the former owner.

E. Maintenance of the log (OSHA No. 200)

In addition to keeping the log on a calendar year basis, employers are required to update this form to include newly discovered cases and to reflect changes which occur in recorded cases after the end of the calendar year. Maintenance or updating of the log is different from the retention of records discussed in the previous section. Although all OSHA injury and illness records must be retained, only the log must be updated by the employer. If, during the 5-year retention period, there is a change in the extent or outcome of an injury or illness which affects an entry on a previous year's log, then the first entry should be lined out and a corrected entry made on that log. Also, new entries should be made for previously unrecorded cases that are discovered or for cases that initially weren't recorded but were found to be recordable after the end of the year in which the case occurred. The entire entry should be lined out for recorded cases that are later found nonrecordable. Log totals should also be modified to reflect these changes.

Chapter IV. Deciding Whether a Case Should Be Recorded and How To Classify It

This chapter presents guidelines for determining whether a case must be recorded under the OSHA recordkeeping requirements. These requirements should not be confused with recordkeeping requirements of various workers' compensation systems, internal industrial safety and health monitoring systems, the ANSI Z.16 standards for recording and measuring work injury and illness experience, and private insurance company rating systems. Reporting a case on the OSHA records should not affect recordkeeping determinations under these or other systems. Also—

Recording an injury or illness under the OSHA system does not necessarily imply that management was at fault, that the worker was at fault, that a violation of an OSHA standard has occurred, or that the injury or illness is compensable under workers' compensation or other systems.

A. Employees vs. other workers on site

Employers must maintain injury and illness records for their own employees at each of their establishments, but they are *not* responsible for maintaining records for employees of other firms or for independent contractors, even though these individuals may be working temporarily in their establishment or on one of their jobsites at the time an injury or illness exposure occurs. Therefore, before deciding whether a case is recordable an employment relationship needs to be determined.

Employee status generally exists for recordkeeping purposes when the employer supervises not only the output, product, or result to be accomplished by the person's work, but also the details, means, methods, and processes by which the work is accomplished. This means the employer who supervises the worker's day-to-day activities is responsible for recording his injuries and illnesses. Independent contractors are not considered employees; they are primarily subject to supervision by the using firm only in regard to the result to be accomplished or end product to be delivered. Independent contractors keep their own injury and illness records.

Other factors which may be considered in determining employee status are: (1) Whom the worker considers to be his or her employer; (2) who pays the worker's wages; (3) who withholds the worker's Social Security taxes; (4) who hired the worker; and (5) who has the authority to terminate the worker's employment.

B. Method used for case analysis

The decisionmaking process consists of five steps:

1. Determine whether a case occurred; that is, whether there was a death, illness, or an injury;
2. Establish that the case was work related; that it resulted from an event or exposure in the work environment;
3. Decide whether the case is an injury or an illness; and
4. If the case is an illness, record it and check the appropriate illness category on the log; or
5. If the case is an injury, decide if it is recordable based on a finding of medical treatment, loss of consciousness, restriction of work or motion, or transfer to another job.

Chart 1 presents this methodology in graphic form.

C. Determining whether a case occurred

The first step in the decisionmaking process is the determination of whether or not an injury or illness has occurred. Employers have nothing to record unless an employee has experienced a work-related injury or illness. In most instances, recognition of these injuries and illnesses is a fairly simple matter. However, some situations have troubled employers over the years. Two of these are:

1. *Hospitalization for observation.* If an employee goes to or is sent to a hospital for a brief period of time for observation, it is not recordable, provided no medical treatment was given, or no illness was recognized. The determining factor is not that the employee went to the hospital, but whether the incident is recordable as a work-related illness or as an injury requiring medical treatment or involving loss of consciousness, restriction of work or motion, or transfer to another job.

2. *Differentiating a new case from the recurrence of a previous injury or illness.* Employers are required to make new entries on their OSHA forms for each new recordable injury or illness. However, new entries should

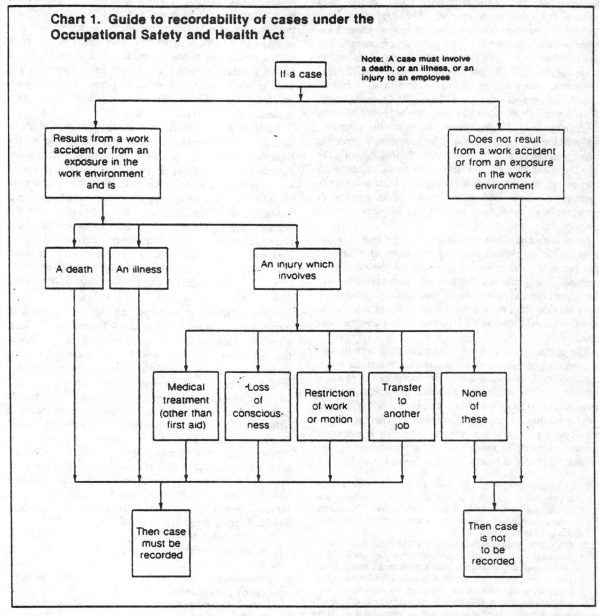

not be made for the recurrence of symptoms from previous cases, and it is sometimes difficult to decide whether or not a situation is a new case or a recurrence. The following guidelines address this problem:

a. *Injuries.* The aggravation of a previous injury almost always results from some new incident involving the employee (such as a slip, trip, fall, sharp twist, etc.). Consequently, when work related, these new incidents should be recorded as new cases.

b. *Illnesses.* Generally, each occupational illness should be recorded with a separate entry on the OSHA No. 200. However, certain illnesses, such as silicosis, may have prolonged effects which recur over time. The recurrence of these symptoms should not be recorded as new cases on the OSHA forms. The recurrence of symptoms of previous illnesses may require adjustment of entries on the log for previously recorded illnesses to reflect possible changes in the extent or outcome of the particular case.

Some occupational illnesses, such as certain dermatitis or respiratory conditions, may recur as the result of new exposures to sensitizing agents, and should be recorded as new cases.

D. **Establishing work relationship**

The Occupational Safety and Health Act of 1970

requires employers to record only those injuries and illnesses that are work related. *Work relationship is established under the OSHA recordkeeping system when the injury or illness results from an event or exposure in the work environment. The work environment is primarily composed of: (1) The employer's premises, and (2) other locations where employees are engaged in work-related activities or are present as a condition of their employment.* When an employee is off the employer's premises, work relationship must be established; when on the premises, this relationship is presumed. The employer's premises encompass the total establishment, including not only the primary work facility, but also such areas as company storage facilities. In addition to physical locations, equipment or materials used in the course of an employee's work are also considered part of the employee's work environment.

1. *Injuries and illnesses resulting from events or exposures on the employer's premises.* Injuries and illnesses that result from an event or exposure on the employer's premises are generally considered work related. The employer's premises consist of the total establishment. They include the primary work facilities and other areas which are considered part of the employer's general work area.

The presumption of work relationship for activities on the employer's premises is rebuttable. Situations where the presumption would not apply include: (1) When a worker is on the employer's premises as a member of the general public and not as an employee, and (2) when employees have symptoms that merely surface on the employer's premises, but are the result of a nonwork-related event or exposure off the premises.

The following subjects warrant special mention:

a. Company restrooms, hallways, and cafeterias are all considered to be *part* of the employer's premises and constitute part of the work environment. Therefore, injuries occurring in these places are generally considered work related.
b. For OSHA recordkeeping purposes, the definition of work premises *excludes* all employer controlled ball fields, tennis courts, golf courses, parks, swimming pools, gyms, and other similar recreational facilities which are often apart from the workplace and used by employees on a voluntary basis for their own benefit, primarily during off-work hours. Therefore, injuries to employees in these recreational facilities are not recordable unless the employee was engaged in some work-related activity, or was required by the employer to participate.
c. Company parking facilities are generally *not* considered part of the employer's premises for OSHA recordkeeping purposes. Therefore, injuries to employees on these parking lots are not presumed to be work related, and are not recordable unless the employee was engaged in some work-related activity.

2. *Injuries and illnesses resulting from events or exposures off the employer's premises.* When an employee is off the employer's premises and suffers an injury or an illness exposure, work relationship must be established; it is not presumed. Injuries and illness exposures off premises are considered work related if the employee is engaged in a work activity or if they occur in the work environment. The work environment in these instances includes locations where employees are engaged in job tasks or work-related activities, or places where employees are present due to the nature of their job or as a condition of their employment.

Employees who travel on company business shall be considered to be engaged in work-related activities all the time they spend in the interest of the company, including, but not limited to, travel to and from customer contacts, and entertaining or being entertained for the purpose of transacting, discussing, or promoting business, etc. However, an injury/illness would not be recordable if it occurred during normal living activities (eating, sleeping, recreation); or if the employee deviates from a reasonably direct route of travel (side trip for vacation or other personal reasons). He would again be in the course of employment when he returned to the normal route of travel.

When a traveling employee checks into a hotel or motel, he establishes a "home away from home." Thereafter, his activities are evaluated in the same manner as for nontraveling employees. For example, if an employee on travel status is to report each day to a fixed worksite, then injuries sustained when traveling to this worksite would be considered off the job. The rationale is that an employee's normal commute from home to office would not be considered work related. However, there are situations where employees in travel status report to, or rotate among several different worksites after they establish their "home away from home" (such as a salesperson traveling to and from different customer contacts). In these situations, the injuries sustained when traveling to and from the sales locations would be considered job related.

Traveling sales personnel may establish only one base of operations (home or company office). A sales person with his home as an office is considered at work when he is in that office and when he leaves his premises in the interest of the company.

Chart 2 provides a guide for establishing the work relationship of cases.

E. Distinguishing between injuries and illnesses

Under the OSH Act, all work-related illnesses must be recorded, while injuries are recordable *only* when they require medical treatment (other than first aid), or involve loss of consciousness, restriction of work or motion, or transfer to another job. The distinction between injuries and illnesses, therefore, has significant recordkeeping implications.

Whether a case involves an injury or illness is determined by the nature of the original event or exposure

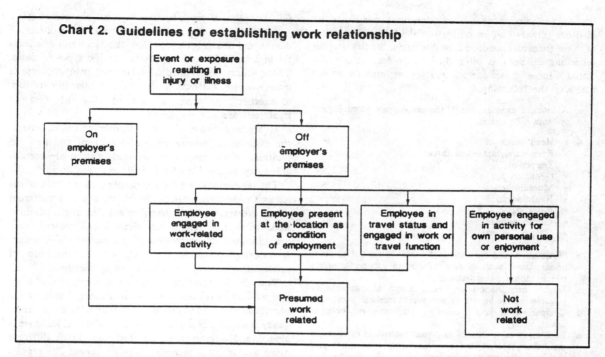

Chart 2. Guidelines for establishing work relationship

which caused the case, not by the resulting condition of the affected employee. Injuries are caused by instantaneous events in the work environment. Cases resulting from anything other than instantaneous events are considered illnesses. This concept of illnesses includes acute illnesses which result from exposures of relatively short duration.

Some conditions may be classified as either an injury or an illness (but not both), depending upon the nature of the event that produced the condition. For example, a loss of hearing resulting from an explosion (an instantaneous event) is classified as an injury; the same condition arising from exposure to industrial noise over a period of time would be classified as an occupational illness.

F. Recording occupational illnesses

Employers are required to record the occurrence of all occupational illnesses, which are defined in the instructions of the log and summary as:

> Any abnormal condition or disorder, other than one resulting from an occupational injury, caused by exposure to environmental factors associated with employment. It includes acute and chronic illnesses or diseases which may be caused by inhalation, absorption, ingestion, or direct contact.

The instructions also refer to recording illnesses which were "diagnosed or recognized." Illness exposures ultimately result in conditions of a chemical, physical, biological, or psychological nature.

Occupational illnesses must be diagnosed to be recordable. However, they do not necessarily have to be diagnosed by a physician or other medical personnel. Diagnosis may be by a physician, registered nurse, or a person who by training or experience is capable to make such a determination. Employers, employees, and others may be able to detect some illnesses such as skin diseases or disorders without the benefit of specialized medical training. However, a case more difficult to diagnose, such as silicosis, would require evaluation by properly trained medical personnel.

In addition to recording the occurrence of occupational illnesses, employers are required to record each illness case in 1 of the 7 categories on the front of the log. The back of the log form contains a listing of types of illnesses or disorders and gives examples for each illness category. These are only examples, however, and should not be considered as a complete list of types of illnesses under each category.

Recording and classifying occupational illnesses may be difficult for employers, especially the chronic and long term latent illnesses. Many illnesses are not easily detected; and once detected, it is often difficult to determine whether an illness is work related. Also, employees may not report illnesses because the symptoms may not be readily apparent, or because they do not think their illness is serious or work related.

The following material is provided to assist in detecting occupational illnesses and in establishing their work relationship:

1. *Detection and diagnosis of occupational illnesses.* An occupational illness is defined in the instructions on

the log as any work-related abnormal condition or disorder (other than an occupational injury). Detection of these abnormal conditions or disorders, the first step in recording illnesses, is often difficult. When an occupational illness is suspected, employers may want to consider the following:

a. A medical examination of the employee's physiological systems. For example:

- Head and neck
- Eyes, ears, nose, and throat
- Endocrine
- Genitourinary
- Musculoskeletal
- Neurological
- Respiratory
- Cardiovascular, and
- Gastrointestinal;

b. Observation and evaluation of behavior related to emotional status, such as deterioration in job performance which cannot be explained;
c. Specific examination for health effects of suspected or possible disease agents by competent medical personnel;
d. Comparison of date of onset of symptoms with occupational history;
e. Evaluation of results of any past biological or medical monitoring (blood, urine, other sample analysis) and previous physical examinations;
f. Evaluation of laboratory tests: Routine (complete blood count, blood chemistry profile, urinalysis) and specific tests for suspected disease agents (e.g., blood and urine tests for specific agents, chest or other X-rays, liver function tests, pulmonary function tests); and
g. Reviewing the literature, such as Material Safety Data Sheets and other reference documents, to ascertain whether the levels to which the workers were exposed could have produced the ill effects.

2. *Determining whether the illness is occupationally related.* The instructions on the back of the log define occupational illnesses as those "caused by environmental factors associated with employment." In some cases, such as contact dermatitis, the relationship between an illness and work-related exposure is easy to recognize. In other cases, where the occupational cause is not direct and apparent, it may be difficult to determine accurately whether an employee's illness is occupational in nature. In these situations, it may help employers to ask the following questions:

a. Has an illness condition clearly been established?
b. Does it appear that the illness resulted from, or was aggravated by, suspected agents or other conditions in the work environment?
c. Are these suspected agents present (or have they been present) in the work environment?
d. Was the ill employee exposed to these agents in the work environment?
e. Was the exposure to a sufficient degree and/or duration to result in the illness condition?
f. Was the illness attributable solely to a nonoccupational exposure?

G. **Deciding if work-related injuries are recordable**

Although the OSH Act requires that all work-related deaths and illnesses be recorded, the recording of nonfatal injuries is limited to certain specific types of cases: Those which require medical treatment or involve loss of consciousness; restriction of work or motion; or transfer to another job. Minor injuries requiring only first aid treatment are *not* recordable.

1. *Medical treatment.* It is important to understand the distinction between medical treatment and first aid treatment since many work-related injuries are recordable only because medical treatment was given.

The regulations and the instructions on the back of the log and summary, OSHA No. 200, define medical treatment as any treatment, other than first aid treatment, administered to injured employees. Essentially, medical treatment involves the provision of medical or surgical care for injuries that are not minor through the application of procedures or systematic therapeutic measures.

The act also specifically states that work-related injuries which involve only first aid treatment should not be recorded. First aid is commonly thought to mean emergency treatment of injuries before regular medical care is available. However, first aid treatment has a different meaning for OSHA recordkeeping purposes. The regulations define first aid treatment as:

> . . . any one-time treatment, and any followup visit for the purpose of observation, of minor scratches, cuts, burns, splinters, and so forth, which do not ordinarily require medical care. Such one-time treatment, and followup visit for the purpose of observation, is considered first aid even though provided by a physician or registered professional personnel.

The distinction between medical treatment and first aid depends not only on the treatment provided, but also on the severity of the injury being treated. First aid is: (1) Limited to one-time treatment and subsequent observation; *and* (2) involves treatment of only minor injuries, *not* emergency treatment of serious injuries. Injuries are *not* minor if:

a. They must be treated only by a physician or licensed medical personnel;
b. They impair bodily function (i.e., normal use of senses, limbs, etc.);
c. They result in damage to the physical structure of a nonsuperficial nature (e.g., fractures); or
d. They involve complications requiring followup medical treatment.

Physicians or registered medical professionals, working under the standing orders of a physician, routinely treat minor injuries. Such treatment may constitute first aid. Also, some visits to a doctor do not involve treatment at all. For example, a visit to a doctor for an examination or other diagnostic procedure to determine whether the employee has an injury does not constitute medical treatment. Conversely, medical treatment can be provided to employees by lay persons; i.e., someone other than a physician or registered medical personnel.

The following classifications list certain procedures as either medical treatment or first aid treatment.

Medical Treatment:

The following are generally considered medical treatment. Work-related injuries for which this type of treatment was provided or should have been provided are almost always recordable:

- Treatment of INFECTION
- Application of ANTISEPTICS **during second or subsequent visit** to medical personnel
- Treatment of SECOND OR THIRD DEGREE BURN(S)
- Application of SUTURES (stitches)
- Application of BUTTERFLY ADHESIVE DRESSING(S) or STERI STRIP(S) in lieu of sutures
- Removal of FOREIGN BODIES EMBEDDED IN EYE
- Removal of FOREIGN BODIES FROM WOUND; if procedure is COMPLICATED because of depth of embedment, size, or location
- Use of PRESCRIPTION MEDICATIONS (except a single dose administered on first visit for minor injury or discomfort)
- Use of hot or cold SOAKING THERAPY **during second or subsequent visit** to medical personnel
- Application of hot or cold COMPRESS(ES) **during second or subsequent visit** to medical personnel
- CUTTING AWAY DEAD SKIN (surgical debridement)
- Application of HEAT THERAPY **during second or subsequent visit** to medical personnel
- Use of WHIRLPOOL BATH THERAPY **during second or subsequent visit** to medical personnel
- POSITIVE X-RAY DIAGNOSIS (fractures, broken bones, etc.)
- ADMISSION TO A HOSPITAL or equivalent medical facility FOR TREATMENT.

First Aid Treatment:

The following are generally considered first aid treatment (e.g., one-time treatment and subsequent observation of minor injuries) and should not be recorded if the work-related injury does not involve loss of consciousness, restriction of work or motion, or transfer to another job:

- Application of ANTISEPTICS **during first visit** to medical personnel
- Treatment of FIRST DEGREE BURN(S)
- Application of BANDAGE(S) during any visit to medical personnel
- Use of ELASTIC BANDAGE(S) **during first visit** to medical personnel
- Removal of FOREIGN BODIES NOT EMBEDDED IN EYE if only irrigation is required
- Removal of FOREIGN BODIES FROM WOUND; if procedure is UNCOMPLICATED, and is, for example, by tweezers or other simple technique
- Use of NONPRESCRIPTION MEDICATIONS AND administration of **single dose of PRESCRIPTION MEDICATION** on first visit for minor injury or discomfort
- SOAKING THERAPY **on initial visit** to medical personnel or removal of bandages by SOAKING
- Application of hot or cold COMPRESS(ES) **during first visit** to medical personnel
- Application of OINTMENTS to abrasions to prevent drying or cracking
- Application of HEAT THERAPY **during first visit** to medical personnel
- Use of WHIRLPOOL BATH THERAPY **during first visit** to medical personnel
- NEGATIVE X-RAY DIAGNOSIS
- OBSERVATION of injury during visit to medical personnel.

The following procedure, by itself, is not considered medical treatment:

- Administration of TETANUS SHOT(S) or BOOSTER(S). However, these shots are often given in conjunction with more serious injuries; consequently, injuries requiring these shots may be recordable for other reasons.

2. *Loss of consciousness.* If an employee loses consciousness as the result of a work-related injury, the case must be recorded no matter what type of treatment was provided. The rationale behind this recording requirement is that loss of consciousness is generally associated with the more serious injuries.

3. *Restriction of work or motion.* Restricted work activity occurs when the employee, because of the impact of a job-related injury, is physically or mentally unable to perform all or any part of his or her normal assignment during all or any part of the workday or shift. The emphasis is on the employee's ability to perform normal job duties. Restriction of work or motion may result in either a lost worktime injury or a nonlost-worktime injury, depending upon whether the restriction extended beyond the day of injury.

4. *Transfer to another job.* Injuries requiring transfer of the employee to another job are also considered serious enough to be recordable regardless of the type of treatment provided. Transfers are seldom the sole criterion for recordability because injury cases are almost always recordable on other grounds, primarily medical treatment or restriction of work or motion.

Chapter V. Categories for Evaluating the Extent of Recordable Cases

Once the employer decides that a recordable injury or illness has occurred, the case must be evaluated to determine its extent or outcome. There are three categories of recordable cases: Fatalities, lost workday cases, and cases without lost workdays. Every recordable case must be placed in only one of these categories.

A. Fatalities

All work-related fatalities must be recorded, regardless of the time between the injury and the death, or the length of the illness.

B. Lost workday cases

Lost workday cases occur when the injured or ill employee experiences either days away from work, days of restricted work activity, or both. In these situations, the injured or ill employee is affected to such an extent that: (1) Days must be taken off from the job for medical treatment or recuperation; or (2) the employee is unable to perform his or her normal job duties over a normal work shift, even though the employee may be able to continue working.

1. Lost workday cases involving days away from work are cases resulting in days the employee would have worked but could not because of the job-related injury or illness. The focus of these cases is on the employee's inability, because of injury or illness, to be present in the work environment during his or her normal work shift.
2. Lost workday cases involving days of restricted work activity are those cases where, because of injury or illness, (1) the employee was assigned to another job on a temporary basis, or (2) the employee worked at a permanent job less than full time, or (3) the employee worked at his or her permanently assigned job but could not perform all the duties normally connected with it. Restricted work activity occurs when the employee, because of the job-related injury or illness, is physically or mentally unable to perform all or any part of his or her normal job duties over all or any part of his or her normal workday or shift. The emphasis is on the employee's inability to perform normal job duties over a normal work shift.

Injuries and illnesses are not considered lost workday cases unless they affect the employee beyond the day of injury or onset of illness. When counting the number of days away from work or days of restricted work activity, do not include the initial day of injury or onset of illness, or any days on which the employee would not have worked even though able to work (holidays, vacations, etc.).

C. Cases not resulting in death or lost workdays

These cases consist of the relatively less serious injuries and illnesses which satisfy the criteria for recordability but which do not result in death or require the affected employee to have days away from work or days of restricted work activity beyond the date of injury or onset of illness.

Chapter VI. Employer Obligations for Reporting Occupational Injuries and Illnesses

This chapter focuses on the requirements of Section 8(c)(2) of the Occupational Safety and Health Act of 1970 and Title 29, Part 1904, of the Code of Federal Regulations for employers to make reports of occupational injuries and illnesses. It does not include the reporting requirements of other standards or regulations of the Occupational Safety and Health Administration (OSHA) or of any other State or Federal agency.

A. The Annual Survey of Occupational Injuries and Illnesses

The survey is conducted on a sample basis, and firms required to submit reports of their injury and illness experience are contacted by BLS or a participating State agency. A firm not contacted by its State agency or BLS need not file a report of its injury and illness experience. Employers should note, however, that even if they are not selected to participate in the annual survey for a given year, they must still comply with the recordkeeping requirements listed in the preceding chapters as well as with the requirements for reporting fatalities and multiple hospitalization cases provided in the next section of this chapter.

Participants in the annual survey consist of two categories of employers: (1) Employers who maintain OSHA records on a regular basis; and (2) a small, rotating sample of employers who are regularly exempt from OSHA recordkeeping. The survey procedure is different for these two groups of employers.

1. *Participation of firms regularly maintaining OSHA records.* When employers regularly maintaining OSHA records are selected to participate in the Annual Survey of Occupational Injuries and Illnesses, they are mailed the survey questionnaire in February of the year following the reference calendar year of the survey. (A firm selected to participate in the 1985 survey would have been contacted in February of 1986.) The survey form, the Occupational Injuries and Illnesses Survey Questionnaire, OSHA No. 200-S, requests information about the establishment(s) included in the report and the injuries and illnesses experienced during the previous year. Information for the injury and illness portion of the report form usually can be copied directly from the totals on the log and summary, OSHA No. 200, which the employer should have completed and posted in the establishment by the time the questionnaire arrives. The survey form also requests summary information about the type of business activity and number of employees and hours worked at the reporting unit during the reference year.

2. *Participation of normally exempt small employers and employers in low-hazard industries.* A few regularly exempt employers (those with fewer than 11 employees in the previous calendar year and those in designated low-hazard industries) are also required to participate in the annual survey. Their participation is necessary for the production of injury and illness statistics that are comparable in coverage to the statistics published in years prior to the exemptions. These employers are notified prior to the reference calendar year of the survey that they must maintain injury and illness records for the coming year. (A firm selected to participate in the 1985 survey would have been contacted in December 1984.) At the time of notification, they are supplied with the necessary forms and instructions. During the reference calendar year, prenotified employers make entries on the log, OSHA No. 200, but are not required to complete a Supplementary Record of Occupational Injuries and Illnesses, OSHA No. 101, or post the summary of the OSHA No. 200 the following February (regularly participating employers do both).

B. Reporting fatalities and multiple hospitalizations

All employers are required to report accidents resulting in one or more fatalities or the hospitalization of five or more employees. (Some States have more stringent catastrophic reporting requirements.)

The report is made to the nearest office of the Area Director of the Occupational Safety and Health Administration, U.S. Department of Labor, unless the State in which the accident occurred is administering an approved State plan under Section 18(b) of the OSH Act. Those 18(b) States designate a State agency to which the report must be made.

The report must contain three pieces of information: (1) Circumstances surrounding the accident(s), (2) the number of fatalities, and (3) the extent of any injuries. If necessary, the OSHA Area Director may require additional information on the accident.

The report should be made within 48 hours after the occurrence of the accident or within 48 hours after the occurrence of the fatality, regardless of the time lapse between the occurrence of the accident and the death of the employee.

Chapter VII. Access to OSHA Records and Penalties for Failure To Comply With Recordkeeping Obligations

The preceding chapters describe recordkeeping and reporting requirements. This chapter covers subjects related to insuring the integrity of the OSH recordkeeping process—access to OSHA records and penalties for recordkeeping violations.

A. Access to OSHA records

All OSHA records, which are being kept by employers for the 5-year retention period, should be available for inspection and copying by authorized Federal and State government officials. Employees, former employees, and their representatives are provided access to only the log, OSHA No. 200.

Government officials with access to the OSHA records include: Representatives of the Department of Labor, including OSHA safety and health compliance officers and BLS representatives; representatives of the Department of Health and Human Services while carrying out that department's research responsibilities; and representatives of States accorded jurisdiction for inspections or statistical compilations. "Representatives" may include Department of Labor officials inspecting a workplace or gathering information, officials of the Department of Health and Human Services, or contractors working for the agencies mentioned above, depending on the provisions of the contract under which they work.

Employee access to the log is limited to the records of the establishment in which the employee currently works or formerly worked. All current logs and those being maintained for the 5-year retention period must be made available for inspection and copying by employees, former employees, and their representatives. An employee representative can be a member of a union representing the employee, or any person designated by the employee or former employee. Access to the log is to be provided in a reasonable manner and at a reasonable time. Redress for failure to comply with the access provisions of the regulations can be obtained through a complaint to OSHA.

B. Penalties for failure to comply with recordkeeping obligations

Employers committing recordkeeping and/or reporting violations are subject to the same sanctions as employers violating other OSHA requirements such as safety and health standards and regulations.

Glossary of Terms

Annual summary. Consists of a copy of the occupational injury and illness totals for the year from the OSHA No. 200, and the following information: The calendar year covered; company name; establishment address; certification signature, title, and date.

Annual survey. Each year, BLS conducts an annual survey of occupational injuries and illnesses to produce national statistics. The OSHA injury and illness records maintained by employers in their establishments serve as the basis for this survey.

Bureau of Labor Statistics (BLS). The Bureau of Labor Statistics is the agency responsible for administering and maintaining the OSHA recordkeeping system, and for collecting, compiling, and analyzing work injury and illness statistics.

Certification. The person who supervises the preparation of the Log and Summary of Occupational Injuries and Illnesses, OSHA No. 200, certifies that it is true and complete by signing the last page of, or by appending a statement to that effect to, the annual summary.

Cooperative program. A program jointly conducted by the States and the Federal Government to collect occupational injury and illness statistics.

Employee. One who is employed in the business of his or her employer affecting commerce.

Employee representative. Anyone designated by the employee for the purpose of gaining access to the employer's log of occupational injuries and illnesses.

Employer. Any person engaged in a business affecting commerce who has employees.

Establishment. A single physical location where business is conducted or where services or industrial operations are performed; the place where the employees report for work, operate from, or from which they are paid.

Exposure. The reasonable likelihood that a worker is or was subject to some effect, influence, or safety hazard; or in contact with a hazardous chemical or physical agent at a sufficient concentration and duration to produce an illness.

Federal Register. The official source of information and notification on OSHA's proposed rulemaking, standards, regulations, and other official matters, including amendments, corrections, insertions, or deletions.

First aid. Any one-time treatment and subsequent observation of minor scratches, cuts, burns, splinters, and so forth, which do not ordinarily require medical care. Such treatment and observation are considered first aid even though provided by a physician or registered professional personnel.

First report of injury. A workers' compensation form which may qualify as a substitute for the supplementary record, OSHA No. 101.

Incidence rate. The number of injuries, illnesses, or lost workdays related to a common exposure base of 100 full-time workers. The common exposure base enables one to make accurate intrindustry comparisons, trend analysis over time, or comparisons among firms regardless of size. This rate is calculated as:

$$N/EH \times 200,000$$

where:

N = number of injuries and/or illnesses or lost workdays

EH = total hours worked by all employees during calendar year

$200,000$ = base for 100 full-time equivalent workers (working 40 hours per week, 50 weeks per year).

Log and Summary (OSHA No. 200). The OSHA recordkeeping form used to list injuries and illnesses and to note the extent of each case.

Lost workday cases. Cases which involve days away from work or days of restricted work activity, or both.

Lost workdays. The number of workdays (consecutive or not), beyond the day of injury or onset of illness, the employee was away from work or limited to restricted work activity because of an occupational injury or illness.

(1) Lost workdays—away from work. The number of workdays (consecutive or not) on which the employee would have worked but could not because of occupational injury or illness.

(2) Lost workdays—restricted work activity. The number of workdays (consecutive or not) on which, because of injury or illness: (1) The employee was assigned to another job on a temporary basis; or (2) the employee worked at a permanent job less than full time; or (3) the employee worked at a permanently assigned job but could not perform all duties normally connected with it.

The number of days away from work or days of restricted work activity does not include the day of injury or onset of illness or any days on which the employee would not have worked even though able to work.

Low-hazard industries. Selected industries in retail trade; finance, insurance, and real estate; and services which are regularly exempt from OSHA recordkeeping. To be included in this exemption, an industry must fall within an SIC not targeted for general schedule inspections and must have an average lost workday case injury rate for a designated 3-year measurement period at or below 75 percent of the U.S. private sector average rate.

Medical treatment. Includes treatment of injuries administered by physicians, registered professional personnel, or lay persons (i.e., nonmedical personnel). Medical treatment does not include first aid treatment (one-time treatment and subsequent observation of minor scratches, cuts, burns, splinters, and so forth, which do not ordinarily require medical care) even though provided by a physician or registered professional personnel.

Occupational illness. Any abnormal condition or disorder, other than one resulting from an occupational injury, caused by exposure to environmental factors associated with employment. It includes acute and chronic illnesses or diseases which may be caused by inhalation, absorption, ingestion, or direct contact. The following categories should be used by employers to classify recordable occupational illnesses on the log in the columns indicated:

Column 7a. *Occupational skin diseases or disorders.*
Examples: Contact dermatitis, eczema, or rash caused by primary irritants and sensitizers or poisonous plants; oil acne; chrome ulcers; chemical burns or inflammations; etc.

Column 7b. *Dust diseases of the lungs (pneumoconioses).*
Examples: Silicosis, asbestosis, and other asbestos-related diseases, coal worker's pneumoconiosis, byssinosis, siderosis, and other pneumoconioses.

Column 7c. *Respiratory conditions due to toxic agents.*
Examples: Pneumonitis, pharyngitis, rhinitis or acute congestion due to chemicals, dusts, gases, or fumes; farmer's lung, etc.

Column 7d. *Poisoning (systemic effects of toxic materials).*
Examples: Poisoning by lead, mercury, cadmium, arsenic, or other metals; poisoning by carbon monoxide, hydrogen sulfide, or other gases; poisoning by benzol, carbon tetrachloride, or other organic solvents; poisoning by insecticide sprays such as parathion, lead arsenate; poisoning by other chemicals such as formaldehyde, plastics, and resins; etc.

Column 7e. *Disorders due to physical agents (other than toxic materials).*
Examples: Heatstroke, sunstroke, heat exhaustion, and other effects of environmental heat; freezing, frostbite, and effects of exposure to low temperatures; caisson disease; effects of ionizing radiation (isotopes, X-Rays, radium); effects of nonionizing radiation (welding flash, ultra-violet rays, microwaves, sunburn); etc.

Column 7f. *Disorders associated with repeated trauma.*
Examples: Noise-induced hearing loss; synovitis, tenosynovitis, and bursitis; Raynaud's phenomena; and other conditions due to repeated motion, vibration, or pressure.

Column 7g. *All other occupational illnesses.*
Examples: Anthrax, brucellosis, infectious hepatitis, malignant and benign tumors, food poisoning, histoplasmosis, coccidioidomycosis, etc.

Occupational injury. Any injury such as a cut, fracture, sprain, amputation, etc., which results from a work accident or from a single instantaneous exposure in the work environment.
Note: Conditions resulting from animal bites, such as insect or snake bites, and from one-time exposure to chemicals are considered to be injuries.

Occupational injuries and illnesses, extent and outcome. All recordable occupational injuries or illnesses result in either:
(1) Fatalities, regardless of the time between the injury, or the length of illness, and death;
(2) Lost workday cases, other than fatalities, that result in lost workdays; or
(3) Nonfatal cases without lost workdays.

Occupational Safety and Health Administration (OSHA). OSHA is responsible for developing, implementing, and enforcing safety and health standards and regulations. OSHA works with employers and employees to foster effective safety and health programs which reduce workplace hazards.

Posting. The annual summary of occupational injuries and illnesses must be posted at each establishment by February 1 and remain in place until March 1 to provide employees with the record of their establishment's injury and illness experience for the previous calendar year.

Premises, employer's. Consist of the employer's total establishment; they include the primary work facility and other areas in the employer's domain such as company storage facilities, cafeterias, and restrooms.

Recordable cases. All work-related deaths and illnesses, and those work-related injuries which result in: Loss of consciousness, restriction of work or motion, transfer to another job, or require medical treatment beyond first aid.

Recordkeeping system. Refers to the nationwide system for recording and reporting occupational injuries and illnesses mandated by the Occupational Safety and Health Act of 1970 and implemented by Title 29, Code of Federal Regulations, Part 1904. This system is the only source of national statistics on job-related injuries and illnesses for the private sector.

Regularly exempt employers. Employers regularly exempt from OSHA recordkeeping include: (A) All employers with no more than 10 full- or part-time employees at any one time in the previous calendar year; and (B) all employers in retail trade; finance, insurance, and real estate; and services industries; i.e., SIC's 52-89 (except building materials and garden supplies, SIC 52; general merchandise and food stores, SIC's 53 and 54; hotels and other lodging places, SIC 70; repair services, SIC's 75 and 76; amusement and recreation services, SIC 79; and health services, SIC 80). (Note: Some State safety and health laws may require these employers to keep OSHA records.)

Report form. Refers to survey form OSHA No. 200-S which is completed and returned by the surveyed reporting unit.

Restriction of work or motion. Occurs when the employee, because of the result of a job-related injury or illness, is physically or mentally unable to perform all or any part of his or her normal assignment during all or any part of the workday or shift.

Single dose (prescription medication). The measured quantity of a therapeutic agent to be taken at one time.

Small employers. Employers with no more than 10 full- and/or part-time employees among all the establishments of their firm at any one time during the previous calendar year.

Standard Industrial Classification (SIC). A classification system developed by the Office of Management and Budget, Executive Office of the President, for use in the classification of establishments by type of activity in which engaged. Each establishment is assigned an industry code for its major activity which is determined by the product manufactured or service rendered. Establishments may be classified in 2-, 3-, or 4-digit industries according to the degree of information available.

State (when mentioned alone). Refers to a State of the United States, the District of Columbia, and U.S. territories and jurisdictions.

State agency. State agency administering the OSHA recordkeeping and reporting system. Many States cooperate directly with BLS in administering the OSHA recordkeeping and reporting programs. Some States have their own safety and health laws which may impose additional obligations.

Supplementary Record (OSHA No. 101). The form (or equivalent) on which additional information is recorded for each injury and illness entered on the log.

Title 29 of the Code of Federal Regulations, Parts 1900-1999. The parts of the Code of Federal Regulations which contain OSHA regulations.

Volunteers. Workers who are not considered to be employees under the act when they serve of their own free will without compensation.

Work environment. Consists of the employer's premises and other locations where employees are engaged in work-related activities or are present as a condition of their employment. The work environment includes not only physical locations, but also the equipment or materials used by the employee during the course of his or her work.

Workers' compensation systems. State systems that provide medical benefits and/or indemnity compensation to victims of work-related injuries and illnesses.

ORDER FORM

Type or print

Please complete this form and mail it to the appropriate BLS regional office or participating State agency.

From:

Name _____

Firm _____

Street Address _____

City, State, Zip Code _____

Please send me the following items at no charge: Quantity

A Brief Guide to Recordkeeping Requirements for Occupational Injuries
and Illnesses (18 pp.)... _____

Recordkeeping Guidelines for Occupational Injuries and Illnesses (84 pp.)........ _____

OSHA No. 200 Forms (Log and Summary of Occupational Injuries and Illnesses) .... _____

OSHA No. 101 Forms (Supplementary Record of Occupational Injuries and Illnesses) .... _____

ADDRESS LABEL

Name _____

Firm _____

Street Address _____

City, State, Zip Code _____

Participating State Agencies

Agencies preceded by an asterisk (*) have a State safety and health plan under section 18(b) of the act in operation and may be contacted directly for information regarding State regulations.

Alabama Department of Labor
651 Administrative Bldg.
Montgomery, AL 36130
Phone: 205-261-3460

*Alaska Department of Labor
Research and Analysis Section
P.O. Box 25501, Juneau, AK 99802
Phone: 907-465-4520

Government of American Samoa
Department of Manpower Resources
Division of Labor
Pago Pago, AS 96799
Phone: 633-5849

*Industrial Commission of Arizona
Division of Administration/Management
Research and Statistics Section
P.O. Box 19070, Phoenix, AZ 85005
Phone: 602-255-3739

Workers' Compensation Commission OSH
Arkansas Department of Labor
OSH Research and Statistics, Suite 219
1515 W. 7th St., Little Rock, AR 72201
Phone: 501-371-2770

*California Department of Industrial Relations, Labor Statistics and Research Division
P.O. Box 6880, San Francisco, CA 94101
Phone: 415-557-1466

Colorado Department of Labor and
 Employment, Division of Labor
1313 Sherman St., Rm. 323
Denver, CO 80203
Phone: 303-866-3748

*Connecticut Department of Labor
200 Folly Brook Blvd.
Wethersfield, CT 06109
Phone: 203-566-4380

Delaware Department of Labor
Division of Industrial Affairs
820 N. French St., 6th Fl.
Wilmington, DE 19801
Phone: 302-571-2888

Florida Department of Labor and
 Employment Security
Division of Worker's Compensation
2551 Executive Center
Circle West, Rm. 204
Tallahassee, FL 32301-5014
Phone: 904-488-3044

Guam Department of Labor
Bureau of Labor Statistics
Occupational Safety and Health Statistics
P.O. Box 23548, Guam Main Facility
Guam, M.I. 96921
Phone: 477-9241

*State of Hawaii
Department of Labor and Industrial
 Relations, Research and Statistics
 Office—OSHA
P.O. Box 3680, Honolulu, HI 96811
Phone: 808-548-7638

*Indiana Department of Labor
Research and Statistics Division
State Office Bldg.—Rm. 1013
100 N. Senate Ave.
Indianapolis, IN 46204
Phone: 317-232-2681

*Iowa Division of Labor Services
307 East 7th St., Des Moines, IA 50319
Phone: 515-281-3606

Kansas Department of Health and Environment
Division of Policy and Planning
Research and Analysis
Topeka, KS 66620
Phone: 913-862-9360 Ext. 280

*Kentucky Labor Cabinet
Occupational Safety and Health Program
U.S. 127 South Bldg., Frankfort, KY 40601
Phone: 502-564-3100

Louisiana Department of Labor
Office of Employment Security—OSH
P.O. Box 94094
Baton Rouge, LA 70804
Phone: 504-342-3126

Maine Department of Labor
Bureau of Labor Standards
Division of Research and Statistics
State Office Bldg., Augusta, ME 04330
Phone: 207-289-4311

*Maryland Department of Licensing and Regulation
Division of Labor and Industry
501 St. Paul Pl., Baltimore, MD 21202
Phone: 301-333-4202

Massachusetts Department of Labor and Industries
Division of Industrial Safety
100 Cambridge St., Boston, MA 02202
Phone: 617-727-3593

*Michigan Department of Labor
7150 Harris Dr., Secondary Complex
P.O. Box 30015, Lansing, MI 48909
Phone: 517-322-1848

*Minnesota Department of Labor and Industry IMSD
444 Lafayette Rd., 5th Fl.
St. Paul, MN 55101
Phone: 612-296-4893

Mississippi State Department of Health
Office of Public Health Statistics
P.O. Box 1700
Jackson, MS 39215-1700
Phone: 601-354-7233

Missouri Department of Labor and Industrial Relations
Division of Workers' Compensation
P.O. Box 58, Jefferson City, MO 65102
Phone: 314-751-4231

Montana Department of Labor and Industry
Workers' Compensation Division
5 South Last Chance Gulch
Helena, MT 59601
Phone: 406-444-6515

Nebraska Workers' Compensation Court
State Capitol, 12th Fl.
Lincoln, NE 68509-4967
Phone: 402-471-3547

*Nevada Department of Industrial Relations
Division of Occupational Safety and Health
1370 South Curry St.
Carson City, NV 89710
Phone: 702-885-5240

New Jersey Department of Labor and Industry
Division of Planning and Research, CN 057
Trenton, NJ 08625
Phone: 609-292-8997

*New Mexico Health and Environment Department
Environmental Improvement Division
Occupational Health and Safety
P.O. Box 968
Sante Fe, NM 87504-0968
Phone: 505-827-2875

New York Department of Labor
Division of Research and Statistics
1 Main St., Rm 907
Brooklyn, NY 11201
Phone: 718-797-7701

*North Carolina Department of Labor
Research and Statistics Division
214 West Jones St.
Raleigh, NC 27603
Phone: 919-733-4940

Ohio Department of Industrial Relations
OSH Survey Office
P.O. Box 12355, Columbus, OH 43212
Phone: 614-466-7520

Oklahoma Department of Labor
Supplemental Data Division
1315 North Broadway Pl.
Oklahoma City, OK 73105
Phone: 405-235-1447

*Oregon Workers' Compensation Department
Research and Statistics Section
21 Labor and Industries Bldg.
Salem, OR 97310
Phone: 503-378-8254

Pennsylvania Department of Labor and Industry
Office of Employment Security
7th and Forster Sts.
Labor and Industry Bldg.
Harrisburg, PA 17121
Phone: 717-787-1918

*Puerto Rico Department of Labor and Human Resources
Bureau of Labor Statistics
505 Munoz Rivera Ave.
Hato Rey, PR 00918
Phone: 809-754-5339

Rhode Island Department of Labor
220 Elmwood Ave., Providence, RI 02907
Phone: 401-457-1852

*South Carolina Department of Labor
Division of Technical Support
P.O. Box 11329, Columbia, SC 29211
Phone: 803-734-9652

*Tennessee Department of Labor
Research and Statistics
501 Union Bldg., 6th Fl.
Nashville, TN 37219
Phone: 615-741-1748

Texas Department of Health
Division of Occupational Safety
1100 West 49th St., Austin, TX 78756
Phone: 512-458-7287

*Utah Industrial Commission
OSH Statistical Section
160 East 300 South
Salt Lake City, UT 84145-0580
Phone: 801-530-6827

*Vermont Department of Labor and Industry
State Office Bldg.
Montpelier, VT 05602
Phone: 802-828-2765

*Virgin Islands Department of Labor
P.O. Box 3359
St. Thomas, VI 00801
Phone: 809-776-3700

*Virginia Department of Labor and Industry
Planning and Research
205 North 4th St.
P.O. Box 12064, Richmond, VA 23241
Phone: 804-786-2384

*State of Washington
Department of Labor and Industries
Division of Industrial Safety and Health
P.O. Box 2589
Olympia, WA 98504
Phone: 206-753-4013

West Virginia Department of Labor
OSH Project Director
Rm. 319, Bldg. 3, Capitol Complex
1800 Washington St. East
Charleston, WV 25305
Phone: 304-348-7890

Wisconsin Department of Industry, Labor and Human Relations
Workers' Compensation Division
Research Section
201 E. Washington Ave.
P.O. Box 7901
Madison, WI 53707
Phone: 608-266-7850

*Wyoming Department of Labor and Statistics
Herschler Bldg., 2 East
Cheyenne, WY 82002
Phone: 307-777-6370

APPENDIX J

OSHA AIDS Inspection Directive, August 15, 1988

APPENDIX J

U.S. Department of Labor Assistant Secretary for
Occupational Safety and Health
Washington, D.C. 20210

OSHA Instruction CPL 2-2.44A
AUG 15 1988
Office of Health Compliance Assistance

SUBJECT: Enforcement Procedures for Occupational Exposure to Hepatitis B Virus (HBV) and Human Immunodeficiency Virus (HIV).

A. **Purpose.** This instruction provides uniform inspection procedures and guidelines to be followed when conducting inspections and issuing citations under section 5(a)(1) of the Act and pertinent standards for health care workers potentially exposed to HBV and HIV.

B. **Scope.** This instruction applies OSHA-wide.

C. **Cancellation.** This instruction cancels OSHA instruction CPL 2-2.44, January 19, 1988.

D. **References.**

1. OSHA Instruction CPL 2.45A, April 18, 1983, the Field Operations Manual (FOM).

2. OSHA Instruction ADM 1-1.12A, April 1, 1984, the Integrated Management Information System (IMIS) Forms Manual.

3. OSHA Instruction CPL 2-2.36, November 30, 1983, Hepatitis B Risks in the Health Care System.

4. Centers for Disease Control. Acquired Immunodeficiency Syndrome, AIDS--Recommendations and Guidelines November 1982-November 1986.

5. Centers for Disease Control Morbidity and Mortality Weekly Report. Update on Hepatitis B Prevention. June 19, 1987; Vol. 36, No.23.

6. Centers for Disease Control Morbidity and Mortality Weekly Report. 1988 Agent Summary Statement for Human Immunodeficiency Virus and Report on Laboratory-Acquired Infection with Human Immunodeficiency Virus. April 1, 1988; Vol. 37, No. S-4.

OSHA Instruction CPL 2-2.44A
Office of Health Compliance Assistance

E. **Action.** Regional Administrators and Area Directors shall ensure that the policies and procedures explained in this instruction are implemented in scheduling and conducting inspections. Regional Administrators shall further ensure that all relevant procedures established in this instruction are adhered to by State 7(c)(1) consultation projects.

F. **Federal Program Change.** This instruction describes a Federal Program change which affects State programs. Each Regional Administrator shall:

   1. Ensure that this change is promptly forwarded to each State designee.

   2. Explain the technical content of this change to the State designee as requested.

   3. Ensure that State designees are asked to acknowledge receipt of this Federal program change in writing, within 30 days of notification, to the Regional Administrator. This acknowledgment must include the State's intention to follow the appropriate procedures in this instruction, or a description of the State's alternative procedures which are "at least as effective" as the Federal guidelines. Any alternative State policies must be submitted as a plan supplement within 6 months.

      a. If a State intends to follow the inspection procedures described in this instruction, the State must submit either a revised version of this instruction, adapted as appropriate to reference State law, regulations and administrative structure, or a cover sheet describing how references in this instruction correspond to the State's structure. The State's acknowledgment letters may fulfill the plan supplement requirement if the appropriate documentation is provided.

      b. If the State adopts an alternative to Federal enforcement policies, the State's submission must identify and provide a rationale for all substantial differences from Federal policies in order for OSHA to judge whether a different State procedure is as effective as comparable Federal policy.

OSHA Instruction CPL 2-2.44A
Office of Health Compliance Assistance

    c. Although States are not required to enforce General Duty provisions, they are strongly urged to do so.

4. Advise State designees of the following:

    a. In order to ensure a sound and consistent national enforcement and litigation strategy in relation to the complex issues addressed by this instruction, State implementation of the procedures in this instruction, or comparable State procedures, must be carefully coordinated with OSHA.

    b. The State is also responsible for extending coverage under its procedures for addressing occupational exposure to HBV and HIV in the public sector, such as police, fire, ambulance and other emergency response workers.

    c. The Directorate of Technical Support is available to assist the States in locating expert witnesses. (See paragraph O., Expert Witnesses.)

    d. In regard to paragraph I., Inspection Scheduling, Goal and Scope, the State's response to the Regional Administrator is to include a projection of the number of inspections the State expects to conduct in both the private and public sector.

5. After Regional review of the State plan supplement and resolution of any comments thereon, forward the State submission to the National Office in accordance with established procedures. The Regional Administrator shall provide a judgment on the relative effectiveness of each substantial difference in the State plan change and an overall assessment thereon with a recommendation for approval or disapproval by the Assistant Secretary.

6. Review policies, instructions and guidelines issued by the State to determine that this change has been communicated to State program personnel.

OSHA Instruction CPL 2-2.44A

Office of Health Compliance Assistance

G. Definitions.

1. Health Care Facility. Those establishments listed under the Standard Industrial Classification (SIC) codes 80** and 7261; and clinics, health units, and nurses' stations at industrial work sites.

2. Health Care Worker. An employee of a health care facility including, but not limited to, nurses, physicians, dentists and other dental workers, optometrists, podiatrists, chiropractors, laboratory and blood bank technologists and technicians, research laboratory scientists, phlebotomists, dialysis personnel, paramedics, emergency medical technicians, medical examiners, morticians, housekeepers, laundry workers and others whose work may involve direct contact with body fluids, as defined below, from living individuals or corpses.

3. Universal Precautions. The term "universal precautions" refers to a system of infectious disease control which assumes that every direct contact with body fluids is infectious and requires every employee exposed to direct contact with body fluids to be protected as though such body fluids were HBV or HIV infected. Therefore, universal precautions are intended to prevent health care workers from parenteral, mucous membrane, and nonintact skin exposures to blood-borne pathogens.

4. Body Fluids. Fluids that have been recognized by CDC as directly linked to the transmission of HIV and/or HBV and/or to which universal precautions apply: blood, semen, blood products, vaginal secretions, cerebrospinal fluid, synovial fluid, pleural fluid, peritoneal fluid, pericardial fluid, amniotic fluid, and concentrated HIV or HBV viruses.

5. Phlebotomist. A phlebotomist is any health care worker who draws blood samples.

6. Infectious Control (IC) Program. An IC program is the establishment's oral or written policy and implementation of procedures relating to the control of infectious disease hazards where employees may be exposed to direct contact with

APPENDIX J J-7

OSHA Instruction CPL 2-2.44A

Office of Health Compliance Assistance

body fluids. An IC Program must address all of he areas outlined in this directive.

7. <u>Joint Advisory Notice</u>. Department of Labor/Department of Health and Human Services-Joint Advisory Notice (Federal Register, Vol. 52, No. 210; October 30, 1987) is a list of recommendations developed to assist employers in implementing the Centers for Disease Control (CDC) guidelines.

H. <u>Background</u>. In September 1986, OSHA was petitioned by various unions representing health care employees to develop a standard to protect workers from occupational exposure to blood-borne diseases. Although the Agency has decided to pursue development of such a standard, as a result of recent rulemaking petitions and OSHA's evaluation of those petitions, the Agency has concluded that the risk of contracting hepatitis B and AIDS among members of various occupations within the health care system requires an immediate response through a variety of existing mechanisms.

1. Occupational exposure may occur in many ways, including needlestick and cut injuries. Health care workers employed in certain occupations are assumed to be at high risk for blood-borne infections due to their routinely increased exposure to body fluids from potentially infected patients. These high risk occupations include but are not limited to physicians, pathologists, dentists and dental technicians, x-ray technicians, phlebotomists, emergency room, intensive care and operating room nurses and technicians, laboratory and blood bank technologists and technicians. Other health care workers who may be directly exposed to such body fluids depending on their exact work assignments include such occupations as housekeeping personnel, laundry workers, orderlies, morticians, research laboratory workers, paramedics, medical examiners. Employees in any occupation where they are directly exposed to body fluids are considered to be at substantial risk of occupational exposure to HIV and/or HBV.

2. Ward clerks and administrators have virtually no increased risk of contact with body fluids as a result of their employment; they are thus at no

OSHA Instruction CPL 2-2.44A

Office of Health Compliance Assistance

       greater risk of contracting blood-borne diseases than other members of the general population.

  3. Neither HBV nor HIV is transmitted by casual contact in the workplace.

  4. The employer's obligations are those set forth in the Occupational Safety and Health Act (OSH Act) of 1970. However, the CDC has published guidelines to protect workers from HBV and HIV (See Appendices A & B). OSHA is relying on these guidelines as reflecting an appropriate and widely recognized and accepted standard of protection to be followed by health care employers in carrying out their responsibilities under the OSH Act.

  5. The same personal protective equipment and work practices used to prevent occupational transmission of HBV are effective in preventing occupational transmission of HIV. The CDC has called for use of "universal precautions" when working with blood and/or body fluids from any patient.

  6. One difference between the two viruses is that there is currently a vaccine to prevent HBV infection, which the CDC has recommended for persons at substantial risk of occupational exposure, but there is no vaccine for HIV.

I. <u>Inspection Scheduling, Goal and Scope</u>.

  1. Inspection scheduling shall be conducted in accordance with the procedures outlined in the FOM, Chapter II, and for Federal agencies, Chapter XIII, except as modified in the following sections.

  2. The National Office shall develop and distribute a list of health care establishments by State to the appropriate Regional Administrators to identify the establishments to be programmed for inspection within the Region.

    a. This list shall be randomly selected from among all establishments identified as being within the health care industries, viz., those industries having primary Standard Industrial Classification (SIC) codes 80** (Health Ser-

OSHA Instruction CPL 2-2.44A
AUG 15 1988
Office of Health Compliance Assistance

vices) and 7261 (Funeral Service and Crematories).

 b. A widely available commercial listing shall be the source of such establishments (Duns Marketing Service).

 c. All establishments listed as having less than 11 employees shall be excluded from the list.

 d. More establisments will be provided to the Regions than will actually be needed to allow for replacement of establishments.

3. The Regional Administrator shall use an additional random selection method to identify the establishments actually to be programmed for inspection within the Region.

 a. The number of health care establishments to be selected from the list provided by the National Office shall be at least equal to the inspection goal identified in the annual program plan developed for the Region.

 b. The names of the establishments selected by the Regional Administrator shall be distributed to the appropriate Area Office and shall constitute the Inspection Register for that office.

 c. An inspection of all establishments listed on the Inspection Register shall be conducted unless they are deleted for some legitimate reason.

4. Any establishment deleted shall be replaced by another establishment selected according to I.3.

 a. Area Directors shall inform the Regional Administrator of any deletions.

 b. The Regional Administrator shall select the next establishment by random number procedure from the list provided by the National Office.

 c. Because of the random number selection procedures, it may happen that the next establishment does not belong to the same

OSHA Instruction CPL 2-2.44A
AUG 15 1988
Office of Health Compliance Assistance

        Area Office jurisdiction and may have to be assigned to an Area Office other than the one deleting the first establishment.

5. When an inspection of an establishment listed on the Health Care Inspection Register is not conducted because the employer has refused entry, a warrant shall be sought in accordance with the current procedure for handling such refusals. Only if the court refuses to grant a warrant or if the Regional Solicitor declines to apply for a warrant, is such an establishment to be replaced according to I.3.

6. Each Region shall conduct the minimum number of health care facility inspections (e.g., hospitals, clinical laboratories, blood donor centers, etc.) annually identified in its program plan.

7. All inspections, programmed or unprogrammed, conducted at health care facilities or at other facilities (such as manufacturing plants) which support an onsite health care unit shall be directed to all areas involving the hazard of direct exposure to body fluids potentially contaminated with HBV or HIV.

    a. Primary areas of concern are emergency rooms, operating rooms, direct patient care areas, laboratories, and x-ray. Secondary areas of concern are laundry and housekeeping.

    b. Records review procedures for the purpose of conducting a records only inspection do not apply.

    c. Expansion to additional units may be appropriate when:

        (1) The IC program shows significant deficiencies in complying with OSHA requirements, as set forth in this instruction, that may indicate the existence of more widespread problems.

        (2) Relevant complaints are received from employees which are specifically related to direct exposures to body fluids.

OSHA Instruction CPL 2-2.44A
AUG 15 1988
Office of Health Compliance Assistance

J. <u>Inspection Procedures</u>. The procedures given in the FOM, Chapter III, shall be followed except as modified in the following sections.

1. When entering a hospital or health care facility, the CSHO shall locate the Hospital Administrator, the Medical Director or other person in charge and present credentials.

2. Health care facilities generally administer internal IC programs. This function may be performed by a committee or an individual. Upon entry the CSHO shall request the presence of the infection control nurses and/or other individuals who will be responsible for providing records pertinent to the inspection.

3. Careful examination of the facility's IC program is the core element of these inspections. Occupational injury and illness records shall be carefully scrutinized, and employees selected from all appropriate areas of the facility shall be interviewed to verify both the accuracy of the OSHA-200 records and the effectiveness of the IC program.

    NOTE: Employers are not required normally to record needlesticks on the OSHA 200. However, if such needlesticks require medical treatment (i.e., gamma globulin) and are identified as causes of diagnosed occupationally related AIDS, AIDS Related Complex, or hepatitis B they shall be recorded.

4. In the event the facility being inspected does not have a formal IC program, employee interviews, combined with an inspection of appropriate areas of the facility shall be used to determine the effectiveness of the establishment's efforts to protect employees from exposure to potential infectious disease sources.

5. CSHOs shall use appropriate caution when entering patient care areas of the facility. When such visits are judged necessary for determining actual conditions in the facility, the privacy of patients shall be respected. Photographs of patients will not normally be necessary and in no event shall

OSHA Instruction CPL 2-2.44A
AUG 15 1988
Office of Health Compliance Assistance

      identifiable photographs be taken without their consent.

  6. The walkaround portion of the inspection shall consist of a spot-check approach. The CSHO shall identify on the basis of professional judgment what areas should be physically checked out and to what extent. It is not expected that a comprehensive walkaround inspection of the workplace will be necessary. The CSHO is to be satisfied that an IC program is in place and judged to be effective.

  7. If an inspection is conducted in an establishment outside of SIC codes 80** and 7261, and a health care unit is on site, the provisions of Sections J through T apply and shall be enforced.

K. <u>Federal Agency Facilities</u>. Health care facilities owned by agencies of the Federal Government are subject to inspection under this instruction.

  1. Where an inspection other than one scheduled on the regular Federal establishment targeting list is to be made at a Federal health care facility in accordance with this instruction, notification of agency headquarters personnel, through the Office of Federal Agency Programs (OFAP), shall be accomplished <u>prior</u> to any facility visits.

  2. Actual onsite visits shall be coordinated between Regional/Area Office and the establishment being inspected.

L. <u>Citation Policy</u>. The provisions of the FOM, Chapters IV and V, shall be followed when issuing citations for hazards related to blood-borne infectious diseases.

  1. The following requirements apply when citing hazards found in health care facilities. Employers must comply with the provisions of these requirements whenever an employee may reasonably be expected to have direct contact with body fluids regardless of whether the patient is known to have been infected with HBV or HIV. This policy is based on the widespread nature of these viruses and the consequent risk to the health care workers described above. It is also based on the need to maintain patient confidentiality and HBV and HIV

OSHA Instruction CPL 2-2.44A
AUG 15 1988
Office of Health Compliance Assistance

testing limitations.

- o 29 CFR 1910.132--Personal protective equipment.

- o 29 CFR 1910.22(a)(1) and (a)(2)--General requirements, Housekeeping.

- o 29 CFR 1910.141(a)(4)(i) and (ii)--Sanitation, Waste disposal.

- o 29 CFR 1910.145(f)--Specifications for accident prevention signs and tags.

- o Section 5(a)(1)--General Duty Clause and Executive Order 12196, Section 1-201(a) for Federal facilities.

2. Whenever a hazardous condition exists that is not covered by one of the standards listed above, and the decision is made not to cite the condition under the general duty clause, the appropriate letter shall be sent advising the employer of the hazardous conditions and suggesting corrective action.

3. Such recommendations made to employers shall be noted in the case file for special attention in subsequent inspections.

4. Multi-Employer Work Site. The following citation guidelines apply in multi-employer worksites.

   (a) Health care facilities shall be cited for standards and section 5(a)(1) violations to which their own employees are exposed.

   (b) They shall also be cited for standards (but not 5(a)(1)) violations to which employees of other employers on their premises are exposed to the extent that they control the hazard. For example, they shall be cited for not providing personal protective equipment to unprotected employees of other employers on their premises.

   (c) Physicians who are members of professional corporations are generally considered to be employees of that corporation. Therefore, the

OSHA Instruction CPL 2-2.44A
AUG 5 1988
Office of Health Compliance Assistance

    corporation may be cited for all violations affecting those physicians. Hospitals where they work may also be cited for standards violations (but not 5(a)(1)) to which they are exposed.

  (d) No citation shall be issued where the only persons exposed are physicians who are sole practitioners or partners, and thus not employees under the Occupational Safety and Health Act.

M. <u>Violations</u>. The IC program shall be carefully evaluated to determine compliance with OSHA requirements, as clarified by those CDC guidelines relating to health care worker safety and health. The description of the OSHA requirements in this section is based upon those guidelines. Violations of OSHA requirements will normally be classified as serious.

  1. <u>29 CFR 1910.132(a) and (c)</u>. The standard provides in pertinent part:

  "(a) <u>Application</u>. Protective equipment, including personal protective equipment for eyes, face, head, and extremities, protective clothing, respiratory devices, and protective shields and barriers, shall be provided, used, and maintained in a sanitary and reliable condition wherever it is necessary by reason of hazards of processes or environment, ... encountered in a manner capable of causing injury or impairment in the function of any part of the body through absorption, inhalation or physical contact.

  (c) <u>Design</u>. All personal protective equipment shall be of safe design and construction for work to be performed."

  The following personal protective measures shall have been addressed by the IC program and verified by interviews and walkaround.

  a. <u>Gloves</u>. The use of gloves will vary according to the procedure involved. The use of disposable gloves is indicated for procedures where body fluids are handled.

OSHA Instruction CPL 2-2.44A
AUG 15 1988
Office of Health Compliance Assistance

(1) The use of gloves is particularly important in the following circumstances:

   <u>a</u> If the health care worker has cuts, abraded skin, chapped hands, dermatitis or the like.

   <u>b</u> During instrumental examination of otopharynx, gastrointestinal tract and genitourinary tract.

   <u>c</u> When examining abraded or non-intact skin or patients with active bleeding.

   <u>d</u> During invasive procedures.

   <u>e</u> During all cleaning of body fluids and decontaminating procedures.

(2) Gloves must be of appropriate material, usually intact latex or intact vinyl, of appropriate quality for the procedures performed, and of appropriate size for each health care worker. Where gloves do not meet these requirements 29 CFR 1910.132(c) shall be cited.

(3) Employers shall not wash or disinfect surgical or examination gloves for reuse.

(4) General purpose utility (rubber) gloves worn by maintenance, housekeeping, laundry or other non medical personnel may be decontaminated and reused.

(5) No gloves shall be used if they are peeling, cracked, or discolored, or if they have punctures, tears, or other evidence of deterioration. Failure to meet these requirements shall be cited under 29 CFR 1910.132(c).

b. <u>Gowns</u>. The use of gowns, aprons, or lab coats is required when splashes to skin or clothing with body fluids are likely to occur. Gowns, including surgical gowns, shall be made of or lined with impervious material and shall

OSHA Instruction CPL 2-2.44A
AUG 15 1988
Office of Health Compliance Assistance

protect all areas of exposed skin.

c. <u>Masks and Eye Protectors</u>. The use of masks and protective eyewear or face shields is required when contamination of mucosal membranes (eyes, mouth or nose) with body fluids such as splashes or aerosolization of such material (e.g., during surgical or dental procedures), is likely to occur. They are <u>not</u> required for routine care.

d. <u>Resuscitation Equipment</u>. Pocket masks, resuscitation bags, or other ventilation devices shall be provided in strategic locations as well as to key personnel (e.g. paramedics) where the need for resuscitation is likely. This will minimize the need for emergency mouth-to-mouth resuscitation.

e. <u>Invasive Procedures</u>. Personal protective equipment as described above shall be used when performing invasive procedures to avoid exposure. When a health care worker's skin or mucous membranes may come in contact with body fluids, gowns, masks, and eye protection shall be worn, as noted above.

f. <u>Phlebotomy</u>. Gloves shall generally be provided to and used by phlebotomists. Employers who do not make them available shall be cited for failure to <u>provide</u> under 29 CFR 1910.132(a). Employers who make gloves available, but discourage or prohibit their use shall be cited for failure to <u>use</u> under 29 CFR 1910.132(a), if in fact the gloves are not being used. However, no citation for failure to use shall be issued where the phlebotomist voluntarily and without the encouragement of the employer does not wear gloves, unless the following circumstances exist:

(1) For performing phlebotomy when the health care worker has cuts, scratches, or other breaks in his/her skin.

(2) In situations where the CSHO and/or health care worker judges that hand contamination with blood may occur, for example, when

OSHA Instruction CPL 2-2.44A
AUG 15 1988
Office of Health Compliance Assistance

        performing phlebotomy on an uncooperative patient.

       (3) For performing finger and/or heel sticks in infants and children.

       (4) When persons are receiving training in phlebotomy.

   g. **Dentistry.** Gloves are required for contact with oral mucous membranes. Surgical mask and protective eyewear or chin-length plastic face shields are required during dental procedures in which splashing, spattering or aerosolization of blood, saliva or gingival fluids is likely. (Saliva and gingival fluids are included because of the likelihood that they contain blood in their setting.)

   h. **Laboratories.** The use of gloves are required for processing body fluid specimens. Masks and protective eyewear are required when the worker's mucosal membranes may come in contact with body fluids.

   i. **Postmortem Procedures.** Persons performing or assisting in postmortem procedures are required to wear personal protective equipment as noted above to avoid exposure to body fluids.

2. **29 CFR 1910.22(a)(1) and (a)(2).** The standard provides in pertinent part:

   "(a) **Housekeeping.** (1) All places of employment, passageways, storerooms, and service rooms shall be kept clean and orderly and in a sanitary condition.

   (2) The floor of every workroom shall be maintained in a clean and, so far as possible, a dry condition...."

The IC program shall have identified housekeeping operations involving substantial risk of direct exposure to body fluids and shall have addressed the proper precautions to be taken while cleaning

OSHA Instruction CPL 2-2.44A
NOV 15 1988
Office of Health Compliance Assistance

rooms and blood spills. The application of these procedures shall be verified by employee interviews and the walkaround.

a. **Room Cleaning Where Body Fluids are Present**. Schedules shall be as frequent as necessary according to the area of the institution, type of surface to be cleaned, and the amount and type of soil present.

b. **Disinfectants**. Chemical germicides that are approved for use as hospital disinfectants and are tuberculocidal when used at recommended dilutions shall be used to decontaminate spills of blood and other body fluids. A solution of 5.25 percent sodium hypochlorite (household bleach) diluted between 1:10 and 1:100 with water or other suitable disinfectant shall be used for disinfection following the initial cleanup.

3. **29 CFR 1910.141(a)(4)(i) and (ii)**. The standard provides in pertinent part:

"(4) **Waste disposal**. (i) Any receptacle used for putrescible solid or liquid waste or refuse shall be so constructed that it does not leak and may be thoroughly cleaned and maintained in a sanitary condition. Such a receptacle shall be equipped with a solid, tight-fitting cover, unless it can be maintained in a sanitary condition without a cover. This requirement does not prohibit the use of receptacles which are designed to permit the maintenance of a sanitary condition without regard to the aforementioned requirements.

(ii) All sweepings, solid or liquid wastes, refuse, and garbage shall be removed in such a manner as to avoid creating a menace to health and as often as necessary or appropriate to maintain the place of employment in a sanitary condition."

The IC program shall have addressed the handling and disposal of the following potentially contaminated items. The effectiveness of the program in this regard shall be verified through employee

OSHA Instruction CPL 2-2.44A
AUG 15 1988
Office of Health Compliance Assistance

interviews and the walkaround.

a. <u>Sharp instruments and disposable items</u>.
Needles shall not be recapped, purposely bent or broken by hand, removed from disposable syringes, or otherwise manipulated by hand.

   (1) After they are used, disposable syringes and needles, scalpel blades, and other sharp items shall be placed in puncture-resistant containers for disposal.

   (2) Such containers shall be easily accessible to personnel needing them and located in all areas where needles are commonly used, including emergency rooms, intensive care units, and surgical suites and shall be so constructed that they will not spill their contents if knocked over and will not themselves allow injuries when handled.

   (3) These containers shall also be located on patient floors and any other setting where blood is drawn and needles are used.

b. <u>Lab specimens</u>. All specimens of body fluids shall be put in a well constructed container with a secure lid to prevent leaking during transport and shall be disposed of in an approved manner. Contaminated materials used in laboratory tests should be decontaminated before reprocessing or be placed in bags and disposed of in accordance with institutional policies for disposal of infectious waste.

4. <u>29 CFR 1910.145(f)</u>. The standard provides in pertinent part:

   "(f)(3) <u>Use</u>. Tags shall be used as means to prevent accidental injury or illness to employees who are exposed to hazardous or potentially hazardous conditions, equipment or operations which are out of the ordinary, unexpected or not readily apparent. Tags shall be used until such time as the identified hazard is eliminated or the hazardous operation is completed. Tags need not be used where signs, guarding or other positive

OSHA Instruction CPL 2-2.44A
AUG 15 1990
Office of Health Compliance Assistance

means of protection are being used.

(4) <u>General tag criteria</u>. All required tags shall meet the following criteria:

(i) Tags shall contain a signal word and a major message.

(a) The signal word shall be ... "BIOHAZARD," or the biological hazard symbol.

(b) The major message shall indicate the specific hazardous condition or the instruction to be communicated to the employee.

(ii) The signal word shall be readable at a minimum distance of five feet (1.52m) or such greater distance as warranted by the hazard.

(iii) The tag's major message shall be presented in either pictographs, written text or both.

(iv) The signal word and the major message shall be understandable to all employees who may be exposed to the identified hazard.

(v) All employees shall be informed as to the meaning of the various tags used throughout the workplace and what special precautions are necessary.

(vi) Tags shall be affixed as close as safely possible to their respective hazards by a positive means such as string, wire, or adhesive that prevents their loss or unintentional removal.

(f)(8) <u>Biological hazard tags</u>. (i) Biological hazard tags shall be used to identify the actual or potential presence of a biological hazard and to identify equipment, containers, rooms, experimental animals, or combinations thereof, that contain or are contaminated with hazardous biological agents."

The IC program shall have addressed the labeling procedures to be followed in the facility. That these procedures are followed shall confirmed by employee interviews and the walkaround.

OSHA Instruction CPL 2-2.44A
Office of Health Compliance Assistance

a. Bags or other receptacles containing articles contaminated with potentially infectious material, including contaminated disposable items, must be tagged or otherwise identified. The tag shall have the signal word "BIOHAZARD" or the biological hazard symbol. The tag shall indicate that the bag could contain infectious wastes and give any additional instructions; e.g., if the outside of the bag is contaminated with body fluids, a second outer bag should be used.

b. If tags are not used, other equally effective means of identification shall be used (e.g., red bagging).

c. Employees shall be informed of the meaning of tags. With respect to tagged material, they shall also be instructed to use double bagging where puncture or outside contamination is likely.

5. Section 5(a)(1). Section 5(a)(1) provides:

"Each employer shall furnish to each of his employees employment and a place of employment which are free from recognized hazards that are causing or are likely to cause death or serious physical harm to his employees;"

a. Section 5(a)(1) citations must meet the requirements outlined in the FOM, Chapter IV, and can be issued only where there is a hazard which cannot be abated by implementing an abatement method required by the standards above. All applicable abatement methods identified as correcting the same hazard shall be issued under a single 5(a)(1) citation.

b. If a citation under 5(a)(1) is justified, the citation, after setting forth the SAVE for section 5(a)(1), shall state:

Health care workers (specify categories, such as doctors, nurses, etc.) (Specify location) were exposed to the hazard of being infected by HBV and/or HIV through

OSHA Instruction CPL 2-2.44A
AUG 15 1988
Office of Health Compliance Assistance

      possible direct contact with blood or other body fluids. Feasible and useful abatement methods for reducing this hazard, among others, are: (List abatement methods not required by the standards which employer is not implementing.)

  c. Recognition for purposes of citing section 5(a)(1) is recognition of the hazard of being infected with HBV and/or HIV through possible direct contact with body fluids. The health care industry generally accepts and, therefore, recognizes the determination of this hazard by the CDC, which is the acknowledged authority in this area. The employer's IC program can also constitute evidence of recognition.

  d. The following are examples of feasible and useful abatement methods. The non-use of any of these methods is likely to result in the continued existence of a serious hazard and, may, therefore, allow citation under 5(a)(1). Consequently, all of these methods shall have been implemented. To determine whether they are being implemented, the CSHO shall evaluate the IC program and verify with employee interviews and the walkaround.

    (1) <u>Hepatitis B Vaccination</u>. The facility's IC policy regarding hepatitis B vaccinations shall address all circumstances warranting such vaccinations and shall identify employees at substantial risk of directly contacting body fluids. All such employees shall be offered hepatitis B vaccinations in amounts and at times prescribed by standard medical practice.

    (2) <u>Linen</u>. The IC program shall have identified all laundry operations involving substantial risk of direct exposure to body fluids. Linen soiled with body fluids shall be handled as little as possible and with minimum agitation to prevent contamination of the person handling the linen. All soiled linen shall be bagged at the location where it was used; it shall not be sorted

OSHA Instruction CPL 2-2.44A
AUG 15 1988
Office of Health Compliance Assistance

or rinsed in patient-care areas. Soiled linen shall be placed and transported in bags that prevent leakage.

(3) <u>Reusable Equipment</u>. Standard sterilization and disinfection procedures currently recommended for hepatitis B in a variety of health care settings are adequate to sterilize or disinfect instruments, devices, or other items contaminated with body fluids. A recommended source of information is the CDC's Guidelines for Hospital Environmental Control: Cleaning, Disinfection and Sterilization of Hospital Equipment.

(4) <u>Bagging of Articles</u>. Objects that are contaminated with potentially infectious materials shall be placed in an impervious bag. If outside contamination of the bag is likely, a second bag shall be added.

(5) <u>Handwashing</u>. After removing gloves, hands or other skin surfaces shall be washed thoroughly and immediately after contact with body fluids.

(6) <u>Follow-up Procedures After Possible Exposure to HIV/HBV</u>:

    (a) If a health care worker has a percutaneous (needlestick or cut) or mucous membrane (splash to eye, nasal mucosa, or mouth) exposure to body fluids or has a cutaneous exposure to blood when the worker's skin is chapped, abraded, or otherwise nonintact, the source patient shall be informed of the incident and tested for HIV and HBV infections, after consent is obtained.

    (b) If patient consent is refused or if the source patient tests postive, the health care worker shall be evaluated clinically and by HIV antibody testing as soon as possible and advised to report and seek medical

OSHA Instruction CPL 2-2.44A
AUG 15 1988
Office of Health Compliance Assistance

        evaluation of any acute febrile illness that occurs within 12 weeks after exposure. HIV seronegative workers shall be retested 6 weeks post-exposure and on a periodic basis thereafter (12 weeks and 6 months after exposure).

    (c) Follow up procedures shall be taken for health care workers exposed or potentially exposed to HBV. The types of procedures depends on the immunization status of the worker (i.e., whether HBV vaccination has been received and antibody response is adequate) and the HBV serologic status of the source patient. The CDC Immunization Practices Advisory Committee has published its recommendations regarding HBV postexposure prophylaxis in table format in the June 7, 1985, Morbidity and Mortality Weekly Report.

    (d) If an employee refuses to submit to the procedures in (b) or (c) above when such procedures are medically indicated, no adverse action can be taken on that ground alone since the procedures are designed for the benefit of the exposed employee.

(7) <u>Training and Education of Health Care Workers</u>. The employer's training program shall be evaluated in accordance with Appendix C.

    (a) All high risk health care workers such as those listed in H.1. shall receive education on precautionary measures, epidemiology, modes of transmission and prevention of HIV/HBV. Health care workers shall be counseled regarding possible risks to the fetus from HIV/HBV and other associated infectious agents.

OSHA Instruction CPL 2-2.44A
AUG
Office of Health Compliance Assistance

    (b) In addition, such high risk workers must receive training regarding the location and proper use of personal protective equipment. They shall be trained concerning proper work practices and, if the facility has implemented them, shall understand the concept of "universal precautions" as it applies to their work practices. They shall be trained about the meaning of color coding or other methods (except tags) used to designate contaminated articles or infectious waste. Where tags are used, training about tags and precautions to be used in handling contaminated articles or infectious waste is governed by 29 CFR 1910.145(f). (See section M.4.) Workers shall receive training about procedures to be used if they are exposed to needlestick or to body fluids.

N. <u>Other Standards</u>.

1. The hazard communication standard, 29 CFR 1910.1200 only applies to hazardous chemicals or physical hazards in the workplace and thus does not apply to biological hazards such as blood borne diseases.

2. A record concerning employee exposure to HIV and/or HBV is an employee exposure record within the meaning of 29 CFR 1910.20. A record about HIV and/or HBV status is also an employee medical record within the meaning of 29 CFR 1910.20. However, under 29 CFR 1913.10, the CSHO may obtain these records for purposes of determining compliance with 29 CFR 1910.20.

3. Generally, 29 CFR 1910.134 does not apply since there are no respirators approved for biohazards. However, placing respirators in areas where they could be contaminated by body fluids constitutes a violation of 29 CFR 1910.134(b)(6).

O. <u>Expert Witnesses</u>. The Directorate of Technical Support will assist the Regional Offices and the States in

OSHA Instruction CPL 2-2.44A
AUG 15 1989
Office of Health Compliance Assistance

locating expert witnesses.

1. In the event that a 5(a)(1) citation is contested, proper expert witness support shall be immediately obtained. Issues which the expert must be prepared to address include:

   a. The risk to workers associated with the exposure circumstances.

   b. Existence, feasibility and utility of abatement measures.

   c. Recognition of the hazard in the industry, by the employer.

2. Expert witnesses may also be necessary in other cases, particularly those involving 29 CFR 1910.132(a).

P. <u>Recording in the IMIS</u>. Current instructions for completing the appropriate inspection classification boxes (Items 24 and 25) on the OSHA-1, Inspection Report, as found in the IMIS Manual shall be applied when recording inspections conducted at health care facilities or at facilities with health care units:

1. Inspections conducted in such facilities shall be coded as "Comprehensive" or "Partial" in Item 35 of the OSHA-1, as appropriate. Such inspections shall not be coded as records only inspections.

2. The OSHA-1 for health care facility or unit inspections scheduled as a result of a complaint shall be marked as "Safety" or "Health" as appropriate (Item 21.), "Complaint" (Item 24.), and "National Emphasis Program" (Item 25d.). Record "BLOOD" in the space in Item 25d.

3. The OSHA-1 for health care facility or unit inspections scheduled from the Safety or the Health Establishment List or from the Health Care Establishment List shall be marked as "Safety" or "Health," as appropriate (Item 21.), "Planned" (Item 24h.), "Safety" or "Health Planning Guide" as appropriate (Item 25b.), "National Emphasis Program" (Item 25d.). Record "BLOOD" in the space in Item 25d.

OSHA Instruction CPL 2-2.44A
Office of Health Compliance Assistance

4. The OSHA-1 for any unprogrammed safety or health inspection conducted in a health care facility or unit shall be marked "Unprogrammed" (Item 24a. through g., as appropriate), "National Emphasis Program" (Item 25d.) and "BLOOD" recorded in the space in Item 25d.

Q. <u>Referrals</u>.

1. When a complaint or inquiry is received from a source in a State plan State regarding occupational exposure to blood-borne disease, the Regional Administator shall refer it to the State plan designee for action.

2. When a complaint or inquiry regarding occupational exposure to blood-borne disease in a State or local government health care facility is received in a State without an OSHA approved State plan, the Regional Administrator shall refer it to the appropriate State public health agency or local health agency with jurisdiction over the health care facilities.

R. <u>Personal Protective Equipment for CSHOs</u>.

1. CSHOs shall not participate in activities that will require them to come into contact with body fluids, needles or other sharp instruments contaminated with blood. To evaluate such activities, CSHOs normally shall establish the existence of hazards and adequacy of work practices through employee interviews and shall observe them at a safe distance.

2. CSHOs shall take necessary precautions to avoid direct contact with body fluids. It will not normally be necessary for CSHOs actually to enter hazardous areas and, therefore, to use personal protective equipment. On the rare occasions when entry into potentially hazardous areas is judged necessary, the CSHO shall be properly equipped as required by the health care facility as well as by his/her own professional judgement, after consultation with the supervisor (FOM, Chapter III).

S. <u>Copies of Citations</u>. Copies of all citations issued pursuant to this program shall be sent expeditiously to

OSHA Instruction CPL 2-2.44A

Office of Health Compliance Assistance

the Office of Health Compliance Assistance through the Director of Field Programs. Such information is necessary for program monitoring and evaluation.

T. **Informal Settlement Agreements and Notices of Contest.** Copies of all settlement agreements and notices of contest shall be sent expeditiously to the Office of Health Compliance Assistance through the Director of Field Programs.

John A. Pendergrass
Assistant Secretary

DISTRIBUTION: National, Regional and Area Offices
Compliance Officers
State Designees
NIOSH Regional Program Directors
7(c)(1) Project Managers

OSHA Instruction CPL 2-2.44A
Office of Health Compliance Assistance

APPENDIX A

CENTERS FOR DISEASE CONTROL

MMWR

MORBIDITY AND MORTALITY WEEKLY REPORT

June 24, 1988 / Vol. 37 / No. 24

377 Update: Universal Precautions for Prevention of Transmission of Human Immunodeficiency Virus, Hepatitis B Virus, and Other Bloodborne Pathogens in Health-Care Settings

Perspectives in Disease Prevention and Health Promotion

Update: Universal Precautions for Prevention of Transmission of Human Immunodeficiency Virus, Hepatitis B Virus, and Other Bloodborne Pathogens in Health-Care Settings

Introduction

The purpose of this report is to clarify and supplement the CDC publication entitled "Recommendations for Prevention of HIV Transmission in Health-Care Settings" (*1*).*

In 1983, CDC published a document entitled "Guideline for Isolation Precautions in Hospitals" (*2*) that contained a section entitled "Blood and Body Fluid Precautions." The recommendations in this section called for blood and body fluid precautions when a patient was known or suspected to be infected with bloodborne pathogens. In August 1987, CDC published a document entitled "Recommendations for Prevention of HIV Transmission in Health-Care Settings" (*1*). In contrast to the 1983 document, the 1987 document recommended that blood and body fluid precautions be consistently used for all patients regardless of their bloodborne infection status. This extension of blood and body fluid precautions to all patients is referred to as "Universal Blood and Body Fluid Precautions" or "Universal Precautions." Under universal precautions, blood and certain body fluids of all patients are considered potentially infectious for human immunodeficiency virus (HIV), hepatitis B virus (HBV), and other bloodborne pathogens.

---

*The August 1987 publication should be consulted for general information and specific recommendations not addressed in this update.

OSHA Instruction CPL 2-2.44A

Office of Health Compliance Assistance

Update: HIV — Continued

Universal precautions are intended to prevent parenteral, mucous membrane, and nonintact skin exposures of health-care workers to bloodborne pathogens. In addition, immunization with HBV vaccine is recommended as an important adjunct to universal precautions for health-care workers who have exposures to blood (*3,4*).

Since the recommendations for universal precautions were published in August 1987, CDC and the Food and Drug Administration (FDA) have received requests for clarification of the following issues: 1) body fluids to which universal precautions apply, 2) use of protective barriers, 3) use of gloves for phlebotomy, 4) selection of gloves for use while observing universal precautions, and 5) need for making changes in waste management programs as a result of adopting universal precautions.

Body Fluids to Which Universal Precautions Apply

Universal precautions apply to blood and to other body fluids containing visible blood. Occupational transmission of HIV and HBV to health-care workers by blood is documented (*4,5*). **Blood is the single most important source of HIV, HBV, and other bloodborne pathogens in the occupational setting. Infection control efforts for HIV, HBV, and other bloodborne pathogens must focus on preventing exposures to blood as well as on delivery of HBV immunization.**

Universal precautions also apply to semen and vaginal secretions. Although both of these fluids have been implicated in the sexual transmission of HIV and HBV, they have not been implicated in occupational transmission from patient to health-care worker. This observation is not unexpected, since exposure to semen in the usual health-care setting is limited, and the routine practice of wearing gloves for performing vaginal examinations protects health-care workers from exposure to potentially infectious vaginal secretions.

Universal precautions also apply to tissues and to the following fluids: cerebrospinal fluid (CSF), synovial fluid, pleural fluid, peritoneal fluid, pericardial fluid, and amniotic fluid. The risk of transmission of HIV and HBV from these fluids is unknown; epidemiologic studies in the health-care and community setting are currently inadequate to assess the potential risk to health-care workers from occupational exposures to them. However, HIV has been isolated from CSF, synovial, and amniotic fluid (*6–8*), and HBsAg has been detected in synovial fluid, amniotic fluid, and peritoneal fluid (*9–11*). One case of HIV transmission was reported after a percutaneous exposure to bloody pleural fluid obtained by needle aspiration (*12*). Whereas aseptic procedures used to obtain these fluids for diagnostic or therapeutic purposes protect health-care workers from skin exposures, they cannot prevent penetrating injuries due to contaminated needles or other sharp instruments.

Body Fluids to Which Universal Precautions Do Not Apply

Universal precautions do not apply to feces, nasal secretions, sputum, sweat, tears, urine, and vomitus unless they contain visible blood. The risk of transmission of HIV and HBV from these fluids and materials is extremely low or nonexistent. HIV has been isolated and HBsAg has been demonstrated in some of these fluids; however, epidemiologic studies in the health-care and community setting have not implicated these fluids or materials in the transmission of HIV and HBV infections (*13,14*). Some of the above fluids and excretions represent a potential source for nosocomial and community-acquired infections with other pathogens, and recommendations for preventing the transmission of nonbloodborne pathogens have been published (*2*).

Update: HIV — Continued

Precautions for Other Body Fluids in Special Settings

Human breast milk has been implicated in perinatal transmission of HIV, and HBsAg has been found in the milk of mothers infected with HBV (*10,13*). However, occupational exposure to human breast milk has not been implicated in the transmission of HIV nor HBV infection to health-care workers. Moreover, the health-care worker will not have the same type of intensive exposure to breast milk as the nursing neonate. Whereas universal precautions do not apply to human breast milk, gloves may be worn by health-care workers in situations where exposures to breast milk might be frequent, for example, in breast milk banking.

Saliva of some persons infected with HBV has been shown to contain HBV-DNA at concentrations 1/1,000 to 1/10,000 of that found in the infected person's serum (*15*). HBsAg-positive saliva has been shown to be infectious when injected into experimental animals and in human bite exposures (*16–18*). However, HBsAg-positive saliva has not been shown to be infectious when applied to oral mucous membranes in experimental primate studies (*18*) or through contamination of musical instruments or cardiopulmonary resuscitation dummies used by HBV carriers (*19,20*). Epidemiologic studies of nonsexual household contacts of HIV-infected patients, including several small series in which HIV transmission failed to occur after bites or after percutaneous inoculation or contamination of cuts and open wounds with saliva from HIV-infected patients, suggest that the potential for salivary transmission of HIV is remote (*5,13,14,21,22*). One case report from Germany has suggested the possibility of transmission of HIV in a household setting from an infected child to a sibling through a human bite (*23*). The bite did not break the skin or result in bleeding. Since the date of seroconversion to HIV was not known for either child in this case, evidence for the role of saliva in the transmission of virus is unclear (*23*). Another case report suggested the possibility of transmission of HIV from husband to wife by contact with saliva during kissing (*24*). However, follow-up studies did not confirm HIV infection in the wife (*21*).

Universal precautions do not apply to saliva. General infection control practices already in existence — including the use of gloves for digital examination of mucous membranes and endotracheal suctioning, and handwashing after exposure to saliva — should further minimize the minute risk, if any, for salivary transmission of HIV and HBV (*1,25*). Gloves need not be worn when feeding patients and when wiping saliva from skin.

Special precautions, however, are recommended for dentistry (*1*). Occupationally acquired infection with HBV in dental workers has been documented (*4*), and two possible cases of occupationally acquired HIV infection involving dentists have been reported (*5,26*). During dental procedures, contamination of saliva with blood is predictable, trauma to health-care workers' hands is common, and blood spattering may occur. Infection control precautions for dentistry minimize the potential for nonintact skin and mucous membrane contact of dental health-care workers to blood-contaminated saliva of patients. In addition, the use of gloves for oral examinations and treatment in the dental setting may also protect the patient's oral mucous membranes from exposures to blood, which may occur from breaks in the skin of dental workers' hands.

Use of Protective Barriers

Protective barriers reduce the risk of exposure of the health-care worker's skin or mucous membranes to potentially infective materials. For universal precautions,

OSHA Instruction CPL 2-2.44A

Office of Health Compliance Assistance

Update: HIV — Continued

protective barriers reduce the risk of exposure to blood, body fluids containing visible blood, and other fluids to which universal precautions apply. Examples of protective barriers include gloves, gowns, masks, and protective eyewear. Gloves should reduce the incidence of contamination of hands, but they cannot prevent penetrating injuries due to needles or other sharp instruments. Masks and protective eyewear or face shields should reduce the incidence of contamination of mucous membranes of the mouth, nose, and eyes.

Universal precautions are intended to supplement rather than replace recommendations for routine infection control, such as handwashing and using gloves to prevent gross microbial contamination of hands (27). Because specifying the types of barriers needed for every possible clinical situation is impractical, some judgment must be exercised.

The risk of nosocomial transmission of HIV, HBV, and other bloodborne pathogens can be minimized if health-care workers use the following general guidelines:[†]

1. Take care to prevent injuries when using needles, scalpels, and other sharp instruments or devices; when handling sharp instruments after procedures; when cleaning used instruments; and when disposing of used needles. Do not recap used needles by hand; do not remove used needles from disposable syringes by hand; and do not bend, break, or otherwise manipulate used needles by hand. Place used disposable syringes and needles, scalpel blades, and other sharp items in puncture-resistant containers for disposal. Locate the puncture-resistant containers as close to the use area as is practical.

2. Use protective barriers to prevent exposure to blood, body fluids containing visible blood, and other fluids to which universal precautions apply. The type of protective barrier(s) should be appropriate for the procedure being performed and the type of exposure anticipated.

3. Immediately and thoroughly wash hands and other skin surfaces that are contaminated with blood, body fluids containing visible blood, or other body fluids to which universal precautions apply.

Glove Use for Phlebotomy

Gloves should reduce the incidence of blood contamination of hands during phlebotomy (drawing blood samples), but they cannot prevent penetrating injuries caused by needles or other sharp instruments. The likelihood of hand contamination with blood containing HIV, HBV, or other bloodborne pathogens during phlebotomy depends on several factors: 1) the skill and technique of the health-care worker, 2) the frequency with which the health-care worker performs the procedure (other factors being equal, the cumulative risk of blood exposure is higher for a health-care worker who performs more procedures), 3) whether the procedure occurs in a routine or emergency situation (where blood contact may be more likely), and 4) the prevalence of infection with bloodborne pathogens in the patient population. The likelihood of infection after skin exposure to blood containing HIV or HBV will depend on the concentration of virus (viral concentration is much higher for hepatitis B than for HIV), the duration of contact, the presence of skin lesions on the hands of the health-care worker, and — for HBV — the immune status of the health-care worker. Although not accurately quantified, the risk of HIV infection following intact skin contact with infective blood is certainly much less than the 0.5% risk following percutaneous

[†] The August 1987 publication should be consulted for general information and specific recommendations not addressed in this update.

OSHA Instruction CPL 2-2.44A

Office of Health Compliance Assistance

Update: HIV — Continued

needlestick exposures (5). In universal precautions, *all* blood is assumed to be potentially infective for bloodborne pathogens, but in certain settings (e.g., volunteer blood-donation centers) the prevalence of infection with some bloodborne pathogens (e.g., HIV, HBV) is known to be very low. Some institutions have relaxed recommendations for using gloves for phlebotomy procedures by skilled phlebotomists in settings where the prevalence of bloodborne pathogens is known to be very low.

Institutions that judge that routine gloving for *all* phlebotomies is not necessary should periodically reevaluate their policy. Gloves should always be available to health-care workers who wish to use them for phlebotomy. In addition, the following general guidelines apply:

1. Use gloves for performing phlebotomy when the health-care worker has cuts, scratches, or other breaks in his/her skin.
2. Use gloves in situations where the health-care worker judges that hand contamination with blood may occur, for example, when performing phlebotomy on an uncooperative patient.
3. Use gloves for performing finger and/or heel sticks on infants and children.
4. Use gloves when persons are receiving training in phlebotomy.

Selection of Gloves

The Center for Devices and Radiological Health, FDA, has responsibility for regulating the medical glove industry. Medical gloves include those marketed as sterile surgical or nonsterile examination gloves made of vinyl or latex. General purpose utility ("rubber") gloves are also used in the health-care setting, but they are not regulated by FDA since they are not promoted for medical use. There are no reported differences in barrier effectiveness between intact latex and intact vinyl used to manufacture gloves. Thus, the type of gloves selected should be appropriate for the task being performed.

The following general guidelines are recommended:

1. Use sterile gloves for procedures involving contact with normally sterile areas of the body.
2. Use examination gloves for procedures involving contact with mucous membranes, unless otherwise indicated, and for other patient care or diagnostic procedures that do not require the use of sterile gloves.
3. Change gloves between patient contacts.
4. Do not wash or disinfect surgical or examination gloves for reuse. Washing with surfactants may cause "wicking," i.e., the enhanced penetration of liquids through undetected holes in the glove. Disinfecting agents may cause deterioration.
5. Use general-purpose utility gloves (e.g., rubber household gloves) for housekeeping chores involving potential blood contact and for instrument cleaning and decontamination procedures. Utility gloves may be decontaminated and reused but should be discarded if they are peeling, cracked, or discolored, or if they have punctures, tears, or other evidence of deterioration.

Waste Management

Universal precautions are not intended to change waste management programs previously recommended by CDC for health-care settings (1). Policies for defining, collecting, storing, decontaminating, and disposing of infective waste are generally determined by institutions in accordance with state and local regulations. Information

OSHA Instruction CPL 2-2.44A

Office of Health Compliance Assistance

382 MMWR June 24, 1988

Update: HIV — Continued

regarding waste management regulations in health-care settings may be obtained from state or local health departments or agencies responsible for waste management.

Reported by: Center for Devices and Radiological Health, Food and Drug Administration. Hospital Infections Program, AIDS Program, and Hepatitis Br, Div of Viral Diseases, Center for Infectious Diseases, National Institute for Occupational Safety and Health, CDC.

Editorial Note: Implementation of universal precautions does not eliminate the need for other category- or disease-specific isolation precautions, such as enteric precautions for infectious diarrhea or isolation for pulmonary tuberculosis (*1,2*). In addition to universal precautions, detailed precautions have been developed for the following procedures and/or settings in which prolonged or intensive exposures to blood occur: invasive procedures, dentistry, autopsies or morticians' services, dialysis, and the clinical laboratory. These detailed precautions are found in the August 21, 1987, "Recommendations for Prevention of HIV Transmission in Health-Care Settings" (*1*). In addition, specific precautions have been developed for research laboratories (*28*).

(Continued on page 387)

TABLE I. Summary — cases of specified notifiable diseases, United States

Disease	24th Week Ending			Cumulative, 24th Week Ending		
	Jun. 18, 1988	Jun. 20, 1987	Median 1983-1987	Jun. 18, 1988	Jun. 20, 1987	Median 1983-1987
Acquired Immunodeficiency Syndrome (AIDS)	198	U*	187	13,918	8,486	3,287
Aseptic meningitis	98	164	123	1,855	2,374	2,102
Encephalitis: Primary (arthropod-borne & unspec)	10	18	17	300	405	405
Post-infectious	1	4	3	44	54	54
Gonorrhea: Civilian	11,071	14,550	17,073	303,455	363,500	383,650
Military	189	282	407	5,531	7,687	9,454
Hepatitis: Type A	419	481	439	10,868	11,471	10,071
Type B	351	479	532	9,614	11,666	11,451
Non A, Non B	51	60	74	1,137	1,461	1,623
Unspecified	23	75	102	930	1,477	2,212
Legionellosis	16	16	16	376	399	314
Leprosy	6	1	3	80	93	121
Malaria	13	17	20	304	341	349
Measles: Total[†]	21	92	92	1,406	2,379	1,620
Indigenous	12	73	73	1,263	2,089	1,436
Imported	9	19	10	143	290	195
Meningococcal infections	44	55	55	1,592	1,648	1,575
Mumps	84	255	93	2,749	9,053	2,000
Pertussis	43	42	58	984	800	865
Rubella (German measles)	15	15	28	115	196	302
Syphilis (Primary & Secondary): Civilian	728	719	566	17,248	15,492	12,764
Military	1	2	2	84	80	93
Toxic Shock syndrome	6	5	5	131	145	178
Tuberculosis	435	442	475	8,999	9,396	9,397
Tularemia	7	9	8	68	64	68
Typhoid Fever	6	6	5	159	136	136
Typhus fever, tick-borne (RMSF)	27	35	35	130	154	177
Rabies, animal	78	85	111	1,874	2,368	2,368

TABLE II. Notifiable diseases of low frequency, United States

	Cum. 1988		Cum. 1988
Anthrax	-	Leptospirosis	13
Botulism: Foodborne (Md. 1)	10	Plague	2
Infant	16	Poliomyelitis, Paralytic	-
Other	2	Psittacosis (Upstate N.Y. 1)	36
Brucellosis (Minn. 1)	26	Rabies, human	-
Cholera	-	Tetanus	20
Congenital rubella syndrome	3	Trichinosis (Alaska 28)	37
Congenital syphilis, ages < 1 year			
Diphtheria	-		

*Because AIDS cases are not received weekly from all reporting areas, comparison of weekly figures may be misleading.
[†]Nine of the 21 reported cases for this week were imported from a foreign country or can be directly traceable to a known internationally imported case within two generations.

OSHA Instruction CPL 2-2.44A
Office of Health Compliance Assistance

Update: HIV — Continued

References

1. Centers for Disease Control. Recommendations for prevention of HIV transmission in health-care settings. MMWR 1987;36(suppl no. 2S).
2. Garner JS, Simmons BP. Guideline for isolation precautions in hospitals. Infect Control 1983:4;245–325.
3. Immunization Practices Advisory Committee. Recommendations for protection against viral hepatitis. MMWR 1985;34:313-24,329–35.
4. Department of Labor, Department of Health and Human Services. Joint advisory notice: protection against occupational exposure to hepatitis B virus (HBV) and human immunodeficiency virus (HIV). Washington, DC:US Department of Labor, US Department of Health and Human Services, 1987.
5. Centers for Disease Control. Update: Acquired immunodeficiency syndrome and human immunodeficiency virus infection among health-care workers. MMWR 1988;37:229–34,239.
6. Hollander H, Levy JA. Neurologic abnormalities and recovery of human immunodeficiency virus from cerebrospinal fluid. Ann Intern Med 1987;106:692–5.
7. Wirthrington RH, Cornes P, Harris JRW, et al. Isolation of human immunodeficiency virus from synovial fluid of a patient with reactive arthritis. Br Med J 1987;294:484.
8. Mundy DC, Schinazi RF, Gerber AR, Nahmias AJ, Randall HW. Human immunodeficiency virus isolated from amniotic fluid. Lancet 1987;2:459–60.
9. Onion DK, Crumpacker CS, Gilliland BC. Arthritis of hepatitis associated with Australia antigen. Ann Intern Med 1971;75:29–33.
10. Lee AKY, Ip HMH, Wong VCW. Mechanisms of maternal-fetal transmission of hepatitis B virus. J Infect Dis 1978;138:668–71.
11. Bond WW, Petersen NJ, Gravelle CR, Favero MS. Hepatitis B virus in peritoneal dialysis fluid: A potential hazard. Dialysis and Transplantation 1982;11:592–600.
12. Oskenhendler E, Harzic M, Le Roux J-M, Rabian C, Clauvel JP. HIV infection with seroconversion after a superficial needlestick injury to the finger [Letter]. N Engl J Med 1986;315:582.
13. Lifson AR. Do alternate modes for transmission of human immunodeficiency virus exist? A review. JAMA 1988;259:1353–6.
14. Friedland GH, Saltzman BR, Rogers MF, et al. Lack of transmission of HTLV-III/LAV infection to household contacts of patients with AIDS or AIDS-related complex with oral candidiasis. N Engl J Med 1986;314:344–9.
15. Jenison SA, Lemon SM, Baker LN, Newbold JE. Quantitative analysis of hepatitis B virus DNA in saliva and semen of chronically infected homosexual men. J Infect Dis 1987;156:299–306.
16. Cancio-Bello TP, de Medina M, Shorey J, Valledor MD, Schiff ER. An institutional outbreak of hepatitis B related to a human biting carrier. J Infect Dis 1982;146:652–6.
17. MacQuarrie MB, Forghani B, Wolochow DA. Hepatitis B transmitted by a human bite. JAMA 1974;230:723–4.
18. Scott RM, Snitbhan R, Bancroft WH, Alter HJ, Tingpalapong M. Experimental transmission of hepatitis B virus by semen and saliva. J Infect Dis 1980;142:67–71.
19. Glaser JB, Nadler JP. Hepatitis B virus in a cardiopulmonary resuscitation training course: Risk of transmission from a surface antigen-positive participant. Arch Intern Med 1985;145:1653–5.
20. Osterholm MT, Bravo ER, Crosson JT, et al. Lack of transmission of viral hepatitis type B after oral exposure to HBsAg-positive saliva. Br Med J 1979;2:1263–4.
21. Curran JW, Jaffe HW, Hardy AM, et al. Epidemiology of HIV infection and AIDS in the United States. Science 1988;239:610–6.
22. Jason JM, McDougal JS, Dixon G, et al. HTLV-III/LAV antibody and immune status of household contacts and sexual partners of persons with hemophilia. JAMA 1986;255:212–5.
23. Wahn V, Kramer HH, Voit T, Brüster HT, Scrampical B, Scheid A. Horizontal transmission of HIV infection between two siblings [Letter]. Lancet 1986;2:694.
24. Salahuddin SZ, Groopman JE, Markham PD, et al. HTLV-III in symptom-free seronegative persons. Lancet 1984;2:1418–20.
25. Simmons BP, Wong ES. Guideline for prevention of nosocomial pneumonia. Atlanta: US Department of Health and Human Services, Public Health Service, Centers for Disease Control, 1982.

OSHA Instruction CPL 2-2.44A
Office of Health Compliance Assistance

Appendix B

Recommendations for Prevention of HIV Transmission in Health-Care Settings

Introduction

Human immunodeficiency virus (HIV), the virus that causes acquired immunodeficiency syndrome (AIDS), is transmitted through sexual contact and exposure to infected blood or blood components and perinatally from mother to neonate. HIV has been isolated from blood, semen, vaginal secretions, saliva, tears, breast milk, cerebrospinal fluid, amniotic fluid, and urine and is likely to be isolated from other body fluids, secretions, and excretions. However, epidemiologic evidence has implicated only blood, semen, vaginal secretions, and possibly breast milk in transmission.

The increasing prevalence of HIV increases the risk that health-care workers will be exposed to blood from patients infected with HIV, especially when blood and body-fluid precautions are not followed for all patients. Thus, this document emphasizes the need for health-care workers to consider **all** patients as potentially infected with HIV and/or other blood-borne pathogens and to adhere rigorously to infection-control precautions for minimizing the risk of exposure to blood and body fluids of all patients.

The recommendations contained in this document consolidate and update CDC recommendations published earlier for preventing HIV transmission in health-care settings: precautions for clinical and laboratory staffs (1) and precautions for health-care workers and allied professionals (2); recommendations for preventing HIV transmission in the workplace (3) and during invasive procedures (4); recommendations for preventing possible transmission of HIV from tears (5); and recommendations for providing dialysis treatment for HIV-infected patients (6). These recommendations also update portions of the "Guideline for Isolation Precautions in Hospitals" (7) and reemphasize some of the recommendations contained in "Infection Control Practices for Dentistry" (8). The recommendations contained in this document have been developed for use in health-care settings and emphasize the need to treat blood and other body fluids from **all** patients as potentially infective. These same prudent precautions also should be taken in other settings in which persons may be exposed to blood or other body fluids.

Definition of Health-Care Workers

Health-care workers are defined as persons, including students and trainees, whose activities involve contact with patients or with blood or other body fluids from patients in a health-care setting.

OSHA Instruction CPL 2-2.44A
Office of Health Compliance Assistance

Health-Care Workers with AIDS

As of July 10, 1987, a total of 1,875 (5.8%) of 32,395 adults with AIDS, who had been reported to the CDC national surveillance system and for whom occupational information was available, reported being employed in a health-care or clinical laboratory setting. In comparison, 6.8 million persons—representing 5.6% of the U.S. labor force—were employed in health services. Of the health-care workers with AIDS, 95% have been reported to exhibit high-risk behavior; for the remaining 5%, the means of HIV acquisition was undetermined. Health-care workers with AIDS were significantly more likely than other workers to have an undetermined risk (5% versus 3%, respectively). For both health-care workers and non-health-care workers with AIDS, the proportion with an undetermined risk has not increased since 1982.

AIDS patients initially reported as not belonging to recognized risk groups are investigated by state and local health departments to determine whether possible risk factors exist. Of all health-care workers with AIDS reported to CDC who were initially characterized as not having an identified risk and for whom follow-up information was available, 66% have been reclassified because risk factors were identified or because the patient was found not to meet the surveillance case definition for AIDS. Of the 87 health-care workers currently categorized as having no identifiable risk, information is incomplete on 16 (18%) because of death or refusal to be interviewed; 38 (44%) are still being investigated. The remaining 33 (38%) health-care workers were interviewed or had other follow-up information available. The occupations of these 33 were as follows: five physicians (15%), three of whom were surgeons; one dentist (3%); three nurses (9%); nine nursing assistants (27%); seven housekeeping or maintenance workers (21%); three clinical laboratory technicians (9%); one therapist (3%); and four others who did not have contact with patients (12%). Although 15 of these 33 health-care workers reported parenteral and/or other non-needlestick exposure to blood or body fluids from patients in the 10 years preceding their diagnosis of AIDS, none of these exposures involved a patient with AIDS or known HIV infection.

Risk to Health-Care Workers of Acquiring HIV in Health-Care Settings

Health-care workers with documented percutaneous or mucous-membrane exposures to blood or body fluids of HIV-infected patients have been prospectively evaluated to determine the risk of infection after such exposures. As of June 30, 1987, 883 health-care workers have been tested for antibody to HIV in an ongoing surveillance project conducted by CDC (9). Of these, 708 (80%) had percutaneous exposures to blood, and 175 (20%) had a mucous membrane or an open wound contaminated by blood or body fluid. Of 396 health-care workers, each of whom had only a convalescent-phase serum sample obtained and tested ≥90 days post-exposure, one—for whom heterosexual transmission could not be ruled out—was seropositive for HIV antibody. For 425 additional health-care workers, both acute- and convalescent-phase serum samples were obtained and tested; none of 74 health-care workers with nonpercutaneous exposures seroconverted, and three (0.9%) of 351

OSHA Instruction CPL 2-2.44A
Office of Health Compliance Assistance

with percutaneous exposures seroconverted. None of these three health-care workers had other documented risk factors for infection.

Two other prospective studies to assess the risk of nosocomial acquisition of HIV infection for health-care workers are ongoing in the United States. As of April 30, 1987, 332 health-care workers with a total of 453 needlestick or mucous-membrane exposures to the blood or other body fluids of HIV-infected patients were tested for HIV antibody at the National Institutes of Health (10). These exposed workers included 103 with needlestick injuries and 229 with mucous-membrane exposures; none had seroconverted. A similar study at the University of California of 129 health-care workers with documented needlestick injuries or mucous-membrane exposures to blood or other body fluids from patients with HIV infection has not identified any seroconversions (11). Results of a prospective study in the United Kingdom identified no evidence of transmission among 150 health-care workers with parenteral or mucous-membrane exposures to blood or other body fluids, secretions, or excretions from patients with HIV infection (12).

In addition to health-care workers enrolled in prospective studies, eight persons who provided care to infected patients and denied other risk factors have been reported to have acquired HIV infection. Three of these health-care workers had needlestick exposures to blood from infected patients (13-15). Two were persons who provided nursing care to infected persons; although neither sustained a needlestick, both had extensive contact with blood or other body fluids, and neither observed recommended barrier precautions (16,17). The other three were health-care workers with non-needlestick exposures to blood from infected patients (18). Although the exact route of transmission for these last three infections is not known, all three persons had direct contact of their skin with blood from infected patients, all had skin lesions that may have been contaminated by blood, and one also had a mucous-membrane exposure.

A total of 1,231 dentists and hygienists, many of whom practiced in areas with many AIDS cases, participated in a study to determine the prevalence of antibody to HIV; one dentist (0.1%) had HIV antibody. Although no exposure to a known HIV-infected person could be documented, epidemiologic investigation did not identify any other risk factor for infection. The infected dentist, who also had a history of sustaining needlestick injuries and trauma to his hands, did not routinely wear gloves when providing dental care (19).

Precautions To Prevent Transmission of HIV

Universal Precautions

Since medical history and examination cannot reliably identify all patients infected with HIV or other blood-borne pathogens, blood and body-fluid precautions should be consistently used for **all** patients. This approach, previously recommended by CDC (3,4), and referred to as "universal blood and body-fluid precautions" or "universal precautions," should be used in the care of **all** patients, especially including those in emergency-care settings in which the risk of blood exposure is increased and the infection status of the patient is usually unknown (20).

OSHA Instruction CPL 2-2.44A
Office of Health Compliance Assistance

1. All health-care workers should routinely use appropriate barrier precautions to prevent skin and mucous-membrane exposure when contact with blood or other body fluids of any patient is anticipated. Gloves should be worn for touching blood and body fluids, mucous membranes, or non-intact skin of all patients, for handling items or surfaces soiled with blood or body fluids, and for performing venipuncture and other vascular access procedures. Gloves should be changed after contact with each patient. Masks and protective eyewear or face shields should be worn during procedures that are likely to generate droplets of blood or other body fluids to prevent exposure of mucous membranes of the mouth, nose, and eyes. Gowns or aprons should be worn during procedures that are likely to generate splashes of blood or other body fluids.
2. Hands and other skin surfaces should be washed immediately and thoroughly if contaminated with blood or other body fluids. Hands should be washed immediately after gloves are removed.
3. All health-care workers should take precautions to prevent injuries caused by needles, scalpels, and other sharp instruments or devices during procedures; when cleaning used instruments; during disposal of used needles; and when handling sharp instruments after procedures. To prevent needlestick injuries, needles should not be recapped, purposely bent or broken by hand, removed from disposable syringes, or otherwise manipulated by hand. After they are used, disposable syringes and needles, scalpel blades, and other sharp items should be placed in puncture-resistant containers for disposal; the puncture-resistant containers should be located as close as practical to the use area. Large-bore reusable needles should be placed in a puncture-resistant container for transport to the reprocessing area.
4. Although saliva has not been implicated in HIV transmission, to minimize the need for emergency mouth-to-mouth resuscitation, mouthpieces, resuscitation bags, or other ventilation devices should be available for use in areas in which the need for resuscitation is predictable.
5. Health-care workers who have exudative lesions or weeping dermatitis should refrain from all direct patient care and from handling patient-care equipment until the condition resolves.
6. Pregnant health-care workers are not known to be at greater risk of contracting HIV infection than health-care workers who are not pregnant; however, if a health-care worker develops HIV infection during pregnancy, the infant is at risk of infection resulting from perinatal transmission. Because of this risk, pregnant health-care workers should be especially familiar with and strictly adhere to precautions to minimize the risk of HIV transmission.

Implementation of universal blood and body-fluid precautions for **all** patients eliminates the need for use of the isolation category of "Blood and Body Fluid Precautions" previously recommended by CDC (7) for patients known or suspected to be infected with blood-borne pathogens. Isolation precautions (e.g., enteric, "AFB" [7]) should be used as necessary if associated conditions, such as infectious diarrhea or tuberculosis, are diagnosed or suspected.

Precautions for Invasive Procedures

In this document, an invasive procedure is defined as surgical entry into tissues, cavities, or organs or repair of major traumatic injuries 1) in an operating or delivery

OSHA Instruction CPL 2-2.44A

Office of Health Compliance Assistance

room, emergency department, or outpatient setting, including both physicians' and dentists' offices; 2) cardiac catheterization and angiographic procedures; 3) a vaginal or cesarean delivery or other invasive obstetric procedure during which bleeding may occur; or 4) the manipulation, cutting, or removal of any oral or perioral tissues, including tooth structure, during which bleeding occurs or the potential for bleeding exists. The universal blood and body-fluid precautions listed above, combined with the precautions listed below, should be the minimum precautions for **all** such invasive procedures.

1. All health-care workers who participate in invasive procedures must routinely use appropriate barrier precautions to prevent skin and mucous-membrane contact with blood and other body fluids of all patients. Gloves and surgical masks must be worn for all invasive procedures. Protective eyewear or face shields should be worn for procedures that commonly result in the generation of droplets, splashing of blood or other body fluids, or the generation of bone chips. Gowns or aprons made of materials that provide an effective barrier should be worn during invasive procedures that are likely to result in the splashing of blood or other body fluids. All health-care workers who perform or assist in vaginal or cesarean deliveries should wear gloves and gowns when handling the placenta or the infant until blood and amniotic fluid have been removed from the infant's skin and should wear gloves during post-delivery care of the umbilical cord.

2. If a glove is torn or a needlestick or other injury occurs, the glove should be removed and a new glove used as promptly as patient safety permits; the needle or instrument involved in the incident should also be removed from the sterile field.

Precautions for Dentistry*

Blood, saliva, and gingival fluid from **all** dental patients should be considered infective. Special emphasis should be placed on the following precautions for preventing transmission of blood-borne pathogens in dental practice in both institutional and non-institutional settings.

1. In addition to wearing gloves for contact with oral mucous membranes of all patients, all dental workers should wear surgical masks and protective eyewear or chin-length plastic face shields during dental procedures in which splashing or spattering of blood, saliva, or gingival fluids is likely. Rubber dams, high-speed evacuation, and proper patient positioning, when appropriate, should be utilized to minimize generation of droplets and spatter.

2. Handpieces should be sterilized after use with each patient, since blood, saliva, or gingival fluid of patients may be aspirated into the handpiece or waterline. Handpieces that cannot be sterilized should at least be flushed, the outside surface cleaned and wiped with a suitable chemical germicide, and then rinsed. Handpieces should be flushed at the beginning of the day and after use with each patient. Manufacturers' recommendations should be followed for use and maintenance of waterlines and check valves and for flushing of handpieces. The same precautions should be used for ultrasonic scalers and air/water syringes.

*General infection-control precautions are more specifically addressed in previous recommendations for infection-control practices for dentistry (8).

APPENDIX J

OSHA Instruction CPL 2-2.44 A

Office of Health Compliance Assistance

3. Blood and saliva should be thoroughly and carefully cleaned from material that has been used in the mouth (e.g., impression materials, bite registration), especially before polishing and grinding intra-oral devices. Contaminated materials, impressions, and intra-oral devices should also be cleaned and disinfected before being handled in the dental laboratory and before they are placed in the patient's mouth. Because of the increasing variety of dental materials used intra-orally, dental workers should consult with manufacturers as to the stability of specific materials when using disinfection procedures.
4. Dental equipment and surfaces that are difficult to disinfect (e.g., light handles or X-ray-unit heads) and that may become contaminated should be wrapped with impervious-backed paper, aluminum foil, or clear plastic wrap. The coverings should be removed and discarded, and clean coverings should be put in place after use with each patient.

Precautions for Autopsies or Morticians' Services

In addition to the universal blood and body-fluid precautions listed above, the following precautions should be used by persons performing postmortem procedures:

1. All persons performing or assisting in postmortem procedures should wear gloves, masks, protective eyewear, gowns, and waterproof aprons.
2. Instruments and surfaces contaminated during postmortem procedures should be decontaminated with an appropriate chemical germicide.

Precautions for Dialysis

Patients with end-stage renal disease who are undergoing maintenance dialysis and who have HIV infection can be dialyzed in hospital-based or free-standing dialysis units using conventional infection-control precautions (21). Universal blood and body-fluid precautions should be used when dialyzing **all** patients.

Strategies for disinfecting the dialysis fluid pathways of the hemodialysis machine are targeted to control bacterial contamination and generally consist of using 500-750 parts per million (ppm) of sodium hypochlorite (household bleach) for 30-40 minutes or 1.5%-2.0% formaldehyde overnight. In addition, several chemical germicides formulated to disinfect dialysis machines are commercially available. None of these protocols or procedures need to be changed for dialyzing patients infected with HIV.

Patients infected with HIV can be dialyzed by either hemodialysis or peritoneal dialysis and do not need to be isolated from other patients. The type of dialysis treatment (i.e., hemodialysis or peritoneal dialysis) should be based on the needs of the patient. The dialyzer may be discarded after each use. Alternatively, centers that reuse dialyzers—i.e., a specific single-use dialyzer is issued to a specific patient, removed, cleaned, disinfected, and reused several times on the same patient only—may include HIV-infected patients in the dialyzer-reuse program. An individual dialyzer must never be used on more than one patient.

Precautions for Laboratories[†]

Blood and other body fluids from **all** patients should be considered infective. To supplement the universal blood and body-fluid precautions listed above, the following precautions are recommended for health-care workers in clinical laboratories.

[*]Additional precautions for research and industrial laboratories are addressed elsewhere (22,23).

OSHA Instruction CPL 2-2.44A

Office of Health Compliance Assistance

1. All specimens of blood and body fluids should be put in a well-constructed container with a secure lid to prevent leaking during transport. Care should be taken when collecting each specimen to avoid contaminating the outside of the container and of the laboratory form accompanying the specimen.
2. All persons processing blood and body-fluid specimens (e.g., removing tops from vacuum tubes) should wear gloves. Masks and protective eyewear should be worn if mucous-membrane contact with blood or body fluids is anticipated. Gloves should be changed and hands washed after completion of specimen processing.
3. For routine procedures, such as histologic and pathologic studies or microbiologic culturing, a biological safety cabinet is not necessary. However, biological safety cabinets (Class I or II) should be used whenever procedures are conducted that have a high potential for generating droplets. These include activities such as blending, sonicating, and vigorous mixing.
4. Mechanical pipetting devices should be used for manipulating all liquids in the laboratory. Mouth pipetting must not be done.
5. Use of needles and syringes should be limited to situations in which there is no alternative, and the recommendations for preventing injuries with needles outlined under universal precautions should be followed.
6. Laboratory work surfaces should be decontaminated with an appropriate chemical germicide after a spill of blood or other body fluids and when work activities are completed.
7. Contaminated materials used in laboratory tests should be decontaminated before reprocessing or be placed in bags and disposed of in accordance with institutional policies for disposal of infective waste (24).
8. Scientific equipment that has been contaminated with blood or other body fluids should be decontaminated and cleaned before being repaired in the laboratory or transported to the manufacturer.
9. All persons should wash their hands after completing laboratory activities and should remove protective clothing before leaving the laboratory.

Implementation of universal blood and body-fluid precautions for <u>all</u> patients eliminates the need for warning labels on specimens since blood and other body fluids from all patients should be considered infective.

Environmental Considerations for HIV Transmission

No environmentally mediated mode of HIV transmission has been documented. Nevertheless, the precautions described below should be taken routinely in the care of <u>all</u> patients.

Sterilization and Disinfection

Standard sterilization and disinfection procedures for patient-care equipment currently recommended for use (25,26) in a variety of health-care settings—including hospitals, medical and dental clinics and offices, hemodialysis centers, emergency-care facilities, and long-term nursing-care facilities—are adequate to sterilize or disinfect instruments, devices, or other items contaminated with blood or other body fluids from persons infected with blood-borne pathogens including HIV (21,23).

OSHA Instruction CPL 2-2.44A

Office of Health Compliance Assistance

Instruments or devices that enter sterile tissue or the vascular system of any patient or through which blood flows should be sterilized before reuse. Devices or items that contact intact mucous membranes should be sterilized or receive high-level disinfection, a procedure that kills vegetative organisms and viruses but not necessarily large numbers of bacterial spores. Chemical germicides that are registered with the U.S. Environmental Protection Agency (EPA) as "sterilants" may be used either for sterilization or for high-level disinfection depending on contact time.

Contact lenses used in trial fittings should be disinfected after each fitting by using a hydrogen peroxide contact lens disinfecting system or, if compatible, with heat (78 C-80 C [172.4 F-176.0 F]) for 10 minutes.

Medical devices or instruments that require sterilization or disinfection should be thoroughly cleaned before being exposed to the germicide, and the manufacturer's instructions for the use of the germicide should be followed. Further, it is important that the manufacturer's specifications for compatibility of the medical device with chemical germicides be closely followed. Information on specific label claims of commercial germicides can be obtained by writing to the Disinfectants Branch, Office of Pesticides, Environmental Protection Agency, 401 M Street, SW, Washington, D.C. 20460.

Studies have shown that HIV is inactivated rapidly after being exposed to commonly used chemical germicides at concentrations that are much lower than used in practice (27-30). Embalming fluids are similar to the types of chemical germicides that have been tested and found to completely inactivate HIV. In addition to commercially available chemical germicides, a solution of sodium hypochlorite (household bleach) prepared daily is an inexpensive and effective germicide. Concentrations ranging from approximately 500 ppm (1:100 dilution of household bleach) sodium hypochlorite to 5,000 ppm (1:10 dilution of household bleach) are effective depending on the amount of organic material (e.g., blood, mucus) present on the surface to be cleaned and disinfected. Commercially available chemical germicides may be more compatible with certain medical devices that might be corroded by repeated exposure to sodium hypochlorite, especially to the 1:10 dilution.

Survival of HIV in the Environment

The most extensive study on the survival of HIV after drying involved greatly concentrated HIV samples, i.e., 10 million tissue-culture infectious doses per milliliter (31). This concentration is at least 100,000 times greater than that typically found in the blood or serum of patients with HIV infection. HIV was detectable by tissue-culture techniques 1-3 days after drying, but the rate of inactivation was rapid. Studies performed at CDC have also shown that drying HIV causes a rapid (within several hours) 1-2 log (90%-99%) reduction in HIV concentration. In tissue-culture fluid, cell-free HIV could be detected up to 15 days at room temperature, up to 11 days at 37 C (98.6 F), and up to 1 day if the HIV was cell-associated.

When considered in the context of environmental conditions in health-care facilities, these results do not require any changes in currently recommended sterilization, disinfection, or housekeeping strategies. When medical devices are contaminated with blood or other body fluids, existing recommendations include the cleaning of these instruments, followed by disinfection or sterilization, depending on the type of medical device. These protocols assume "worst-case" conditions of

OSHA Instruction CPL 2-2.44A

Office of Health Compliance Assistance

extreme virologic and microbiologic contamination, and whether viruses have been inactivated after drying plays no role in formulating these strategies. Consequently, no changes in published procedures for cleaning, disinfecting, or sterilizing need to be made.

Housekeeping

Environmental surfaces such as walls, floors, and other surfaces are not associated with transmission of infections to patients or health-care workers. Therefore, extraordinary attempts to disinfect or sterilize these environmental surfaces are not necessary. However, cleaning and removal of soil should be done routinely.

Cleaning schedules and methods vary according to the area of the hospital or institution, type of surface to be cleaned, and the amount and type of soil present. Horizontal surfaces (e.g., bedside tables and hard-surfaced flooring) in patient-care areas are usually cleaned on a regular basis, when soiling or spills occur, and when a patient is discharged. Cleaning of walls, blinds, and curtains is recommended only if they are visibly soiled. Disinfectant fogging is an unsatisfactory method of decontaminating air and surfaces and is not recommended.

Disinfectant-detergent formulations registered by EPA can be used for cleaning environmental surfaces, but the actual physical removal of microorganisms by scrubbing is probably at least as important as any antimicrobial effect of the cleaning agent used. Therefore, cost, safety, and acceptability by housekeepers can be the main criteria for selecting any such registered agent. The manufacturers' instructions for appropriate use should be followed.

Cleaning and Decontaminating Spills of Blood or Other Body Fluids

Chemical germicides that are approved for use as "hospital disinfectants" and are tuberculocidal when used at recommended dilutions can be used to decontaminate spills of blood and other body fluids. Strategies for decontaminating spills of blood and other body fluids in a patient-care setting are different than for spills of cultures or other materials in clinical, public health, or research laboratories. In patient-care areas, visible material should first be removed and then the area should be decontaminated. With large spills of cultured or concentrated infectious agents in the laboratory, the contaminated area should be flooded with a liquid germicide before cleaning, then decontaminated with fresh germicidal chemical. In both settings, gloves should be worn during the cleaning and decontaminating procedures.

Laundry

Although soiled linen has been identified as a source of large numbers of certain pathogenic microorganisms, the risk of actual disease transmission is negligible. Rather than rigid procedures and specifications, hygienic and common-sense storage and processing of clean and soiled linen are recommended (26). Soiled linen should be handled as little as possible and with minimum agitation to prevent gross microbial contamination of the air and of persons handling the linen. All soiled linen should be bagged at the location where it was used; it should not be sorted or rinsed in patient-care areas. Linen soiled with blood or body fluids should be placed and transported in bags that prevent leakage. If hot water is used, linen should be washed

OSHA Instruction CPL 2-2.44A

Office of Health Compliance Assistance

with detergent in water at least 71 C (160 F) for 25 minutes. If low-temperature($\leq$70 C [158 F]) laundry cycles are used, chemicals suitable for low-temperature washing at proper use concentration should be used.

Infective Waste

There is no epidemiologic evidence to suggest that most hospital waste is any more infective than residential waste. Moreover, there is no epidemiologic evidence that hospital waste has caused disease in the community as a result of improper disposal. Therefore, identifying wastes for which special precautions are indicated is largely a matter of judgment about the relative risk of disease transmission. The most practical approach to the management of infective waste is to identify those wastes with the potential for causing infection during handling and disposal and for which some special precautions appear prudent. Hospital wastes for which special precautions appear prudent include microbiology laboratory waste, pathology waste, and blood specimens or blood products. While any item that has had contact with blood, exudates, or secretions may be potentially infective, it is not usually considered practical or necessary to treat all such waste as infective (*23,26*). Infective waste, in general, should either be incinerated or should be autoclaved before disposal in a sanitary landfill. Bulk blood, suctioned fluids, excretions, and secretions may be carefully poured down a drain connected to a sanitary sewer. Sanitary sewers may also be used to dispose of other infectious wastes capable of being ground and flushed into the sewer.

Implementation of Recommended Precautions

Employers of health-care workers should ensure that policies exist for:

1. Initial orientation and continuing education and training of all health-care workers—including students and trainees—on the epidemiology, modes of transmission, and prevention of HIV and other blood-borne infections and the need for routine use of universal blood and body-fluid precautions for **all** patients.
2. Provision of equipment and supplies necessary to minimize the risk of infection with HIV and other blood-borne pathogens.
3. Monitoring adherence to recommended protective measures. When monitoring reveals a failure to follow recommended precautions, counseling, education, and/or re-training should be provided, and, if necessary, appropriate disciplinary action should be considered.

Professional associations and labor organizations, through continuing education efforts, should emphasize the need for health-care workers to follow recommended precautions.

OSHA Instruction CPL 2-2.44A

Office of Health Compliance Assistance

Serologic Testing for HIV Infection

Background

A person is identified as infected with HIV when a sequence of tests, starting with repeated enzyme immunoassays (EIA) and including a Western blot or similar, more specific assay, are repeatedly reactive. Persons infected with HIV usually develop antibody against the virus within 6-12 weeks after infection.

The sensitivity of the currently licensed EIA tests is at least 99% when they are performed under optimal laboratory conditions on serum specimens from persons infected for ≥12 weeks. Optimal laboratory conditions include the use of reliable reagents, provision of continuing education of personnel, quality control of procedures, and participation in performance-evaluation programs. Given this performance, the probability of a false-negative test is remote except during the first several weeks after infection, before detectable antibody is present. The proportion of infected persons with a false-negative test attributed to absence of antibody in the early stages of infection is dependent on both the incidence and prevalence of HIV infection in a population (Table 1).

The specificity of the currently licensed EIA tests is approximately 99% when repeatedly reactive tests are considered. Repeat testing of initially reactive specimens by EIA is required to reduce the likelihood of laboratory error. To increase further the specificity of serologic tests, laboratories must use a supplemental test, most often the Western blot, to validate repeatedly reactive EIA results. Under optimal laboratory conditions, the sensitivity of the Western blot test is comparable to or greater than that of a repeatedly reactive EIA, and the Western blot is highly specific when strict criteria are used to interpret the test results. The testing sequence of a repeatedly reactive EIA and a positive Western blot test is highly predictive of HIV infection, even in a population with a low prevalence of infection (Table 2). If the Western blot test result is indeterminant, the testing sequence is considered equivocal for HIV infection.

TABLE 1. Estimated annual number of patients infected with HIV not detected by HIV-antibody testing in a hypothetical hospital with 10,000 admissions/year*

Beginning prevalence of HIV infection	Annual incidence of HIV infection	Approximate number of HIV-infected patients	Approximate number of HIV-infected patients not detected
5.0%	1.0%	550	17-18
5.0%	0.5%	525	11-12
1.0%	0.2%	110	3-4
1.0%	0.1%	105	2-3
0.1%	0.02%	11	0-1
0.1%	0.01%	11	0-1

*The estimates are based on the following assumptions: 1) the sensitivity of the screening test is 99% (i.e., 99% of HIV-infected persons with antibody will be detected); 2) persons infected with HIV will not develop detectable antibody (seroconvert) until 6 weeks (1.5 months) after infection; 3) new infections occur at an equal rate throughout the year; 4) calculations of the number of HIV-infected persons in the patient population are based on the mid-year prevalence, which is the beginning prevalence plus half the annual incidence of infections.

OSHA Instruction CPL 2-2.44A
Office of Health Compliance Assistance

When this occurs, the Western blot test should be repeated on the same serum sample, and, if still indeterminant, the testing sequence should be repeated on a sample collected 3-6 months later. Use of other supplemental tests may aid in interpreting of results on samples that are persistently indeterminant by Western blot.

Testing of Patients

Previous CDC recommendations have emphasized the value of HIV serologic testing of patients for: 1) management of parenteral or mucous-membrane exposures of health-care workers, 2) patient diagnosis and management, and 3) counseling and serologic testing to prevent and control HIV transmission in the community. In addition, more recent recommendations have stated that hospitals, in conjunction with state and local health departments, should periodically determine the prevalence of HIV infection among patients from age groups at highest risk of infection (32).

Adherence to universal blood and body-fluid precautions recommended for the care of all patients will minimize the risk of transmission of HIV and other blood-borne pathogens from patients to health-care workers. The utility of routine HIV serologic testing of patients as an adjunct to universal precautions is unknown. Results of such testing may not be available in emergency or outpatient settings. In addition, some recently infected patients will not have detectable antibody to HIV (Table 1).

Personnel in some hospitals have advocated serologic testing of patients in settings in which exposure of health-care workers to large amounts of patients' blood may be anticipated. Specific patients for whom serologic testing has been advocated include those undergoing major operative procedures and those undergoing treatment in critical-care units, especially if they have conditions involving uncontrolled bleeding. Decisions regarding the need to establish testing programs for patients should be made by physicians or individual institutions. In addition, when deemed appropriate, testing of individual patients may be performed on agreement between the patient and the physician providing care.

In addition to the universal precautions recommended for all patients, certain additional precautions for the care of HIV-infected patients undergoing major surgical operations have been proposed by personnel in some hospitals. For example, surgical procedures on an HIV-infected patient might be altered so that hand-to-hand passing of sharp instruments would be eliminated; stapling instruments rather than

TABLE 2. Predictive value of positive HIV-antibody tests in hypothetical populations with different prevalences of infection

	Prevalence of infection	Predictive value of positive test*
Repeatedly reactive enzyme immunoassay (EIA)[†]	0.2%	28.41%
	2.0%	80.16%
	20.0%	98.02%
Repeatedly reactive EIA followed by positive Western blot (WB)[§]	0.2%	99.75%
	2.0%	99.97%
	20.0%	99.99%

*Proportion of persons with positive test results who are actually infected with HIV.
[†]Assumes EIA sensitivity of 99.0% and specificity of 99.5%.
[§]Assumes WB sensitivity of 99.0% and specificity of 99.9%.

OSHA Instruction CPL 2-2.44A

Office of Health Compliance Assistance

hand-suturing equipment might be used to perform tissue approximation; electrocautery devices rather than scalpels might be used as cutting instruments; and, even though uncomfortable, gowns that totally prevent seepage of blood onto the skin of members of the operative team might be worn. While such modifications might further minimize the risk of HIV infection for members of the operative team, some of these techniques could result in prolongation of operative time and could potentially have an adverse effect on the patient.

Testing programs, if developed, should include the following principles:

- Obtaining consent for testing.
- Informing patients of test results, and providing counseling for seropositive patients by properly trained persons.
- Assuring that confidentiality safeguards are in place to limit knowledge of test results to those directly involved in the care of infected patients or as required by law.
- Assuring that identification of infected patients will not result in denial of needed care or provision of suboptimal care.
- Evaluating prospectively 1) the efficacy of the program in reducing the incidence of parenteral, mucous-membrane, or significant cutaneous exposures of health-care workers to the blood or other body fluids of HIV-infected patients and 2) the effect of modified procedures on patients.

Testing of Health-Care Workers

Although transmission of HIV from infected health-care workers to patients has not been reported, transmission during invasive procedures remains a possibility. Transmission of hepatitis B virus (HBV) — a blood-borne agent with a considerably greater potential for nosocomial spread — from health-care workers to patients has been documented. Such transmission has occurred in situations (e.g., oral and gynecologic surgery) in which health-care workers, when tested, had very high concentrations of HBV in their blood (at least 100 million infectious virus particles per milliliter, a concentration much higher than occurs with HIV infection), and the health-care workers sustained a puncture wound while performing invasive procedures or had exudative or weeping lesions or microlacerations that allowed virus to contaminate instruments or open wounds of patients (*33,34*).

The hepatitis B experience indicates that only those health-care workers who perform certain types of invasive procedures have transmitted HBV to patients. Adherence to recommendations in this document will minimize the risk of transmission of HIV and other blood-borne pathogens from health-care workers to patients during invasive procedures. Since transmission of HIV from infected health-care workers performing invasive procedures to their patients has not been reported and would be expected to occur only very rarely, if at all, the utility of routine testing of such health-care workers to prevent transmission of HIV cannot be assessed. If consideration is given to developing a serologic testing program for health-care workers who perform invasive procedures, the frequency of testing, as well as the issues of consent, confidentiality, and consequences of test results — as previously outlined for testing programs for patients — must be addressed.

OSHA Instruction CPL 2-2.44A

Office of Health Compliance Assistance

Management of Infected Health-Care Workers

Health-care workers with impaired immune systems resulting from HIV infection or other causes are at increased risk of acquiring or experiencing serious complications of infectious disease. Of particular concern is the risk of severe infection following exposure to patients with infectious diseases that are easily transmitted if appropriate precautions are not taken (e.g., measles, varicella). Any health-care worker with an impaired immune system should be counseled about the potential risk associated with taking care of patients with any transmissible infection and should continue to follow existing recommendations for infection control to minimize risk of exposure to other infectious agents (7,35). Recommendations of the Immunization Practices Advisory Committee (ACIP) and institutional policies concerning requirements for vaccinating health-care workers with live-virus vaccines (e.g., measles, rubella) should also be considered.

The question of whether workers infected with HIV—especially those who perform invasive procedures—can adequately and safely be allowed to perform patient-care duties or whether their work assignments should be changed must be determined on an individual basis. These decisions should be made by the health-care worker's personal physician(s) in conjunction with the medical directors and personnel health service staff of the employing institution or hospital.

Management of Exposures

If a health-care worker has a parenteral (e.g., needlestick or cut) or mucous-membrane (e.g., splash to the eye or mouth) exposure to blood or other body fluids or has a cutaneous exposure involving large amounts of blood or prolonged contact with blood—especially when the exposed skin is chapped, abraded, or afflicted with dermatitis—the source patient should be informed of the incident and tested for serologic evidence of HIV infection after consent is obtained. Policies should be developed for testing source patients in situations in which consent cannot be obtained (e.g., an unconscious patient).

If the source patient has AIDS, is positive for HIV antibody, or refuses the test, the health-care worker should be counseled regarding the risk of infection and evaluated clinically and serologically for evidence of HIV infection as soon as possible after the exposure. The health-care worker should be advised to report and seek medical evaluation for any acute febrile illness that occurs within 12 weeks after the exposure. Such an illness—particularly one characterized by fever, rash, or lymphadenopathy—may be indicative of recent HIV infection. Seronegative health-care workers should be retested 6 weeks post-exposure and on a periodic basis thereafter (e.g., 12 weeks and 6 months after exposure) to determine whether transmission has occurred. During this follow-up period—especially the first 6-12 weeks after exposure, when most infected persons are expected to seroconvert—exposed health-care workers should follow U.S. Public Health Service (PHS) recommendations for preventing transmission of HIV (36,37).

No further follow-up of a health-care worker exposed to infection as described above is necessary if the source patient is seronegative unless the source patient is at high risk of HIV infection. In the latter case, a subsequent specimen (e.g., 12 weeks following exposure) may be obtained from the health-care worker for antibody

OSHA Instruction CPL 2-2.44A

Office of Health Compliance Assistance

testing. If the source patient cannot be identified, decisions regarding appropriate follow-up should be individualized. Serologic testing should be available to all health-care workers who are concerned that they may have been infected with HIV.

If a patient has a parenteral or mucous-membrane exposure to blood or other body fluid of a health-care worker, the patient should be informed of the incident, and the same procedure outlined above for management of exposures should be followed for both the source health-care worker and the exposed patient.

References
1. CDC. Acquired immunodeficiency syndrome (AIDS): Precautions for clinical and laboratory staffs. MMWR 1982;31:577-80.
2. CDC. Acquired immunodeficiency syndrome (AIDS): Precautions for health-care workers and allied professionals. MMWR 1983;32:450-1.
3. CDC. Recommendations for preventing transmission of infection with human T-lymphotropic virus type III/lymphadenopathy-associated virus in the workplace. MMWR 1985;34:681-6, 691-5.
4. CDC. Recommendations for preventing transmission of infection with human T-lymphotropic virus type III/lymphadenopathy-associated virus during invasive procedures. MMWR 1986;35:221-3.
5. CDC. Recommendations for preventing possible transmission of human T-lymphotropic virus type III/lymphadenopathy-associated virus from tears. MMWR 1985;34:533-4.
6. CDC. Recommendations for providing dialysis treatment to patients infected with human T-lymphotropic virus type III/lymphadenopathy-associated virus infection. MMWR 1986;35:376-8, 383.
7. Garner JS, Simmons BP. Guideline for isolation precautions in hospitals. Infect Control 1983;4 (suppl) :245-325.
8. CDC. Recommended infection control practices for dentistry. MMWR 1986;35:237-42.
9. McCray E, The Cooperative Needlestick Surveillance Group. Occupational risk of the acquired immunodeficiency syndrome among health care workers. N Engl J Med 1986;314:1127-32.
10. Henderson DK, Saah AJ, Zak BJ, et al. Risk of nosocomial infection with human T-cell lymphotropic virus type III/lymphadenopathy-associated virus in a large cohort of intensively exposed health care workers. Ann Intern Med 1986;104:644-7.
11. Gerberding JL, Bryant-LeBlanc CE, Nelson K, et al. Risk of transmitting the human immunodeficiency virus, cytomegalovirus, and hepatitis B virus to health care workers exposed to patients with AIDS and AIDS-related conditions. J Infect Dis 1987;156:1-8.
12. McEvoy M, Porter K, Mortimer P, Simmons N, Shanson D. Prospective study of clinical, laboratory, and ancillary staff with accidental exposures to blood or other body fluids from patients infected with HIV. Br Med J 1987;294:1595-7.
13. Anonymous. Needlestick transmission of HTLV-III from a patient infected in Africa. Lancet 1984;2:1376-7.
14. Oksenhendler E, Harzic M, Le Roux JM, Rabian C, Clauvel JP. HIV infection with seroconversion after a superficial needlestick injury to the finger. N Engl J Med 1986;315:582.
15. Neisson-Vernant C, Arfi S, Mathez D, Leibowitch J, Monplaisir N. Needlestick HIV seroconversion in a nurse. Lancet 1986;2:814.
16. Grint P, McEvoy M. Two associated cases of the acquired immune deficiency syndrome (AIDS). PHLS Commun Dis Rep 1985;42:4.
17. CDC. Apparent transmission of human T-lymphotropic virus type III/lymphadenopathy-associated virus from a child to a mother providing health care. MMWR 1986;35:76-9.
18. CDC. Update: Human immunodeficiency virus infections in health-care workers exposed to blood of infected patients. MMWR 1987;36:285-9.
19. Kline RS, Phelan J, Friedland GH, et al. Low occupational risk for HIV infection for dental professionals [Abstract]. In: Abstracts from the III International Conference on AIDS, 1-5 June 1985. Washington, DC: 155.
20. Baker JL, Kelen GD, Sivertson KT, Quinn TC. Unsuspected human immunodeficiency virus in critically ill emergency patients. JAMA 1987;257:2609-11.
21. Favero MS. Dialysis-associated diseases and their control. In: Bennett JV, Brachman PS, eds. Hospital infections. Boston: Little, Brown and Company, 1985:267-84.

OSHA Instruction CPL 2-2.44A

Office of Health Compliance Assistance

22. Richardson JH, Barkley WE, eds. Biosafety in microbiological and biomedical laboratories, 1984. Washington, DC: US Department of Health and Human Services, Public Health Service. HHS publication no. (CDC) 84-8395.
23. CDC. Human T-lymphotropic virus type III/lymphadenopathy-associated virus: Agent summary statement. MMWR 1986;35:540-2, 547-9.
24. Environmental Protection Agency. EPA guide for infectious waste management. Washington, DC :U.S. Environmental Protection Agency, May 1986 (Publication no. EPA/530-SW-86-014).
25. Favero MS. Sterilization, disinfection, and antisepsis in the hospital. In: Manual of clinical microbiology. 4th ed. Washington, DC: American Society for Microbiology, 1985;129-37.
26. Garner JS, Favero MS. Guideline for handwashing and hospital environmental control, 1985. Atlanta: Public Health Service, Centers for Disease Control, 1985. HHS publication no. 99-1117.
27. Spire B, Montagnier L, Barré-Sinoussi F, Chermann JC. Inactivation of lymphadenopathy associated virus by chemical disinfectants. Lancet 1984;2:899-901.
28. Martin LS, McDougal JS, Loskoski SL. Disinfection and inactivation of the human T lymphotropic virus type III/lymphadenopathy-associated virus. J Infect Dis 1985; 152:400-3.
29. McDougal JS, Martin LS, Cort SP, et al. Thermal inactivation of the acquired immunodeficiency syndrome virus-III/lymphadenopathy-associated virus, with special reference to antihemophilic factor. J Clin Invest 1985;76:875-7.
30. Spire B, Barré-Sinoussi F, Dormont D, Montagnier L, Chermann JC. Inactivation of lymphadenopathy-associated virus by heat, gamma rays, and ultraviolet light. Lancet 1985;1:188-9.
31. Resnik L, Veren K, Salahuddin SZ, Tondreau S, Markham PD. Stability and inactivation of HTLV-III/LAV under clinical and laboratory environments. JAMA 1986;255:1887-91.
32. CDC. Public Health Service (PHS) guidelines for counseling and antibody testing to prevent HIV infection and AIDS. MMWR 1987;3:509-15..
33. Kane MA, Lettau LA. Transmission of HBV from dental personnel to patients. J Am Dent Assoc 1985;110:634-6.
34. Lettau LA, Smith JD, Williams D, et. al. Transmission of hepatitis B with resultant restriction of surgical practice. JAMA 1986;255:934-7.
35. Williams WW. Guideline for infection control in hospital personnel. Infect Control 1983;4 (suppl) :326-49.
36. CDC. Prevention of acquired immune deficiency syndrome (AIDS): Report of inter-agency recommendations. MMWR 1983;32:101-3.
37. CDC. Provisional Public Health Service inter-agency recommendations for screening donated blood and plasma for antibody to the virus causing acquired immunodeficiency syndrome. MMWR 1985;34:1-5.

OSHA Instruction CPL 2-2.44A
Office of Health Compliance Assistance

APPENDIX C

Evaluation of Employer Training and Education Programs

Training programs must be evaluated through program review and discussion with management and employees.

1. Training programs shall normally include epidemiology, clinical presentation, modes of transmission and prevention of HBV and HIV as well as protective measures to be taken to prevent exposure.

2. The following questions provide a general outline of training topics to be reviewed when conducting an inspection at a health care facility. Responses shall be documented in the case file. Areas of interest include, but are not limited to, direct patient care areas, emergency room, operating rooms, clinical laboratories, x-ray, housekeeping and laundry.

    a. Has a training and information program been established for employees actually or potentially exposed to blood and/or body fluids?

    b. How often is training provided and does it cover:

       (1) Universal precautions?

       (2) Personal protective equipment?

       (3) Workplace practices including blood drawing, room cleaning, laundry handling, cleanup of blood spills?

       (4) Needlestick exposure/management?

       (5) Hepatitis B Vaccination?

    c. Does new employee orientation cover infectious disease control?

    d. Does the employer evaluate the effectiveness of the training program through monitoring of employee compliance with the guidelines?

OSHA Instruction CPL 2-2.44A

Office of Health Compliance Assistance

e. Have employees been informed of the precautionary measures outlined in the CDC guidelines?

f. Is personal protective equipment provided to employees? In all appropriate locations? (Specifically, ask about gloves, masks, eye protection, gowns (as appropriate).)

g. Is the necessary equipment (i.e., mouthpieces, resuscitation bags, or other ventilation devices) provided for administering mouth-to-mouth resuscitation on potentially infected patients?

h. Does training identify the specific procedures implemented by the employer to provide protection, such as proper use of personal protective equipment?

i. Are facilities available to comply with workplace practices, such as handwashing sinks, needle containers, detergents and disinfectants to clean up spills?

j. Are employees aware of specific workplace practices to follow when appropriate? Specifically ask about:

- Handwashing.
- Handling sharp instruments.
- Routine examinations.
- Blood spills.
- Handling of laundry.
- Disposal of contaminated materials.
- Reusable equipment.

k. Are workers aware of procedures to follow after a needlestick or blood exposure? Have they had such experiences, and are the guidelines followed?

l. Are employees aware of the Hepatitis B vaccination program? Do they take advantage of it?

*U.S. Government Printing Office: 1988-202-109/94917

APPENDIX K

Employee Health and Safety Whistleblower Protection Act (S 436)

APPENDIX K

Employee Health and Safety Whistleblower Protection Act (S. 436)

101ST CONGRESS
1ST SESSION
S. 436

To strengthen the protections available to employees against reprisals for disclosing information, to protect the public health and safety, and for other purposes.

IN THE SENATE OF THE UNITED STATES

FEBRUARY 23 (legislative day, JANUARY 3), 1989

Mr. METZENBAUM (for himself, Mr. GRASSLEY, Mr. SIMON, Mr. PELL, Mr. DODD, Ms. MIKULSKI, Mr. HARKIN, Mr. LEVIN, and Mr. KENNEDY) introduced the following bill; which was read twice and referred to the Committee on Labor and Human Resources

---

A BILL

To strengthen the protections available to employees against reprisals for disclosing information, to protect the public health and safety, and for other purposes.

1 *Be it enacted by the Senate and House of Representa-*
2 *tives of the United States of America in Congress assembled,*

3 **SECTION 1. SHORT TITLE.**

4 This Act may be cited as the "Employee Health and
5 Safety Whistleblower Protection Act".

6 **SEC. 2. FINDINGS AND PURPOSES.**

7 (a) FINDINGS.—Congress finds that—

(1) employees who report unlawful or hazardous activities are performing an important public service by helping ensure the health and safety of workers and the general public;

(2) employees should be able to report unlawful or hazardous activities without fear of reprisal; and

(3) employees who make such reports are not adequately protected against reprisal by the current Federal laws designed to protect private sector whistleblowers.

(b) PURPOSE.—The purpose of this Act is to strengthen and improve protection for employees who report unlawful or hazardous activities, to prevent retaliation against employees, and to improve the health and safety of workers and the general public by—

(1) extending whistleblower protection regarding health and safety matters to all employees not currently protected by Federal law;

(2) creating a uniform statute of limitations for all whistleblower complaints by private sector employees;

(3) ensuring that the Secretary of Labor has subpoena power to investigate whistleblower complaints;

(4) providing for a limited private right of action for certain whistleblower complaints; and

(5) ensuring that reports of unlawful or hazardous activities are referred to the appropriate Federal agencies for investigation.

SEC. 3. DEFINITIONS.

As used in this Act—

(1) COMMERCE.—The term "commerce" means trade, traffic, commerce, transportation, or communication among the several States, or between a State and any place outside thereof, or within the District of Columbia, or a possession of the United States, or between points in the same State but through a point outside thereof.

(2) EMPLOYEE.—The term "employee" means any individual who is employed by an employer, or who is a former employee of an employer, to the extent that—

(A) such individual is discharged or discriminated against for any of the conduct described in section 4; and

(B) no remedy is available to such individual for such discharge or discrimination under an existing Federal whistleblower protection law.

(3) EMPLOYER.—The term "employer"—

(A) means a person engaged in a business affecting commerce who has employees;

(B) includes—

(i) a State or political subdivision of a State, or an interstate governmental agency; and

(ii) an agent of an employer and an employer of an employer; and

(C) does not include the United States, except for a government corporation (as defined in section 103(1) of title 5, United States Code) and the United States Postal Service.

(4) EXISTING FEDERAL WHISTLEBLOWER PROTECTION LAW.—The term "existing Federal whistleblower protection law" means those provisions of Federal law that protect employees against discharge or discrimination for any of the conduct described in section 4, that are contained in—

(A) section 942 of Department of Defense Authorization Act of 1987 (10 U.S.C. 2409);

(B) section 23 of the Toxic Substances Control Act (15 U.S.C. 2622);

(C) section 211 of the Asbestos Hazard Emergency Response Act of 1986 (15 U.S.C. 2651);

(D) section 11(c) of the Occupational Safety and Health Act of 1970 (29 U.S.C. 660(c));

(E) section 505(a) of the Migrant and Seasonal Agricultural Worker Protection Act (29 U.S.C. 1855(a));

(F) section 105(c) of the Federal Mine Safety and Health Act of 1977 (30 U.S.C. 815(c));

(G) section 703 of the Surface Mining Control and Reclamation Act of 1977 (30 U.S.C. 1293);

(H) section 507 of the Federal Water Pollution Control Act (33 U.S.C. 1367);

(I) section 1450 of the Public Health Service Act (42 U.S.C. 300j–9);

(J) section 210 of the Energy Reorganization Act of 1974 (42 U.S.C. 5851);

(K) section 7001 of the Solid Waste Disposal Act (42 U.S.C. 6971);

(L) section 322 of the Clean Air Act (42 U.S.C. 7622);

(M) section 110 of the Comprehensive Environmental Response, Compensation, and Liability Act of 1980 (42 U.S.C. 9610);

(N) section 212 of the Federal Railroad Safety Act of 1970 (45 U.S.C. 441);

(O) section 7 of the International Safe Container Act (46 U.S.C. 1506); or

(P) section 405 of the Surface Transportation Assistance Act of 1982 (49 U.S.C. App. 2305).

(5) FEDERAL AGENCY.—The term "Federal agency" means each authority of the Government of the United States, whether or not it is within or subject to review by another agency, or any interstate governmental agency or commission. For purposes of section 4, the term includes each authority of the Government of a State or a political subdivision of a State that administers or enforces a Federal health or safety law.

(6) FEDERAL HEALTH OR SAFETY LAW.—The term "Federal health or safety law" means any Federal law, or any portion thereof, including any Federal regulation, rule, or order implementing such law or portion thereof, a primary purpose or effect of which is to protect health or safety. The term includes a State law, regulation, rule, or order, enacted as part of a State plan established to carry out a Federal health or safety law.

(7) INVESTIGATION OFFICE.—The term "investigation office" means the office or unit created or designated within the Department of Labor pursuant to section 5(c)(1).

(8) PERSON.—The term "person" means one or more individuals, partnerships, associations, corporations, business trusts, legal representatives, or any organized group of persons.

(9) PROCEEDING.—The term "proceeding" means a trial, hearing, investigation, inquiry, inspection, administrative rulemaking, or adjudication involving a Federal agency.

(10) RESPONDENT.—The term "respondent" means a person named in a complaint as a respondent.

(11) SECRETARY.—The term "Secretary" means the Secretary of Labor.

(12) STATE.—The term "State" means a State of the United States, the District of Columbia, Commonwealth of Puerto Rico, the Virgin Islands, American Samoa, Guam, the Trust Territory of the Pacific Islands, or any other Territory or possession of the United States.

SEC. 4. EMPLOYEE PROTECTION.

(a) IN GENERAL.—No person shall discharge or in any other manner discriminate against any employee because the employee—

(1) discloses, or demonstrates an intent to disclose, an activity, policy, or practice that the employee

believes evidences a violation of a Federal health or safety law;

 (2) initiates, assists, participates, or demonstrates an intent to initiate, assist, or participate, in a proceeding with respect to—

  (A) an activity, policy, or practice that the employee believes evidences a violation of a Federal health or safety law;

  (B) enactment, adoption, promulgation, issuance, or amendment of a Federal health or safety law; or

  (C) administration or enforcement of a Federal health or safety law or of this Act; or

 (3) refuses to participate in an activity, policy, or practice when—

  (A) such participation constitutes a violation of a Federal health or safety law; or

  (B)(i) the employee has a reasonable apprehension of serious injury to the employee, other employees, or the public due to such participation;

  (ii) the activity, policy, or practice causing the employee's apprehension of such injury is of such a nature that a reasonable person, under the circumstances then confronting the employee, would conclude there is a bona fide danger of an accident, injury, or serious impairment of health or safety, resulting from participation in the activity, policy, or practice; and

(iii) before refusing to participate, the employee, where possible, sought, and was unable to obtain, a correction of the dangerous activity, policy, or practice.

(b) WAIVER.—Except as provided in section 5(l) or pursuant to any court supervised settlement agreement, an agreement by an employee, express or implied, written or oral, purporting to waive or modify the rights of the employee under this Act shall be void as contrary to public policy.

(c) EXEMPTION.—This Act shall not apply under circumstances where an employee acted without direction from the employee's employer, deliberately causing a violation of a Federal health or safety law.

SEC. 5. REMEDY.

(a) STATUTE OF LIMITATIONS.—The statute of limitations for filing a complaint pursuant to this Act or an existing Federal whistleblower protection law shall be 180 days after the later of the date—

(1) an alleged violation occurs;

(2) the discharge or other discrimination has taken effect; or

(3) the employee first learns or reasonably should have learned of the violation.

(b) COMPLAINT.—

(1) FILING.—Any employee who believes the employee has been discharged or otherwise discriminated against by any person in violation of section 4 may, within the period prescribed in subsection (a), file (or have filed by any person on the employee's behalf) a complaint with the Secretary alleging such violation. Except as provided in section 4(a)(3)(B)(iii), the complaint may be filed without regard to exhaustion of any other remedies.

(2) COPY.—On receipt of a complaint, the Secretary shall cause a copy of the complaint to be served on the respondent and on any Federal agency that may have jurisdiction over the activity, policy, or practice alleged in the complaint.

(c) INVESTIGATION.—

(1) ADMINISTRATIVE PROVISIONS.—The Secretary shall create or designate an identifiable office or unit within the Department of Labor that shall be responsible for investigation of complaints filed under subsection (b). The Secretary shall refer all such complaints immediately to the investigation office.

(2) IN GENERAL.—Within 60 days of the receipt of a complaint filed under subsection (b), the investigation office shall conduct an investigation, determine whether there is reasonable cause to believe the complaint has merit, and make findings of fact. If there is reasonable cause to believe a violation of this Act has occurred, the investigation office shall accompany the findings of fact with a preliminary order providing the relief prescribed by subsection (i), including immediate reinstatement of the employee pending a final order on the complaint.

(3) HEARING.—The findings and any preliminary order issued by the investigation office shall be served on all parties to the investigation and shall include a statement of the right of any party to request a hearing pursuant to subsection (d). Where a hearing is not timely requested, the findings and any preliminary order issued by the investigation office shall be deemed the final order of the Secretary and shall not be subject to judicial review.

(4) HEARING PROCEDURE.—In all hearings conducted under this Act, once the complainant has established by a preponderance of the evidence that an activity protected by this Act was a factor in the alleged discharge or discrimination, the burden of proof shall

be on the respondent to prove by clear and convincing evidence that the alleged discharge or discrimination would have occurred even if the employee had not engaged in activities protected by this Act.

(d) HEARING REQUEST.—

(1) TIMING.—The complainant or the respondent may request a hearing before an administrative law judge—

    (A) within 30 days of receipt of the findings or preliminary order of the investigation office; or

    (B) if no findings have been issued on or before the 60th day of receipt of the complaint, within 30 days thereafter.

(2) JURISDICTION.—The jurisdiction of the investigation office to make findings or issue a preliminary order shall cease on the filing of a request by the complainant for a hearing or the expiration of the period for filing such a request, whichever occurs first. Once the investigation office has lost jurisdiction to make findings or issue a preliminary order, any investigative material compiled by the investigating office in connection with the investigation shall be made available to the complainant or the respondent on request, subject to section 552a of title 5, United States Code.

(3) Filing with department.—Any request for a hearing shall be filed with the investigation office. The investigation office shall file the request with the Office of Administrative Law Judges in the Department of Labor within 5 days of receipt.

(4) No stay of reinstatement.—Such a request shall not operate to stay any reinstatement remedy contained in a preliminary order issued by the investigation office.

(5) No findings.—If the investigation office has failed to make findings within the time period prescribed, the complainant may include in the hearing request a request to the administrative law judge for a preliminary order for immediate reinstatement.

(e) Intervention.—

(1) Violation determined by investigation office.—If the investigation office has determined there has been a violation of section 4, the investigation office shall intervene on behalf of the complainant in the proceedings authorized under this section.

(2) Violation not determined.—In all other cases, within 15 days after receipt of a request for a hearing, the investigation office shall make a determination in writing whether, in its discretion, to intervene on behalf of the complainant.

(3) REPRESENTATION.—Such determination or the intervention by the investigation office shall not preclude the complainant or any other person on behalf of the complainant from representing the complainant in the proceedings authorized under this section.

(f) SUBPOENA AND RULEMAKING.—

(1) SUBPOENA POWER.—For purposes of any hearing or investigation provided for in this Act, or any hearing or investigation conducted pursuant to an existing Federal whistleblower protection law where the Secretary has authority to conduct investigations or hearings, the provisions of sections 9 and 10 (relating to attendance of witnesses and the production of books, papers, and documents) of the Federal Trade Commission Act (15 U.S.C. 49 and 50) shall be applicable to the jurisdiction, powers, and duties of the Secretary.

(2) RULEMAKING AUTHORITY.—The Secretary shall promulgate such rules and regulations as may be appropriate to administer this Act.

(g) HEARING PROCEDURE.—

(1) TIME FOR COMPLETION.—Any hearing requested pursuant to subsection (d) shall be conducted by an administrative law judge and completed within

180 days of receipt of the request for a hearing by the Office of Administrative Law Judges.

(2) NOTICE.—The parties shall have a reasonable opportunity for prehearing discovery and shall be given written notice of the time and place of the hearing at least 15 days prior to the hearing.

(3) ADMINISTRATIVE PROCEDURE.—The hearing shall be subject to the requirements of section 554 of title 5, United States Code.

(4) TRIAL DE NOVO.—The hearing shall be de novo, except that any preliminary order directing immediate reinstatement shall be enforceable.

(5) DECISION.—Within 60 days after the conclusion of the hearing on the record, or within 240 days after receipt of the hearing request by the investigation office, whichever is sooner, the administrative law judge shall issue a decision containing findings of fact, conclusions of law, and the remedy ordered, if any. The decision shall be served on the Secretary and on all parties to the hearing and shall include a statement to the parties of their right to appeal.

(h) ADMINISTRATIVE APPEAL.—

(1) REVIEW.—Any decision issued pursuant to subsection (g) shall be the final order of the Secretary subject to judicial review pursuant to subsection (k)

unless a party petitions the Secretary for review or the Secretary issues an order of review within 30 days after the issuance of the decision.

(2) TIME FOR REVIEW.—Within 60 days after receipt of the petition for review or issuance of an order of review, the Secretary shall issue a final order, based on the record and the decision of the administrative law judge, which shall be served on all parties. If the Secretary does not issue a final order within the period prescribed by this subsection, the decision of the administrative law judge shall become the final order of the Secretary.

(i) REMEDIES.—

(1) IN GENERAL.—If, in response to a complaint filed under subsection (b), the Secretary, investigation office, or administrative law judge determines that a violation of this Act occurred, the Secretary, investigation office, or administrative law judge shall order—

    (A) the person who committed the violation to take affirmative action to abate the violation;

    (B) such person to reinstate the complainant to the complainant's former position, if reinstatement is sought, together with compensation, including back pay and lost benefits, terms, conditions, and privileges of the complainant's employment;

  (C) compensatory damages; and

  (D) such equitable relief as is necessary to correct any violation of this Act or to make the complainant whole.

 (2) COSTS.—On the issuance or designation of a final order of the Secretary sustaining the complainant's complaint, the Secretary, at the request of the complainant, may assess against the respondent a sum equal to aggregate amount of all costs and expenses (including attorney's fees) reasonably incurred, as determined by the Secretary, by the complainant for or in connection with, the bringing of the complaint on which the order was issued.

(j) PRIVATE RIGHT OF ACTION.—

 (1) IN GENERAL.—If no decision by an administrative law judge is issued within the period prescribed by subsection (g), the complainant may file a civil action against the respondent within 60 days for damages and declaratory and equitable relief in any district court of the United States having jurisdiction over the parties, without regard to the amount in controversy, the citizenship of the parties, or further exhaustion of any administrative remedies provided in this Act,

except that the court shall not take jurisdiction if the decision of the administrative law judge is issued before the filing of any civil action. The complainant may not file a civil action against the respondent under this subsection once the filing period has expired.

(2) LOSS OF JURISDICTION.—The jurisdiction of the Administrative Law Judge and the Secretary shall cease upon the filing of such civil action. The filing of such civil action shall not operate to stay any preliminary orders, including any orders for immediate reinstatement, issued by the investigation office or by the administrative law judge, unless otherwise ordered by the court.

(3) NOTICE.—On expiration of the period prescribed by subsection (g), the Secretary shall notify the complainant of the opportunity to file a civil action pursuant to this subsection.

(4) COSTS.—On the issuance or designation of a final order of the court sustaining the complainant's complaint, the court, at the request of the complainant, may assess against the respondent a sum equal to aggregate amount of all costs and expenses (including attorney's fees) reasonably incurred, as determined by the court, by the complainant for or in connection with,

the bringing of the complaint on which the order was issued.

(5) EVIDENTIARY RECORD.—If the complainant files a civil action within the period prescribed in this subsection, either party to that action may request that the Administrative Law Judge provide the complete evidentiary record compiled during the administrative hearing to the district court. Provision of the evidentiary record shall not preclude either party from supplementing the record before the district court.

(k) JUDICIAL REVIEW.—Any person adversely affected or aggrieved by a final order issued pursuant to subsection (h) may obtain review of the order in the United States court of appeals for the circuit in which the violation allegedly occurred. The petition for review must be filed within 60 days after the issuance or designation of the final order. The review shall be conducted in accordance with chapter 7 of title 5, United States Code. The filing of a petition for review under this subsection shall not, unless ordered in accordance with the Federal Rules of Appellate Procedure, operate as a stay of the order. A final order of the Secretary with respect to which review could have been obtained under this paragraph shall not be subject to judicial review in any criminal or other civil proceeding.

(l) SETTLEMENT OR ALTERNATIVE DISPUTE RESOLUTION.—

(1) SETTLEMENT.—On receipt of a complaint pursuant to subsection (b), the investigation office shall attempt to reach a settlement between the complainant and the respondent. The Secretary, investigation office, or administrative law judge shall not enter into a settlement terminating a proceeding without the participation and written consent of the complainant. Any settlement order issued pursuant to this subsection shall be a final order of the Secretary and shall include findings that the settlement is fair, adequate, reasonable, and in the public interest.

(2) ALTERNATIVE DISPUTE RESOLUTION.—While the complaint is pending within the Department of Labor, the Secretary may use alternative means of dispute resolution to resolve the complaint, but the parties must consent, in writing, before the Secretary may employ such means.

(m) TIME COMPUTATIONS.—Except as otherwise provided in this Act, for purposes of a proceeding on a complaint, all time computations shall be made in accordance with the Federal Rules of Civil Procedure.

(n) POSTING.—Each employer subject to this Act shall post and keep posted in conspicuous places on its premises a notice to be prepared or approved by the Secretary setting forth such information as the Secretary considers appropriate to effectuate the purposes of this Act.

SEC. 6. ENFORCEMENT.

(a) BY THE SECRETARY.—Whenever a person has failed to comply with an order issued under this Act, the Secretary may file a civil action in the United States district court for the district in which the violation is alleged to have occurred to enforce such order. In an action brought under this subsection, the district court shall have jurisdiction to grant all appropriate relief, including injunctive and declaratory relief and compensatory and exemplary damages.

(b) BY THE PARTIES.—Any person on whose behalf an order is issued pursuant to this Act may commence a civil action against the person named in the order to require compliance with such order. The appropriate United States district court shall have jurisdiction, without regard to the amount in controversy or the citizenship of the parties, to enforce such order. On issuance of an order on behalf of any person seeking enforcement under this subsection, the court may award costs of litigation (including reasonable attorney and expert witness fees) to the person seeking enforcement.

SEC. 7. COMPLAINT REFERRAL.

(a) REFERRAL.—On issuance of a final order (including an order approving a settlement pursuant to section 5(l)) by the Secretary, investigation office, or administrative law judge, the Secretary shall cause a copy of such order, but only the relevant portions of a settlement order, to be served on each Federal agency that may have jurisdiction over the activity, policy, or practice alleged in the complaint. Any such Federal agency shall take appropriate action with respect to the activity, policy, or practice. Within 30 days after taking such action any Federal agency shall inform the parties to the complaint filed under section 5(b) of any action taken and shall inform the relevant committees of Congress. If appropriate, the Federal agency shall inform the parties and Congress in the form of a report.

(b) OTHER INVESTIGATIVE AUTHORITY.—Subject to section 8, nothing in this section shall limit the authority of any Federal agency under any other law.

SEC. 8. EFFECT ON OTHER RIGHTS AND REMEDIES.

(a) FEDERAL RIGHTS AND REMEDIES.—It is the express intent of Congress to supersede an existing Federal whistleblower protection law only insofar as such law—

 (1) may provide a statute of limitations for filing complaints that is less than 180 days; and

 (2) does not authorize subpoena power for the Secretary as provided in section 5(f).

(b) STATE RIGHTS AND REMEDIES.—The rights and remedies provided to employees by this Act are in addition to, and not in lieu of, any other rights and remedies of the employees provided under State law, and are not intended to alter or effect such rights and remedies.

(c) CONTRACTUAL RIGHTS AND REMEDIES.—The rights and remedies provided to employees by this Act are in addition to, and not in lieu of, any other contractual rights and remedies of the employees, and are not intended to alter or effect such rights and remedies.

SEC. 9. EFFECTIVE DATE.

(a) EFFECTIVE DATE.—This Act shall become effective on the date of enactment.

(b) PENDING CASES.—This Act shall apply only to those complaints filed on or after the effective date of this Act.

APPENDIX L

Connnecticut Right To Act (House Bill No. 7346)

File No. 271

Substitute House Bill No. 7346

State of Connecticut
House of Representatives

House of Representatives, April 6, 1989. The Committee on Labor and Public Employees reported through REP. ADAMO, 116th DIST., Chairman of the Committee on the part of the House, that the substitute bill ought to pass.

AN ACT CONCERNING AN EMPLOYEE'S RIGHT TO ACT.

Be it enacted by the Senate and House of Representatives in General Assembly convened:

1 Section 1. (NEW) (a) As used in this act:
2 (1) "Person" means one or more individuals,
3 partnerships, associations, corporations, business
4 trusts, legal representatives or any organized
5 group of persons;
6 (2) "Employer" means a person engaged in
7 business who has employees, excluding the state
8 and any political subdivision of the state;
9 (3) "Employee" means any person engaged in
10 service to an employer in a business of his
11 employer;
12 (4) "Hazardous condition" means a condition
13 which causes or creates a substantial risk of
14 death, disease or serious physical harm, whether
15 imminent or in the future, and which is beyond the
16 ordinary expected risks inherent in a job after
17 all feasible safety and health precautions have
18 been taken.
19 (b) No employer shall discharge, discipline or
20 otherwise penalize any employee because the
21 employee (1) informs another employee that such
22 other employee is working in or exposed to a
23 hazardous condition or (2) refuses in good faith

24 to expose himself to a hazardous condition in the
25 workplace, provided (A) the condition causing the
26 employee's apprehension of death, disease or
27 serious physical harm is of such a nature that a
28 reasonable person, under the circumstances
29 confronting the employee, would conclude that
30 there is a hazardous condition, (B) there is
31 insufficient time, due to the urgency of the
32 situation, to eliminate or abate the hazardous
33 condition through resort to regular statutory
34 enforcement procedures, (C) the employee notifies
35 the employer of the hazardous condition and asks
36 the employer to correct or abate the hazardous
37 condition and (D) the employer is unable or
38 refuses to abate or correct such condition. No
39 employee shall be penalized while any dispute over
40 the existence of a hazardous condition is being
41 resolved or while a hazardous condition is being
42 abated. Any employee found to have refused to
43 work without a reasonable belief that a hazardous
44 condition existed shall be subject to disciplinary
45 action by his employer up to and including
46 dismissal.
47 (c) Any employee who believes that there is a
48 violation by the employer of such employee of any
49 provision of this act may file a written complaint
50 with the labor commissioner. The complaint shall
51 be signed and shall set forth with reasonable
52 particularity the grounds for the complaint.
53 Within thirty days after receipt of such
54 complaint, the labor commissioner shall notify the
55 employer in writing of the complaint. The
56 commissioner, or his authorized representative,
57 upon presenting appropriate credentials to the
58 employer, operator or agent in charge, may
59 inspect, at reasonable times, the employer's
60 workplace and all conditions pertinent to the
61 grounds of the complaint and shall, in a
62 reasonable manner, make any additional
63 investigation deemed necessary by the commissioner
64 or his representative for full and effective
65 determination of any complaint he receives.
66 (d) If, upon inspection or investigation of a
67 complaint, the labor commissioner or his
68 authorized representative believes that an
69 employer has violated any provisions of this act,
70 he shall hold a hearing and shall, at least ten
71 days prior to the date of such hearing, mail a

File No. 271

72 notice of such hearing to the employer and the
73 employee. The commissioner shall resolve all
74 issues relating to any dispute arising under the
75 provisions of this act.
76   (e) The labor commissioner shall promulgate
77 regulations, adopted in accordance with the
78 provisions of chapter 54 of the general statutes,
79 to implement the provisions of this act.
80   (f) Nothing in this act shall be construed to
81 diminish or impair the rights of any person under
82 any collective bargaining agreement.
83   Sec. 2. The department of labor shall
84 institute an action for a declaratory judgment in
85 federal court on whether the provisions of this
86 act are preempted by the Occupational Safety and
87 Health Act of 1970.
88   Sec. 3. This act shall take effect upon the
89 issuance of a decision, pursuant to the provisions
90 of section 2 of this act, that its provisions are
91 not preempted by the Occupational Safety and
92 Health Act of 1970 and if any section, clause or
93 provision of this act is preempted in whole or in
94 part, to the extent that it is not preempted it
95 shall be valid and effective and no other section,
96 clause or provision shall on account thereof be
97 deemed invalid or ineffective.

98 STATEMENT OF LEGISLATIVE COMMISSIONERS: Subsection
99 (c) of section 1 was reworded in part for clarity.

100 Committee Vote: Yea 9 Nay 2

4   File No. 271

* * * * *

"THE FOLLOWING FISCAL IMPACT STATEMENT AND BILL ANALYSIS ARE PREPARED FOR THE BENEFIT OF MEMBERS OF THE GENERAL ASSEMBLY, SOLELY FOR PURPOSES OF INFORMATION, SUMMARIZATION AND EXPLANATION AND DO NOT REPRESENT THE INTENT OF THE GENERAL ASSEMBLY OR EITHER HOUSE THEREOF FOR ANY PURPOSE."

* * * * *

FISCAL IMPACT STATEMENT - BILL NUMBER SHB 7346

STATE IMPACT Potential Cost, see explanation below

MUNICIPAL IMPACT None

STATE AGENCY(S) Department of Labor

EXPLANATION OF ESTIMATES:

STATE IMPACT: If the bill becomes effective upon issuance of a decision by a federal court that the bill's provisions are not preempted by OSHA (which is uncertain at this time), the Department of Labor could need an additional staff person at a cost of approximately $24,000 per year if there were any activity generated by the law. The Department would be responsible for enforcing the provisions of the bill which allows employees to file complaints with the Labor Commissioner, allows the Commissioner to investigate an employer's workplace and requires a public hearing in the case of a violation. The Department does not have the resources to handle the added responsibilities. Depending upon the level of activity additional personnel could be needed as well.

* * * * *

OLR BILL ANALYSIS

sHB 7346

AN ACT CONCERNING AN EMPLOYEE'S RIGHT TO ACT

SUMMARY: This bill prohibits an employer from penalizing an employee who refuses, in good faith and

File No. 271   5

within certain limits, to work in hazardous conditions. It defines a hazardous condition as one that creates a substantial immediate or long-term risk of death, disease, or serious physical harm. The bill's protection also extends to employees who tell a fellow employee that he is working in or exposed to such conditions. It covers private sector employers only; the state and its political subdivisions are covered by another state law.

The bill gives the labor commissioner authority to investigate and adjudicate employee complaints of violations by their employers. The investigatory power includes authority to inspect workplaces and all conditions pertinent to alleged violations.

The bill does not impair or diminish anyone's rights under a collective bargaining agreement.

Finally, the bill requires the labor commissioner to obtain a declaratory judgment in federal court on the issue of whether the federal Occupational Safety and Health Act (OSHA) preempts any or all of this bill's provisions. The bill does not become effective until the court has ruled. If the court rules that only part of the bill is preempted, the parts not invalidated will go into effect.

EFFECTIVE DATE: Upon issuance of a decision by a federal court that the bill's provisions are not preempted by OSHA.

FURTHER EXPLANATION

<u>Limits on Employee's Refusal to Work</u>

In order to be covered by the bill's protections, the employee's fear of harm must be reasonable and the risk confronting him must be beyond that which is ordinary, expected, and inherent in the job after taking all feasible safety and health precautions. In addition, there must be insufficient time to mitigate the hazardous condition using regular statutory enforcement procedures, the employee must tell his employer about the hazard and ask him to abate it, and the employer must either refuse or be unable to do so.

File No. 271

The bill prohibits an employer from penalizing an employee while any dispute over the existence of a hazardous condition continues. But if an employee is found to have refused to work without a reasonable belief that a hazardous condition existed, the employer may dismiss or penalize him in other ways.

Enforcement

The bill gives the labor commissioner authority to enforce its provisions and requires her to adopt regulations to do so. It allows any employee who believes his employer has violated the bill to file a signed complaint with the labor commissioner, giving particulars of the grounds for suspecting a violation.

The labor commissioner must notify the employer of the complaint within 30 days of receiving it. The bill allows the commissioner or her representative, at reasonable times and after presenting proper credentials, to inspect the employer's workplace and all conditions relevant to the complaint. It also requires her to conduct any other reasonable investigation needed to enable her to resolve the complaint.

If, after investigating the complaint, the commissioner believes there has been a violation, she must hold a hearing, providing at least 10 days advance notice to both the employer and the employee. The bill requires her to resolve all issues raised in any dispute arising under it.

COMMITTEE ACTION

Labor and Public Employees Committee

Joint Favorable Substitute
Yea 9   Nay 2